STRENGTH OF MATERIALS

PART I
Elementary Theory and Problems

By

S. TIMOSHENKO

Professor Emeritus of Engineering Mechanics
Stanford University

THIRD EDITION

CBS Publishers & Distributors Pvt. Ltd.

New Delhi • Bengaluru • Chennai • Kochi • Kolkata • Mumbai
Hyderabad • Nagpur • Patna • Pune • Vijayawada

ISBN: 81-239-1030-4

First Indian Edition: 1986
Reprint: 2000, 2002

This edition has been published in India by arrangement with
Wadsworth Publishing Company, USA

Published by **Satish Kumar Jain** and produced by **Varun Jain** for
CBS Publishers & Distributors Pvt. Ltd.,
4819/XI Prahlad Street, 24 Ansari Road, Daryaganj, New Delhi - 110002
delhi@cbspd.com, cbspubs@airtelmail.in • www.cbspd.com
Ph.: 23289259, 23266861, 23266867 • Fax: 011-23243014

Corporate Office: 204 FIE, Industrial Area, Patparganj, Delhi - 110 092
Ph: 49344934 • Fax: 011-49344935
E-mail: publishing@cbspd.com • publicity@cbspd.com

Branches:
• *Bengaluru:* 2975, 17th Cross, K.R. Road, Bansankari 2nd Stage,
Bengaluru - 70 • Ph: +91-80-26771678/79 • Fax: +91-80-26771680
E-mail: cbsbng@gmail.com, bangalore@cbspd.com
• *Chennai:* No. 7, Subbaraya Street, Shenoy Nagar, Chennai - 600030
Ph: +91-44-26681266, 26680620 • Fax: +91-44-42032115
E-mail: chennai@cbspd.com
• *Kochi:* Ashana House, 39/1904, A.M. Thomas Road, Valanjambalam,
Ernakulum, Kochi • Ph: +91-484-4059061-65
Fax: +91-484-4059065 • E-mail: cochin@cbspd.com
• *Kolkata:* 6-B, Ground Floor, Rameshwar Shaw Road, Kolkata - 700014
Ph: +91-33-22891126/7/8 • E-mail: kolkata@cbspd.com
• *Mumbai:* 83-C, Dr. E. Moses Road, Worli, Mumbai - 400018
Ph: +91-9833017933, 022-24902340/41 • E-mail: mumbai@cbspd.com

Representatives:
• Bhubaneswar 0-9911037372 • Hyderabad 0-9885175004 • Jharkhand 0-9811541605
• Nagpur 0-9021734563 • Patna 0-9334159340 • Pune 0-9623451994
• Uttarakhand 0-9716462459 • Dhaka (Bangladesh) 01912-003485

Printed at:
Neekunj Print Process, Delhi (India)

PREFACE TO THE THIRD EDITION

In the preparation of the third edition of this book a considerable number of new problems were added, and answers to many of the old problems inserted. The book was expanded by the addition of two new chapters; namely, Chapter VIII which deals with bending of beams in a plane which is not a plane of symmetry, and Chapter XII on the bending of curved bars. In Chapter VIII the notion of shear center, which is of great practical importance in the case of thin walled structures, is introduced. In Chapter XII is presented the material on curved bars which previously appeared in the second volume of this book. That material has been entirely rewritten and new material added. It is hoped with these major changes, as well as the innumerable minor changes throughout the entire text, that the volume will be not only more complete, but also more satisfactory as a textbook in elementary courses in strength of materials. The author wishes to thank Professor James M. Gere of Stanford University, who assisted in revising the volume and in reading the proofs.

S. Timoshenko

Stanford University

PREFACE TO THE SECOND EDITION

In preparing the second edition of this volume, an effort has been made to adapt the book to the teaching requirements of our engineering schools.

With this in view, a portion of the material of a more advanced character which was contained in the previous edition of this volume has been removed and will be included in the new edition of the second volume. At the same time, some portions of the book, which were only briefly discussed in the first edition, have been expanded with the intention of making the book easier to read for the beginner. For this reason, chapter II, dealing with combined stresses, has been entirely rewritten. Also, the portion of the book dealing with shearing force and bending moment diagrams has been expanded, and a considerable amount of material has been added to the discussion of deflection curves by the integration method. A discussion of column theory and its application has been included in chapter VIII, since this subject is usually required in undergraduate courses of strength of materials. Several additions have been made to chapter X dealing with the application of strain energy methods to the solution of statically indetermined problems. In various parts of the book there are many new problems which may be useful for class and home work.

Several changes in the notations have been made to conform to the requirements of American Standard Symbols for Mechanics of Solid Bodies recently adopted by The American Society of Mechanical Engineers.

It is hoped that with the changes made the book will be found more satisfactory for teaching the undergraduate course of strength of materials and that it will furnish a better foundation for the study of the more advanced material discussed in the second volume.

S. Timoshenko

Palo Alto, California
June 13, 1940

PREFACE TO THE FIRST EDITION

At the present time, a decided change is taking place in the attitude of designers towards the application of analytical methods in the solution of engineering problems. Design is no longer based principally upon empirical formulas. The importance of analytical methods combined with laboratory experiments in the solution of technical problems is becoming generally accepted.

Types of machines and structures are changing very rapidly, especially in the new fields of industry, and usually time does not permit the accumulation of the necessary empirical data. The size and cost of structures are constantly increasing, which consequently creates a severe demand for greater reliability in structures. The economical factor in design under the present conditions of competition is becoming of growing importance. The construction must be sufficiently strong and reliable, and yet it must be designed with the greatest possible saving in material. Under such conditions, the problem of a designer becomes extremely difficult. Reduction in weight involves an increase in working stresses, which can be safely allowed only on a basis of careful analysis of stress distribution in the structure and experimental investigation of the mechanical properties of the materials employed.

It is the aim of this book to present problems such that the student's attention will be focussed on the practical applications of the subject. If this is attained, and results, in some measure, in increased correlation between the studies of strength of materials and engineering design, an important forward step will have been made.

The book is divided into two volumes. The first volume contains principally material which is usually covered in required courses of strength of materials in our engineering

schools.   The more advanced portions of the subject are of
interest chiefly to graduate students and research engineers,
and are incorporated in the second volume of the book.   This
contains also the new developments of practical importance in
the field of strength of materials.

In writing the first volume of strength of materials, attention was given to simplifying all derivations as much as
possible so that a student with the usual preparation in mathematics will be able to read it without difficulty.   For example,
in deriving the theory of the deflection curve, the *area moment
method* was extensively used.   In this manner, a considerable
simplification was made in deriving the deflections of beams for
various loading and supporting conditions.   In discussing
statically indeterminate systems, the *method of superposition*
was applied, which proves very useful in treating such problems
as continuous beams and frames.   For explaining combined
stresses and deriving principal stresses, use was made of the
Mohr's circle, which represents a substantial simplification in
the presentation of this portion of the theory.

Using these methods of simplifying the presentation, the
author was able to condense the material and to discuss some
problems of a more advanced character.   For example, in
discussing torsion, the twist of rectangular bars and of rolled
sections, such as angles, channels, and I beams, is considered.
The deformation and stress in helical springs are discussed in
detail.   In the theory of bending, the case of non-symmetrical
cross sections is discussed, the *center of twist* is defined and
explained, and the effect of shearing force on the deflection of
beams is considered.   The general theory of the bending of
beams, the materials of which do not follow Hooke's law, is
given and is applied in the bending of beams beyond the yielding
point.   The bending of reinforced concrete beams is given
consideration.   In discussing combinations of direct and bending stress, the effect of deflections on the bending moment is
considered, and the limitation of the method of superposition
is explained.   In treating combined bending and torsion,
the cases of rectangular and elliptical cross sections are dis-

cussed, and applications in the design of crankshafts are given. Considerable space in the book is devoted to methods for solving elasticity problems based on the consideration of the strain energy of elastic bodies. These methods are applied in discussing statically indeterminate systems. The stresses produced by impact are also discussed. All these problems of a more advanced character are printed in small type, and may be omitted during the first reading of the book.

The book is illustrated with a number of problems to which solutions are presented. In many cases, the problems are chosen so as to widen the field covered by the text and to illustrate the application of the theory in the solution of design problems. It is hoped that these problems will be of interest for teaching purposes, and also useful for designers.

The author takes this opportunity of thanking his friends who have assisted him by suggestions, reading of manuscript and proofs, particularly Messrs. W. M. Coates and L. H. Donnell, teachers of mathematics and mechanics in the Engineering College of the University of Michigan, and Mr. F. L. Everett of the Department of Engineering Research of the University of Michigan. He is indebted also to Mr. F. C. Wilharm for the preparation of drawings, to Mrs. E. D. Webster for the typing of the manuscript, and to the Van Nostrand Company for its care in the publication of the book.

S. TIMOSHENKO

ANN ARBOR, MICHIGAN
May 1, 1930

NOTATIONS

α . . . . Angle, coefficient of thermal expansion, numerical coefficient

β . . . . Angle, numerical coefficient

γ . . . . Shearing strain, weight per unit volume

Δ . . . . Unit volume expansion, distance

δ . . . . Total elongation, total deflection, distance

ϵ . . . . Unit strain

$\epsilon_x, \epsilon_y, \epsilon_z$ . . . . . Unit strains in x, y and z directions

θ . . . . Angle, angle of twist per unit length of shaft

μ . . . . Poisson's ratio

σ . . . . Unit normal stress

σ_1, σ_2 . . . . Principal stresses

σ_n . . . . Unit normal stress on plane perpendicular to the direction n

$\sigma_x, \sigma_y, \sigma_z$ . . . . Unit normal stresses on planes perpendicular to the x, y and z axes

σ_U . . . . Ultimate stress

σ_W . . . . Working stress

$\sigma_{Y.P.}$ . . . . Yield point stress

τ . . . . Unit shear stress

$\tau_{xy}, \tau_{yz}, \tau_{zx}$ . . Unit shear stresses on planes perpendicular to the x, y and z axes, and parallel to the y, z and x axes

τ_W . . . . Working stress in shear

$\tau_{Y.P.}$ . . . . Yield point stress in shear

φ . . . . Angle

ω . . . . Angular velocity

A . . . . Cross sectional area

a, b, c, d . . . . . Distances

C . . . . Torsional rigidity, constant of integration

D, d . . . . Diameters

E . . . . Modulus of elasticity

G.......... Modulus of elasticity in shear

H.......... Horizontal force, horsepower

h.......... Height, thickness

I_p.......... Polar moment of inertia of a plane area

I_y, I_z....... Moments of inertia of a plane area with respect to the y and z axes

I_{yz}.......... Product of inertia of a plane area with respect to the y and z axes

K.......... Bulk modulus of elasticity

k.......... Spring constant, numerical factor

k_y, k_z....... Radii of gyration of a plane area with respect to the y and z axes

l.......... Length, span

M.......... Bending moment

M_t.......... Torque

n.......... Factor of safety, revolutions per minute, normal to a plane

P, Q....... Concentrated forces

p.......... Pressure, steel ratio for reinforced concrete beams

q.......... Load per unit length, pressure

R.......... Reaction, force, radius

r.......... Radius, radius of curvature

S.......... Axial force in a bar

t.......... Temperature, thickness

U.......... Strain energy

u.......... Deflection, distance

V.......... Volume, shearing force

v.......... Velocity, deflection, distance

W.......... Total load, weight

w.......... Weight per unit length, strain energy per unit volume

w_1.......... Strain energy per unit weight

X, Y, Z..... Axial forces in bars, unknown reactions

x, y, z....... Rectangular coordinates

Z.......... Section modulus

CONTENTS

CONTENTS

PART I

CHAPTER I

TENSION AND COMPRESSION WITHIN THE ELASTIC LIMIT

1. Elasticity.—A material body consists of small particles, or molecules, between which forces are acting. These molecular forces resist the change in the shape of the body which external forces tend to produce. Under the action of external forces the particles of the body are displaced and the displacements continue until equilibrium is established between the external and internal forces. The body is then in a *state of strain*. During deformation the external forces acting upon the body do work, and this work is transformed completely or partially into *potential energy of strain*. A watch spring is an example of such an accumulation of potential energy in a strained body. If the forces which produced the deformation of the body are now gradually diminished, the body will return wholly or partly to its original shape and during this reversed deformation the potential energy of strain which was accumulated in the body may be recovered in the form of external work.

Consider, for instance, a prismatic bar loaded at the end as shown in Fig. 1.[1] Under the action of this

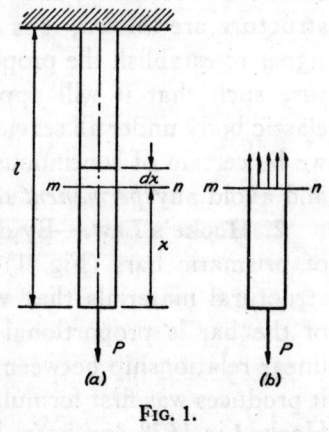

Fig. 1.

load a certain elongation of the bar will take place. The point of application of the load will then move in a downward direction and positive work will be done by the load during this

---

[1] It is assumed that the load is acting along the axis of the bar, i.e., along the line passing through the centroids of the cross sections.

1

motion. When the load is diminished, the elongation of the bar diminishes also, the loaded end of the bar moves upward and the potential energy of strain will be transformed into the work of moving the load in the upward direction.

The property by which a body returns to its original shape after removal of the load is called *elasticity*. The body is *per fectly elastic* if it recovers its original shape completely after unloading; it is *partially elastic* if the deformation produced by the external forces does not disappear completely after unloading. In the case of a perfectly elastic body the work done by the external forces during deformation is completely transformed into potential energy of strain.[2] In the case of a partially elastic body, part of the work done by the external forces during deformation is dissipated in the form of heat, which is developed in the body during the non-elastic deformation. Experiments show that such structural materials as steel, wood and stone may be considered as perfectly elastic within certain limits, depending upon the properties of the material. Assuming that the external forces acting upon the structure are known, it is a fundamental problem for the designer to establish the proportions of the members of the structure such that it will approach the condition of a perfectly elastic body under all service conditions. Only in this way can we be certain of continuous reliable service from the structure and avoid any *permanent set* in its members.

2. Hooke's Law.—By direct experiment with the extension of prismatic bars (Fig. 1) it has been established for many structural materials that within certain limits the elongation of the bar is proportional to the tensile force. This simple linear relationship between the force and the elongation which it produces was first formulated by the English scientist Robert Hooke[3] in 1678 and bears his name. Using the notation:

P = force producing extension of bar,
l = length of bar,

---

[2] The small temperature changes which usually accompany elastic deformation and the corresponding heat exchange with the surroundings are neglected in this consideration (see Part II).

[3] Robert Hooke, *De Potentia restitutiva*, London, 1678.

A = cross-sectional area of bar,
δ = total elongation of bar,
E = elastic constant of the material, called the *Modulus of Elasticity*,

Hooke's experimental law may be given by the following equation:

$$\delta = \frac{Pl}{AE}. \tag{1}$$

The elongation of the bar is proportional to the tensile force and to the length of the bar and inversely proportional to the cross-sectional area and to the modulus of elasticity. In making tensile tests precautions are usually taken to ensure central application of the tensile force. In Fig. 2 is shown a method of fixing the ends of a circular tensile test specimen in a tensile test machine. In this manner any bending of the bar will be prevented. Excluding from consideration those portions of the bar in the vicinity of the applied forces,[4] it may be assumed that during tension all longitudinal fibers of the prismatic bar have the same elongation and that cross sections of the bar originally plane and perpendicular to the axis of the bar remain so after extension.

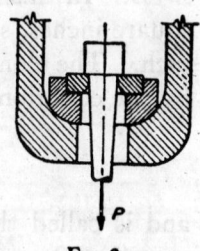

Fig. 2.

In discussing the magnitude of internal forces let us imagine the bar cut into two parts by a cross section mn and let us consider the equilibrium of the lower portion of the bar (Fig. 1b). At the lower end of this portion the tensile force P is applied.

On the upper end the forces represent the action of the particles of the upper portion of the strained bar on the particles of the lower portion. These forces are continuously distributed over the cross section. Familiar examples of such a continuous distribution of forces over a surface are hydrostatic pressure and steam pressure. In handling such continuously distributed forces *the intensity of force*, i.e., the force per unit area, is of great importance. In the present case of axial tension, in

---

[4] The more complicated stress distribution near the points of application of the forces is discussed in Part II.

which all fibers have the same elongation, the distribution of forces over the cross section mn will be *uniform*. The resultant of these forces will pass through the centroid of the cross section and will act along the axis of the bar. Taking into account that the sum of these forces, from the condition of equilibrium (Fig. 1b), must be equal to P and denoting the force per unit of cross-sectional area by σ, we obtain

$$\sigma = \frac{P}{A}. \tag{2}$$

This force per unit area is called *unit tensile stress* or simply *stress*. In this book, force is measured in pounds and area in square inches, so that stress is measured in pounds per square inch. The elongation of the bar per unit length is determined by the equation

$$\epsilon = \frac{\delta}{l} \tag{3}$$

and is called the *unit elongation* or the *tensile strain*. Using eqs. (1), (2) and (3), Hooke's law may also be written in the following form:

$$E = \frac{\sigma}{\epsilon} \tag{4}$$

and we see that *modulus of elasticity is equal to unit stress divided by unit strain* and may be easily calculated provided the stress and corresponding unit elongation are found from a tensile test. The unit elongation ϵ is a pure number representing the ratio of two lengths (see eq. 3); therefore, from eq. (4) it may be concluded that modulus of elasticity is measured in the same units as stress σ, i.e., in pounds per square inch. Average values of the modulus E for several materials are given in the first column of Table 1.[5]

Eqs. (1)–(4) may be used also in the case of compression of prismatic bars. Then δ denotes the total longitudinal contraction, ϵ the *compressive strain* and σ the *compressive stress*.

---

[5] More details on the mechanical properties of materials are given in Part II.

TABLE 1: MECHANICAL PROPERTIES OF MATERIALS

Material	E lb/in.^{2}	Yield Point lb/in.^{2}	Ultimate Strength lb/in.^{2}
Structural carbon steel 0.15 to 0.25% carbon.........	30×10^6	$30 \times 10^3 - 40 \times 10^3$	$55 \times 10^3 - 65 \times 10^3$
Nickel steel 3 to 3.5% nickel	29×10^6	$40 \times 10^3 - 50 \times 10^3$	$78 \times 10^3 - 100 \times 10^3$
Duraluminum..............	10×10^6	$35 \times 10^3 - 45 \times 10^3$	$54 \times 10^3 - 65 \times 10^3$
Copper, cold rolled.........	16×10^6		$28 \times 10^3 - 40 \times 10^3$
Glass....................	10×10^6		3.5×10^3
Pine, with the grain........	1.5×10^6		$8 \times 10^3 - 20 \times 10^3$
Concrete, in compression....	4×10^6		3×10^3

For most structural materials the modulus of elasticity for compression is the same as for tension. In calculations, tensile stress and tensile strain are considered as positive, and compressive stress and strain as negative.

Problems

1. Determine the total elongation of a steel bar 25 in. long, if the tensile stress is equal to 15×10^3 lb per sq in.

Answer. $\delta = \frac{1}{80}$ in.

2. Determine the tensile force on a cylindrical steel bar of 1 in. diameter, if the unit elongation is equal to 0.7×10^{-3}.

Solution. The tensile stress in the bar, from eq. (4), is

$$\sigma = \epsilon \cdot E = 21 \times 10^3 \text{ lb per sq in.}$$

The tensile force, from eq. (2), is

$$P = \sigma \cdot A = 21 \times 10^3 \times \frac{\pi}{4} = 16,500 \text{ lb.}$$

3. What is the ratio of the moduli of elasticity of the materials of two bars of the same size if under the action of equal tensile forces the unit elongations of the bars are in the ratio $1 : \frac{1.5}{8}$? Determine these elongations if one of the bars is of steel, the other of copper, and the tensile stress is 10,000 lb per sq in.

Solution. The moduli are inversely proportional to the unit elongations. For steel

$$\epsilon = \frac{10,000}{30 \times 10^6} = \frac{1}{3,000},$$

for copper

$$\epsilon = \frac{1}{1,600}.$$

4. A prismatic steel bar 25 in. long is elongated $\frac{1}{40}$ in. under the action of a tensile force P. Find the magnitude of the force if the volume of the bar is 25 in.[3]

Answer. $P = 30,000$ lb.

5. A piece of wire 100 ft long subjected to a tensile force $P = 1,000$ lb elongates by 1 in. Find the modulus of elasticity of the material if the cross-sectional area of the wire is 0.04 sq in.

Answer. $E = 30 \times 10^6$ lb per sq in.

6. Determine the total elongation of the steel bar AB having a cross-sectional area $A = 1$ in.[2] and submitted to the action of forces $Q = 10,000$ lb and $P = 5,000$ lb (Fig. 3).

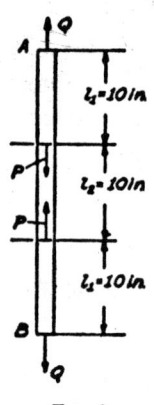

Solution. The tensile force in the upper and lower portions of the bar is equal to Q and that in the middle portion is $Q - P$. Then the total elongation will be

$$\delta = 2\frac{Ql_1}{AE} + \frac{(Q-P)l_2}{AE} = 2\frac{10,000 \times 10}{1 \times 30 \times 10^6}$$

$$+ \frac{5,000 \times 10}{1 \times 30 \times 10^6} = \frac{1}{150} + \frac{1}{600} = \frac{1}{120} = 0.00833 \text{ in.}$$

Fıg. 3.

7. Solve Prob. 6, assuming that the material is duraluminum and that $P = Q = 10,000$ lb.

3. The Tensile Test Diagram.

The proportionality between the tensile force and the corresponding elongation holds only up to a certain limiting value of the tensile stress, called the *proportional limit*, which depends upon the properties of the material. Beyond this limit, the relationship between the elongation and the tensile stress becomes more complicated. For such a material as structural steel the proportionality between the load and elongation holds within a considerable range and the proportional limit is as high as 25×10^3–30×10^3 lb per sq in. For materials such as cast iron or soft copper the proportional limit is very low, i.e., deviations from Hooke's law may be noticed at a low tensile stress.

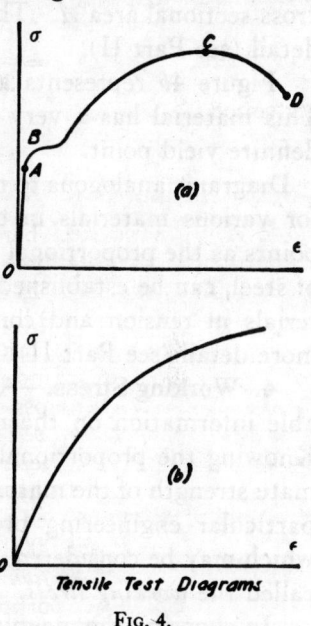

Tensile Test Diagrams

FIG. 4.

In investigating the mechanical properties of materials beyond the proportional limit, the relationship between the strain and the corresponding stress is usually presented graphically by a *tensile test diagram*. Fig. 4a presents a typical diagram for structural steel. Here the elongations are plotted along the horizontal axis and the corresponding stresses are given by the ordinates of the curve OABCD. From O to A the stress and the strain are proportional; beyond A the deviation from Hooke's law becomes marked; hence the stress at A is the *proportional limit*. Upon loading beyond this limit the elongation increases more rapidly and the diagram becomes curved. At B a sudden elongation of the bar takes place without an appreciable increase in the tensile force. This phenomenon, called *yielding* of the metal, is shown in the diagram by an almost horizontal portion of the curve. The stress corresponding to the point B is called the *yield point*. Upon further stretching of the bar, the material recovers its resistance and, as is seen from the diagram, the tensile force increases with the elongation up to the point C, where the force attains its maximum value. The corresponding stress is called the *ultimate strength* of the material. Beyond the point C, elongation of the bar takes place with a diminution of the load and fracture finally occurs at a load corresponding to point D on the diagram.

It should be noted that stretching of the bar is accompanied by lateral contraction but it is established practice in calculating.the yield point and the ultimate strength to use the initial cross-sectional area A. This question is discussed later in more detail (see Part II).

Figure 4*b* represents a tensile test diagram for cast iron. This material has a very low proportional limit [6] and has no definite yield point.

Diagrams analogous to those in tension may also be obtained for various materials in compression and such characteristic points as the proportional limit, and the yield point in the case of steel, can be established. The mechanical properties of materials in tension and compression will be discussed later in more detail (see Part II).

4. Working Stress.—A tensile test diagram gives very valuable information on the mechanical properties of a material. Knowing the proportional limit, the yield point, and the ultimate strength of the material, it is possible to establish for each particular engineering problem the magnitude of the stress which may be considered as a *safe stress*. This stress is usually called the *working stress*.

In choosing the magnitude of the working stress for steel it must be noted that at stresses below the proportional limit the material may be considered as perfectly elastic, whereas beyond this limit a part of the strain usually remains after unloading the bar, i.e., *permanent set* occurs. In order to have the structure in an elastic condition and to eliminate any possibility of permanent set, it is common practice to keep the working stress well below the proportional limit. In the experimental determination of this limit, sensitive measuring instruments (extensometers) are necessary and the position of the limit depends to some extent upon the accuracy with which the measurements are made. In order to eliminate this difficulty one usually takes the *yield point* or the *ultimate strength* of the material as a basis for determining the magnitude of the working stress. Denoting by σ_w, $\sigma_{Y.P.}$ and σ_U, respectively, the working stress, the yield point and the ultimate strength of the material, the magnitude of the working stress

---

[6] This limit can be established only by using very sensitive extensometers in measuring elongations. See Grüneisen, *Ber. deut. physik. Ges.*, 1906.

is determined by one of the following equations:

$$\sigma_W = \frac{\sigma_{Y.P.}}{n} \quad \text{or} \quad \sigma_W = \frac{\sigma_U}{n_1}. \tag{5}$$

Here n and n_1 are constants called *factors of safety*, which determine the magnitude of the working stress. In the case of structural steel, it is logical to take the yield point as the basis for calculating the working stress because at yield stress considerable permanent set occurs, which is not permissible in engineering structures. In this case a factor of safety $n = 2$ gives a conservative value for the working stress, provided that only constant or static loads act upon the structure. In the case of suddenly applied loads or variable loads, such as occur frequently in machine parts, a larger factor of safety becomes necessary. For brittle materials such as cast iron, concrete or various kinds of stone, and for material such as wood, the ultimate strength is usually taken as the basis for determining the working stress.

The magnitude of the factor of safety depends upon the accuracy with which the external forces acting upon a structure are known, upon the accuracy with which the stresses in the members of a structure can be calculated, and also upon the homogeneity of the materials used. This important question of working stresses will be discussed in more detail later (see Part II). Here we will give several simple examples of the determination of safe cross-sectional dimensions of bars, assuming that the working stress is given.

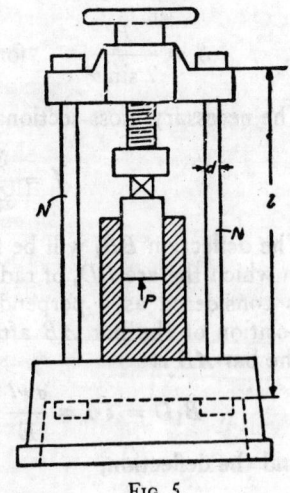

FIG. 5.

Problems

1. Determine the diameter d of the steel bolts N of a press for a maximum compressive force $P = 100,000$ lb (Fig. 5), if the working stress for steel in this case is $\sigma_W = 10,000$ lb per sq in. Determine the total elongation of the bolts at the maximum load, if the length between their heads is $l = 50$ in.

Solution. The necessary cross-sectional area, from eq. (2), is

$$A = \frac{\pi d^2}{4} = \frac{P}{2\sigma_W} = \frac{50,000}{10,000} = 5 \text{ in.}^2$$

Then

$$d = \sqrt{\frac{20}{\pi}} = 2.52 \text{ in.}$$

Total elongation, from eqs. (3) and (4),

$$\delta = \epsilon l = \frac{\sigma l}{E} = \frac{10^4 \cdot 50}{30 \cdot 10^6} = \frac{1}{60} \text{ in.}$$

2. A structure ABC consisting of two equal steel bars (Fig. 6) 15 ft long and with hinged ends is subjected to the action of a vertical load P. Determine the necessary cross-sectional areas of the bars and the deflection of the point B when $P = 5,000$ lb, $\sigma_W = 10,000$ lb per sq in. and the initial angle of inclination of the bars $\theta = 30°$. Neglect the weight of the bars as a small quantity in comparison with the load P.

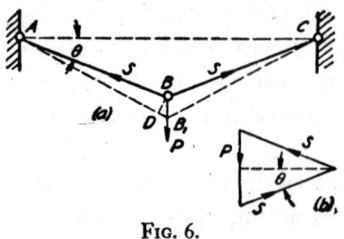

FIG. 6.

Solution. From Fig. 6b, representing the condition for equilibrium of the hinge B, the tensile force in the bars is

$$S = \frac{P}{2 \sin \theta}, \quad \text{for } \theta = 30°, \quad S = P = 5,000 \text{ lb.}$$

The necessary cross-sectional area is

$$A = \frac{S}{\sigma_W} = \frac{5,000}{10,000} = \frac{1}{2} \text{ in.}^2$$

The deflection BB_1 will be found from the small right triangle DBB_1 in which the arc BD, of radius equal to the initial length of the bars, is considered as a perpendicular dropped upon AB_1, which is the position of the bar AB after deformation. Then the elongation of the bar AB is

$$B_1 D = \epsilon \cdot l = \frac{\sigma_W l}{E} = \frac{10,000 \times 15 \times 12}{30 \times 10^6} = 0.06 \text{ in..}$$

and the deflection,

$$BB_1 = \frac{B_1 D}{\sin \theta} = 0.12 \text{ in.}$$

It is seen that the change of the angle due to the deflection BB_1 is very small and the previous calculation of S, based upon the assumption that $\theta = 30°$, is accurate enough.

3. Determine the cross-sectional dimensions of the wooden beam BC and of the steel bar AB of the structure ABC, loaded at B, if the working stress for pine wood is taken as $\sigma_W = 160$ lb per sq in. and for steel $\sigma_W = 10,000$ lb per sq in. The load $P = 6,000$ lb. The dimensions of the structure are shown in Fig. 7. Determine the vertical and the horizontal components of the displacement of the point B due to deformation of the bars. Neglect the weight of the structure.

Solution. From the triangle in Fig. 7b, giving the condition for equilibrium of hinge B and similar to the triangle ABC of Fig. 7a, we have

$$S = \frac{P \cdot 15}{9} = 10,000 \text{ lb,}$$

$$S_1 = \frac{P \cdot 12}{9} = 8,000 \text{ lb.}$$

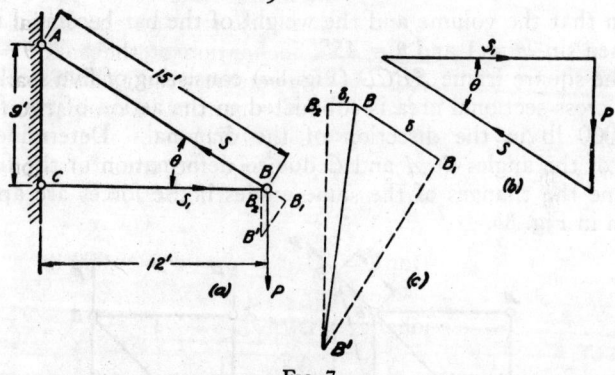

Fig. 7.

The cross-sectional areas of the steel bar and of the wooden beam are

$$A = \frac{S}{\sigma_W} = \frac{10,000}{10,000} = 1 \text{ in.}^2, \qquad A_1 = \frac{S_1}{\sigma_W} = \frac{8,000}{160} = 50 \text{ in.}^2$$

The total elongation of the steel bar and the total compression of the wooden beam are

$$\delta = \frac{S \cdot l}{E_s A} = \frac{10,000 \cdot 15 \cdot 12}{30 \times 10^6} = 0.060 \text{ in.,}$$

$$\delta_1 = \frac{S_1 l_1}{E_w A_1} = \frac{160 \times 12 \times 12}{1.5 \times 10^6} = 0.0154 \text{ in.}$$

To determine the displacement of the hinge B, due to deformation, arcs are drawn with centers A and C (Fig. 7a) and radii equal to the lengths of the elongated bar and of the compressed beam, respectively. They intersect in the new position B' of the hinge B. This is shown on a larger scale in Fig. 7c, where BB_1 is the elongation of the steel bar and BB_2, the compression of the wooden beam. The dotted perpendiculars replace the arcs mentioned above. Then BB' is the displacement of the hinge B. The components of this displacement may be easily obtained from the figure.

4. Determine in the previous problem the angle of inclination θ of the bar AB so as to make its weight a minimum.

Solution. If θ denotes the angle between the bar and the horizontal beam and l_1 is the length of the beam, then the length of the bar is $l = l_1/\cos \theta$, the tensile force in the bar is $S = P/\sin \theta$ and the necessary cross-sectional area is $A = P/\sigma_W \sin \theta$. The volume of the bar will be

$$l \cdot A = \frac{l_1 P}{\sigma_W \sin \theta \cos \theta} = \frac{2l_1 P}{\sigma_W \sin 2\theta}.$$

It is seen that the volume and the weight of the bar become a minimum when $\sin 2\theta = 1$ and $\theta = 45°$.

5. The square frame $ABCD$ (Fig. 8a) consisting of five steel bars of 1 in.^{2} cross-sectional area is submitted to the action of two forces $P = 10,000$ lb in the direction of the diagonal. Determine the changes of the angles at A and C due to deformation of the frame. Determine the changes of the same angles if the forces are applied as shown in Fig. 8b.

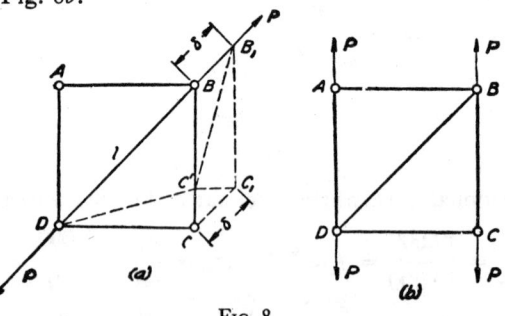

Fig. 8.

Solution. In the case shown in Fig. 8a the diagonal will take the complete load P. Assuming that the hinge D and the direction of the diagonal are stationary, the displacement of the hinge B in the direction of the diagonal will be equal to the elongation of the diagonal $\delta = Pl/AE$. The determination of the new position C' of the hinge C is indicated in the figure by dotted lines. It is seen from the small right triangle CC_1C' that $CC' = \delta/\sqrt{2}$. Then the angle of rotation of the bar DC due to deformation of the frame is equal to

$$\frac{CC'}{DC} = \frac{\delta\sqrt{2}}{\sqrt{2}l} = \frac{\delta}{l} = \frac{P}{1 \cdot E} = \frac{1}{3,000} \text{ radian.}$$

Then the increase of the angle at C will be

$$2 \times \frac{1}{3,000} = \frac{1}{1,500} \text{ radian.}$$

The solution of the problem shown in Fig. 8b is left to the student.

6. Determine the position of the load P on the beam ABD so that the force in the bar BC becomes a maximum. Determine the angle θ to make the volume of the bar BC a minimum (Fig. 9).

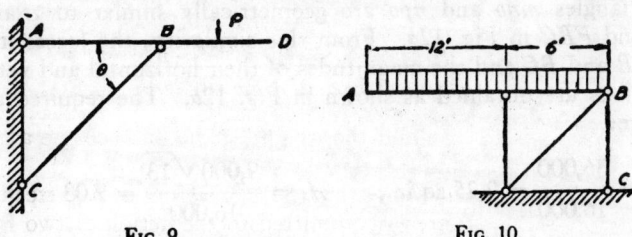

FIG. 9.   FIG. 10.

Answer. The force in the bar BC becomes maximum when the load P has its extreme position on the right at point D. The volume of the bar will be a minimum when $\theta = 45°$.

7. Determine the necessary cross-sectional area of the steel bar BC (Fig. 10) if the working stress $\sigma_W = 15,000$ lb per sq in. and the uniformly distributed vertical load per ft of beam AB is $q = 1,000$ lb. per ft.

Answer. $A = 0.6$ sq in.

8. Determine the necessary cross-sectional areas A and A_1 of the bars AB and BC of the structure shown in Fig. 11 if $\sigma_W = 16,000$ lb per sq in.

Answer. $A = 2.5$ sq in., $A_1 = 2$ sq in.

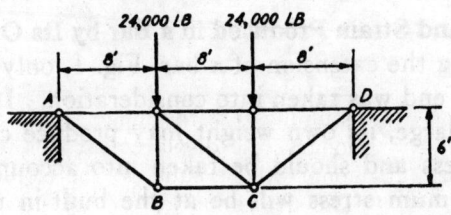

FIG. 11.

9. Determine the necessary cross-sectional areas A and A_1 of the bars AB and BC of the structure shown in Fig. 12a if $\sigma_W = 16,000$ lb per sq in.

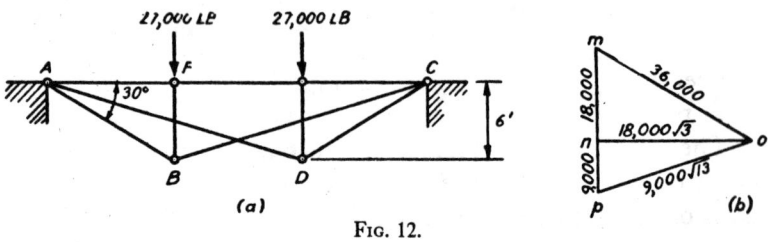

Fig. 12.

Solution. Let Δmpo in Fig. 12b represent the triangle of forces acting on hinge B. Then, drawing line on horizontally, we conclude that triangles mno and npo are geometrically similar to triangles BFA and FBC in Fig. 12a. From this similarity, the forces in the bars AB and BC and the magnitudes of their horizontal and vertical projections are obtained as shown in Fig. 12b. The required areas then are:

$$A = \frac{36,000}{16,000} = 2.25 \text{ sq in.}, \qquad A_1 = \frac{9,000\sqrt{13}}{16,000} = 2.03 \text{ sq in.}$$

10. Find the cross-sectional area of the bar CD in Fig. 11 and the total elongation of the bar if the material is structural steel and $\sigma_W = 16,000$ lb per sq in.

Answer. $A = 2.5$ sq in., $\delta = 0.064$ in.

11. Solve Prob. 8, assuming that the load is applied at only one joint of the upper chord at a distance 8 ft from support A.

Answer. $A = 1.67$ sq in., $A_1 = 1.33$ sq in.

12. A square steel frame is loaded as shown in Fig. 13. Find the total elongation of each bar if the cross-sectional area of each bar is 1 sq in.

Answer. Elongation of all the bars except AB is zero. For bar AB, $\delta = 0.12$ in.

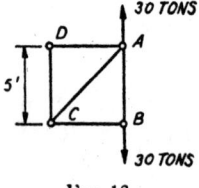

Fig. 13.

5. Stress and Strain Produced in a Bar by Its Own Weight. —In discussing the extension of a bar, Fig. 1, only the load P applied at the end was taken into consideration. If the length of the bar is large, its own weight may produce considerable additional stress and should be taken into account. In this case the maximum stress will be at the built-in upper cross section. Denoting by γ the weight per unit volume of the bar, the total weight becomes $A\gamma l$ and the maximum stress is given by the equation:

$$\sigma_{max} = \frac{P + A\gamma l}{A} = \frac{P}{A} + \gamma l. \tag{6}$$

The second term on the right side of eq. (6) represents the stress produced by the weight of the bar.

The weight of the portion of the bar below cross section mn at distance x from the lower end (Fig. 1) is $A\gamma x$ and the stress on that cross section is given by the equation:

$$\sigma = \frac{P + A\gamma x}{A}. \tag{7}$$

Substituting the working stress σ_W for σ_{max} in eq. (6), the equation for calculating the safe cross-sectional area will be

$$A = \frac{P}{\sigma_W - \gamma l}. \tag{8}$$

It is interesting to note that with increasing length l the weight of the bar becomes more and more important, the denominator of the right side of eq. (8) diminishes and the necessary cross-sectional area A increases. When $\gamma l = \sigma_W$, i.e., the stress due to the weight of the bar alone becomes equal to the working stress, the right side of eq. (8) becomes infinite. Under such circumstances it is impossible to use a prismatic design and a bar of variable cross section must be used.

In calculating the total elongation of a prismatic bar submitted to the action of its own weight and a tensile force P at the end, let us consider first the elongation of an element of length dx cut from the bar by two adjacent cross sections (see Fig. 1). It may be assumed that along the very short length dx the tensile stress is constant and is given by eq. (7). Then the elongation $d\delta$ of the element will be

$$d\delta = \frac{\sigma dx}{E} = \frac{P + A\gamma x}{AE} dx.$$

The total elongation of the bar will be obtained by summing

the elongations of all the elements. Then

$$\delta = \int_0^l \frac{P + A\gamma x}{AE} \, dx = \frac{l}{AE}(P + \tfrac{1}{2}A\gamma l). \qquad (9)$$

Comparing this with eq. (1) it is seen that the total elongation produced by the bar's own weight is equal to that produced by a load of half its weight applied at the end.

Problems

1. Determine the cross-sectional area of a vertical prismatic steel bar (Fig. 1) carrying at its lower end a load $P = 70,000$ lb, if the length of the bar is 720 ft, the working stress $\sigma_w = 10,000$ lb per sq in. and the weight of a cu ft of steel is 490 lb. Determine the total elongation of the bar.

Solution. The cross-sectional area, from eq. (8), is

$$A = \frac{70,000}{10,000 - \dfrac{490 \times 720 \times 12}{12^3}} = 9.27 \text{ in.}^2$$

The total elongation, from eq. (9), is

$$\delta = \frac{720 \times 12}{30 \times 10^6}\left(7,550 + \frac{2,450}{2}\right) = 2.53 \text{ in.}$$

2. Determine the elongation of a conical bar under the action of its own weight (Fig. 14) if the length of the bar is l, the diameter of the base is d and the weight per unit volume of the material is γ.

Solution. The weight of the bar is

$$Q = \frac{\pi d^2}{4}\frac{l\gamma}{3}.$$

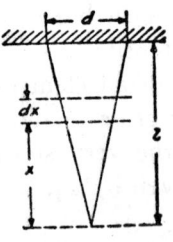

FIG. 14.

For any cross section at distance x from the lower end of the bar the tensile force, equal to the weight of the lower portion of the bar, is

$$\frac{Qx^3}{l^3} = \frac{\pi d^2}{4}\frac{\gamma x^3}{3l^2}.$$

Assuming that the tensile force is uniformly distributed over the cross section [7] and considering the element of length dx as a pris-

---

[7] Such an assumption is justifiable when the angle of the cone is small.

matic bar, the elongation of this element will be

$$d\delta = \frac{\gamma x}{3E} dx,$$

and the total elongation of the bar is

$$\delta = \frac{\gamma}{3E} \int_0^l x\,dx = \frac{\gamma l^2}{6E}.$$

This elongation is one third that of a prismatic bar of the same length (see eq. 9).

3. The vertical prismatic rod of a mine pump is moved up and down by a crankshaft (Fig. 15). Assuming that the material is steel and the working stress is $\sigma_W = 7{,}000$ lb per sq in., determine the cross-sectional area of the rod if the resistance of the piston during motion downward is 200 lb and during motion upward is 2,000 lb. The length of the rod is 320 ft. Determine the necessary length of the radius r of the crank if the stroke of the pump is equal to 8 in.

Fig. 15.

Solution. The necessary cross-sectional area of the rod may be found from eq. (8) by substituting $P = 2{,}000$ lb. Then

$$A = \frac{2{,}000}{7{,}000 - \dfrac{490\cdot320\cdot12}{12^3}} = 0.338 \text{ in.}^2$$

The difference in total elongation of the rod when it moves up and when it moves down is due to the resistance of the piston and will be equal to

$$\Delta\delta = \frac{(2{,}000 + 200)\cdot320\cdot12}{30\cdot10^6 \times 0.338} = 0.833 \text{ in.}$$

The radius of the crank should be

$$r = \frac{8 + 0.833}{2} = 4.42 \text{ in.}$$

4. Lengths of steel and aluminum wire are suspended vertically. Determine for each the length at which the stress due to the weight of the wire equals the ultimate strength if for steel wire $\sigma_U = 300{,}000$ lb per sq in. and $\gamma = 490$ lb per cu ft, and for aluminum wire $\sigma_U = 50{,}000$ lb per sq in. and $\gamma = 170$ lb per cu ft.

Answer. For steel $l = 88{,}200$ ft, for aluminum $l = 42{,}300$ ft.

5. In what proportion will the maximum stress produced in a prismatic bar by its own weight increase if all the dimensions of the bar are increased in the proportion $n:1$ (Fig. 1)?

Answer. The stress will increase in the ratio $n:1$.

6. A pillar consisting of two prismatic portions of equal length (Fig. 16) is centrally loaded at the upper end by a compressive force $P = 600,000$ lb. Determine the volume of masonry if the height of the pillar is 120 ft, its weight per cu ft is 100 lb, and the maximum compressive stress in each portion is 150 lb per sq in. Compare this volume with that of a single prismatic pillar designed for the same condition.

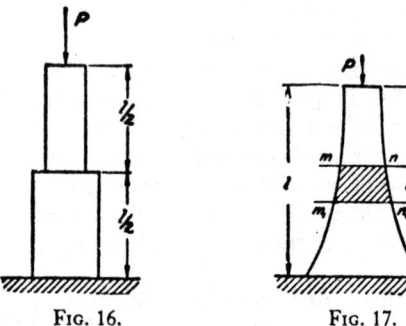

FIG. 16.   FIG. 17.

Solution. The cross-sectional area of the upper portion of the pillar, from eq. (8), is

$$A = \frac{600,000}{150 - \dfrac{100 \times 60 \times 12}{1,728}} = 5,540 \text{ sq in.} = 38.5 \text{ sq ft.}$$

For the lower portion,

$$A_1 = \frac{150 \times 5,540}{150 - \dfrac{100 \times 60 \times 12}{1,728}} = 7,680 \text{ sq in.} = 53.3 \text{ sq ft.}$$

Total volume of masonry $V = (38.5 + 53.3)60 = 5,500$ cu ft. For a prismatic pillar,

$$A = \frac{600,000}{150 - \dfrac{100 \times 120 \times 12}{1,728}} = 9,000 \text{ sq in.} = 62.5 \text{ sq ft,}$$

$$V = 62.5 \times 120 = 7,500 \text{ cu ft.}$$

7. Solve the preceding problem assuming three prismatic portions of equal length.

Answer. $A = 34.1$ sq ft, $A_1 = 41.8$ sq ft, $A_2 = 51.3$ sq ft, $V = 5,090$ cu ft.

8. Determine the form of the pillar in Fig. 17 such that the stress in each cross section is just equal to σ_W. The form satisfying this condition is called the *form of equal strength*.

Solution. Considering a differential element, shaded in the figure, it is evident that the compressive force on the cross section m_1n_1 is larger than that on the cross section mn by the magnitude of the weight of the element. Thus since the stress in both cross sections is to be the same and equal to σ_W, the difference dA in the cross-sectional area must be such as to compensate for the difference in the compressive force. Hence

$$dA\sigma_W = \gamma A dx, \qquad (a)$$

where the right side of the equation represents the weight of the element. Dividing this equation by $A\sigma_W$ and integrating, we find

$$\int \frac{dA}{A} = \int \frac{\gamma dx}{\sigma_W},$$

from which

$$\log_e A = \frac{\gamma x}{\sigma_W} + C_1$$

and

$$A = Ce^{\gamma x/\sigma_W}, \qquad (b)$$

where e is the base of natural logarithms and $C = e^{C_1}$. At $x = 0$ this equation gives for the cross-sectional area at the top of the pillar

$$(A)_{x=0} = C.$$

But the cross-sectional area at the top is equal to P/σ_W; hence $C = P/\sigma_W$ and eq. (b) becomes

$$A = \frac{P}{\sigma_W} e^{\gamma x/\sigma_W}. \qquad (c)$$

The cross-sectional area at the bottom of the pillar is obtained by substituting $x = l$ in eq. (c), which gives

$$A_{\max} = \frac{P}{\sigma_W} e^{\gamma l/\sigma_W}. \qquad (d)$$

9. Find the volume of the masonry for a pillar of equal strength designed to meet the conditions of Prob. 6.

Solution. By using eq. (*d*) the difference of the cross-sectional areas at the bottom of the pillar and at its top is found to be

$$\frac{P}{\sigma_W} e^{\gamma l/\sigma_W} - \frac{P}{\sigma_W} = \frac{P}{\sigma_W} (e^{\gamma l/\sigma_W} - 1).$$

This difference multiplied by the working stress σ_W evidently gives the weight of the pillar; its volume is thus

$$V = \frac{P}{\gamma} (e^{\gamma l/\sigma_W} - 1) = 4{,}440 \text{ cu ft.}$$

6. Statically Indeterminate Problems in Tension and Compression.—There are cases in which the axial forces acting in the bars of a structure cannot be determined from the equations of statics alone and the deformation of the structure must be taken into consideration. Such structures are called *statically indeterminate systems.*

A simple example of such a system is shown in Fig. 18. The load P produces extension in the bars OB, OC and OD, which are in the same plane. The conditions for equilibrium of the hinge O give two equations of statics which are not sufficient to determine the three unknown tensile forces in the bars, and for a third equation a consideration of the deformation of the system becomes necessary. Let us assume, for simplicity, that the system is symmetrical with respect to the vertical axis OC, that the vertical bar is of steel with A_s and E_s as the cross-sectional area and the modulus of elasticity for the material, and that the inclined bars are of copper with A_c and E_c as area and modulus. The length of the vertical bar is l and that of the inclined bars is $l/\cos \alpha$. Denoting by X the tensile force in the vertical bar and by Y the forces in the inclined bars, the only equation of equilibrium for the hinge O in this case of symmetry will be

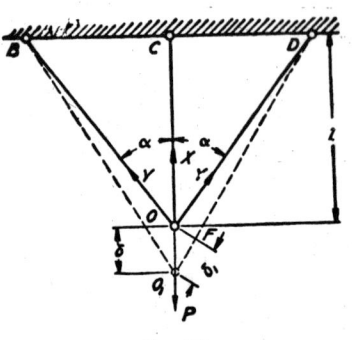

Fig. 18.

$$X + 2Y \cos \alpha = P. \tag{a}$$

In order to derive the second equation necessary for determining the unknown quantities X and Y, the deformed configuration of the system indicated in the figure by dotted lines must be considered. Let δ be the total elongation of the vertical bar under the action of the load P; then the elongation δ_1 of the inclined bars will be found from the triangle OFO_1. Assuming that these elongations are very small, the circular arc OF from the center D may be replaced by a perpendicular line and the angle at O_1 may be taken equal to the initial angle α. Then

$$\delta_1 = \delta \cos \alpha.$$

The unit elongations and the stresses for the vertical and the inclined bars will be

$$\epsilon_s = \frac{\delta}{l}, \quad \sigma_s = \frac{E_s \delta}{l} \quad \text{and} \quad \epsilon_c = \frac{\delta \cos^2 \alpha}{l}, \quad \sigma_c = \frac{E_c \delta \cos^2 \alpha}{l},$$

respectively. Then the forces in the bars will be obtained by multiplying the stresses by the cross-sectional areas as follows:

$$X = \sigma_s A_s = \frac{A_s E_s \delta}{l}, \quad Y = \sigma_c A_c = \frac{A_c E_c \delta \cos^2 \alpha}{l}, \quad (b)$$

from which

$$Y = X \cos^2 \alpha \cdot \frac{A_c E_c}{A_s E_s}.$$

Substituting in eq. (a), we obtain

$$X = \frac{P}{1 + 2 \cos^3 \alpha \dfrac{A_c E_c}{A_s E_s}}. \quad (10)$$

It is seen that the force X depends not only upon the angle of inclination α but also upon the cross-sectional areas and the mechanical properties of the materials of the bars. In the particular case in which all bars have the same cross section and the same modulus we obtain, from eq. (10),

$$X = \frac{P}{1 + 2 \cos^3 \alpha}.$$

When α approaches zero, cos α approaches unity, and the force in the vertical bar approaches $\frac{1}{3}P$. When α approaches 90°, the inclined bars become very long and the complete load will be taken by the middle bar.

As another example of a statically indeterminate system let us consider a prismatic bar with built-in ends, loaded axially at an intermediate cross section mn (Fig. 19). The load P will be in equilibrium with the reactions R and R_1 at the ends and we have

$$P = R + R_1. \tag{c}$$

In order to derive the second equation for determining the forces R and R_1 the deformation of the bar must be considered. The load P with the force R produces shortening of the lower portion of the bar and with the force R_1 elongation of the upper portion. The total shortening of one part must be equal to the total elongation of the other. Then, by using eq. (1), we obtain

$$\frac{R_1 a}{AE} = \frac{Rb}{AE}.$$

Hence

$$\frac{R}{R_1} = \frac{a}{b}. \tag{d}$$

Fig. 19.

i.e., the forces R and R_1 are inversely proportional to the distances of their points of application from the loaded cross section mn. Now from eqs. (c) and (d) the magnitudes of these forces and the stresses in the bar may be readily calculated.

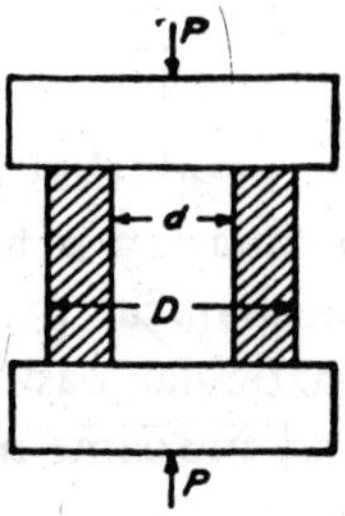

Problems

1. A steel cylinder of diameter d and a copper tube of outer diameter D are compressed between the plates of a press (Fig. 20). Determine the stresses in steel and copper and also the unit compression if $P = 100,000$ lb, $d = 4$ in. and $D = 8$ in.

Solution. Here again static conditions are inadequate, and the deformation of cylinder and tube must be considered in order to obtain the load carried by each material. The unit shortening in the steel and in the copper must

Fig. 20.

be equal; therefore the stresses in each material will be in the same ratio as their moduli (see eq. 4), i.e., the compressive stress in the steel will be $\frac{15}{8}$ the compressive stress in the copper. Then the magnitude of the stress σ_c in the copper may be found from the equation of statics,

$$P = \frac{\pi d^2}{4}\frac{15}{8}\sigma_c + \frac{\pi}{4}(D^2 - d^2)\sigma_c.$$

Substituting numerical values, we obtain

$$\sigma_c = 1{,}630 \text{ lb per sq in.}, \qquad \sigma_s = \tfrac{15}{8}\cdot\sigma_c = 3{,}060 \text{ lb per sq in.},$$

and unit compression

$$\epsilon = \frac{\sigma_c}{E_c} = 102 \times 10^{-6}.$$

2. A square column of reinforced concrete is compressed by an axial force $P = 60{,}000$ lb. What part of this load will be taken by the concrete and what part by the steel if the cross-sectional area of the steel is $\frac{1}{10}$ of the cross-sectional area of the concrete? Assume that the steel bars are symmetrically located with respect to the axis of the column.

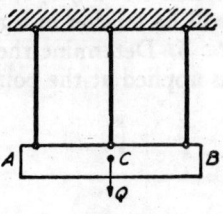

Answer. The load taken by the steel is $\frac{3}{4}$ of the load taken by the concrete.

Fig. 21.

3. A rigid body AB of weight Q hangs on three vertical wires symmetrically situated with respect to the center of gravity C of the body (Fig. 21). Determine the tensile forces in the wires if the middle wire is of steel and the two others of copper. Cross-sectional areas of all wires are equal.

Suggestion. Use method of Prob. 1.

4. Determine the forces in the four legs of a square table, Fig. 22, produced by the load P acting at point A. The top of the table and the floor are assumed absolutely rigid and the legs are attached to the floor so that they can undergo tension as well as compression.

Solution. Assuming that the new position of the top of the table is that indicated by the dotted line mn, the compression of legs 2 and 4 will be the average of that of legs 1 and 3. Hence

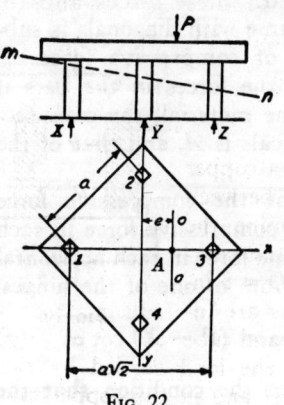

Fig. 22.

$$2Y = X + Z,$$

and since $2Y + X + Z = P$, we obtain

$$2Y = X + Z = \tfrac{1}{2}P. \qquad (e)$$

An additional equation for determining X and Z is obtained by taking the moment of all the forces with respect to the horizontal axis O–O parallel to y and passing through A. Then

$$X(\tfrac{1}{2}a\sqrt{2} + \text{·}) + \tfrac{1}{2}P \cdot e = Z(\tfrac{1}{2}a\sqrt{2} - e).$$

From (e) and (f) we obtain

$$X = P\left(\frac{1}{4} - \frac{e}{a\sqrt{2}}\right), \qquad Y = \frac{P}{4}, \qquad Z = P\left(\frac{1}{4} + \frac{e}{a\sqrt{2}}\right).$$

When $e > a\sqrt{2}/4$, X becomes negative. This indicates that there will be tension in leg 1.

5. Determine the forces in the legs of the above table if the load is applied at the point with the coordinates

$$x = \frac{a}{4}, \qquad y = \frac{a}{5}.$$

Hint. In solving this problem it should be noted that when the point of application of the load P is not on the diagonal of the table, this load may be replaced by two loads statically equivalent to the load P and applied at points on the two diagonals. The forces produced in the legs by each of these two loads are found as explained

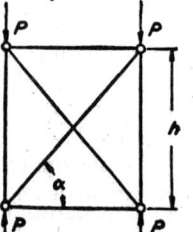

above. Adding the effects of the two component loads, the forces in the legs for any position of the load P may be found.

6. A rectangular frame with diagonals is submitted to the action of compressive forces P (Fig. 23). Determine the forces in the bars if they are all of the same material, the cross-sectional area of the verticals is A, and that of the remaining bars A_1.

Solution. Let X be the compressive force in each vertical, Y the compressive force in each diagonal and Z the tensile force in each horizontal

Fig. 23.

bar. Then from the condition of equilibrium of one of the hinges,

$$Y = (P - X)/\sin \alpha, \qquad Z = Y \cos \alpha = (P - X) \cot \alpha. \qquad (g)$$

The third equation will be obtained from the condition that the frame after deformation remains rectangular by virtue of symmetry; therefore

$$(a^2 + h^2)\left(1 - \frac{Y}{A_1E}\right)^2 = h^2\left(1 - \frac{X}{AE}\right)^2 + a^2\left(1 + \frac{Z}{A_1E}\right)^2.$$

From this, neglecting the small quantities of higher order, we get

$$\frac{(a^2 + h^2)Y}{A_1E} = \frac{h^2X}{AE} - \frac{a^2Z}{A_1E}. \qquad (h)$$

Solving eqs. (g) and (h), the following value of the force in a diagonal is obtained:

$$Y = \frac{P}{\dfrac{a^2 + h^2}{h^2} \cdot \dfrac{A}{A_1} + \dfrac{a^2}{h^2} \cdot \dfrac{A}{A_1} \cos \alpha + \sin \alpha}.$$

The forces in the other bars may now be easily determined from eqs. (g).

7. Solve the above problem, assuming $a = h$, $A = 5A_1$ and $P = 50,000$ lb.

8. What stresses will be produced in a steel bolt and a copper tube (Fig. 24) by $\frac{1}{4}$ of a turn of the nut if the length of the bolt $l = 30$ in., the pitch of the bolt thread $h = \frac{1}{8}$ in., the area of the cross section of the bolt $A_s = 1$ sq in., the area of the cross section of the tube $A_c = 2$ sq in.?

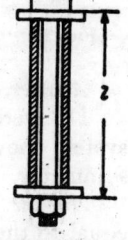

FIG. 24.

Solution. Let X denote the unknown tensile force in the bolt and the compressive force in the tube. The magnitude of X will be found from the condition that the extension of the bolt plus the shortening of the tube is equal to the displacement of the nut along the bolt. In our case, assuming the length of the tube equal to the length of the bolt, we obtain

$$\frac{Xl}{A_sE_s} + \frac{Xl}{A_cE_c} = \frac{1}{4}h,$$

from which

$$X = \frac{hA_sE_s}{4l\left(1 + \dfrac{A_sE_s}{A_cE_c}\right)} = \frac{30 \times 10^6}{32 \times 30\left(1 + \dfrac{15}{16}\right)} = 16,100 \text{ lb.}$$

The tensile stress in the bolt is $\sigma_s = X/A_s = 16,100$ lb per sq in. The compressive stress in the tube is $\sigma_c = X/A_c = 8,050$ lb per sq in.

9. What change in the stresses calculated in the above problem will be produced by tensile forces $P = 5{,}000$ lb applied to the ends of the bolt?

Solution. Let X denote the increase in the tensile force in the bolt and Y the decrease in the compressive force in the tube. Then from the condition of equilibrium,

$$X + Y = P. \qquad (i)$$

A second equation is obtained from the consideration that the unit elongations of the bolt and tube under the application of the forces P must be equal, i.e.,

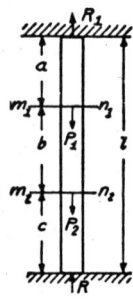

FIG. 25.

$$\frac{X}{A_s E_s} = \frac{Y}{A_c E_c}. \qquad (j)$$

From eqs. (i) and (j) the forces X and Y and the corresponding stresses are easily calculated.

10. A prismatic bar with built-in ends is loaded axially at two intermediate cross sections (Fig. 25) by forces P_1 and P_2. Determine the reactions R and R_1.

Hint. Use eq. (d) on p. 22, calculating the reactions produced by each load separately and then summing these reactions. Determine the reactions when $a = 0.3l$, $b = 0.3l$ and $P_1 = 2P_2 = 1{,}000$ lb.

Answer. $R = 600$ lb, $R_1 = 900$ lb.

11. Determine the forces in the bars of the system shown in Fig. 26, where OA is an axis of symmetry.

Answer. The tensile force in the bar OB is equal to the compressive force in the bar OC and is $P/2 \sin \alpha$. The force in the horizontal bar OA is equal to zero.

12. Solve Prob. 10 assuming that the lower portion of length c of the bar has a cross-sectional area two times larger than the cross-sectional area of the two upper parts of lengths a and b.

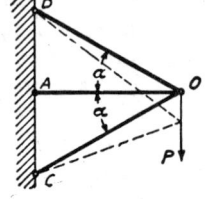

FIG. 26.

Answer. $R = \dfrac{2aP_1 + 2P_2(l - c)}{2l - c}$, $R_1 = \dfrac{P_1(2b + c) + cP_2}{2l - c}$.

7. **Assembly and Thermal Stresses.**—In a statically indeterminate system it is possible to have initial stresses produced in the bars during assembly. These stresses may be due to unavoidable inaccuracies in the lengths of the bars or to intentional deviations from the correct lengths, and are called

assembly stresses. Such stresses will exist even when external loads are absent, and depend only upon the geometric proportions of the system, on the mechanical properties of the materials and on the magnitude of the inaccuracies. Assume, for example, that the system represented in Fig. 18 has, by mistake, $l + a$ as the length of the vertical bar instead of l. Then after assembling the bars BO and DO, the vertical bar can be put into place only after being initially compressed, and consequently some tensile force will be produced in the inclined bars. Let X denote the compressive force which exists after assembly in the vertical bar. Then the corresponding tensile force in the inclined bars will be $X/2 \cos \alpha$ and the displacement of the hinge O due to the extension of these bars will be (see eq. b, p. 21)

$$\delta = \frac{Xl}{2A_c E_c \cos^3 \alpha}. \qquad (a)$$

The shortening of the vertical bar will be

$$\delta_1 = \frac{Xl}{A_s E_s}. \qquad (b)$$

From elementary geometrical considerations, the displacement of the hinge O, together with the shortening of the vertical bar, must be equal to the error a in the length of the vertical bar. This gives the following equation for determining X:

$$\frac{Xl}{2A_c E_c \cos^3 \alpha} + \frac{Xl}{A_s E_s} = a.$$

Hence

$$X = \frac{a A_s E_s}{l \left(1 + \dfrac{A_s E_s}{2A_c E_c \cos^3 \alpha} \right)}. \qquad (11)$$

The initial stresses in all the bars may now be calculated.

Expansion of the bars of a system due to changes in temperature may have the same effect as inaccuracies in lengths. Consider a bar with built-in ends. If the temperature of the bar is raised from t_0 to t and thermal expansion is prevented by the reactions at the ends, there will be produced in the bar compressive stresses, whose magnitude may be calculated from the condition that the length remains unchanged. Let α denote the coefficient of thermal expansion and σ the compressive stress produced by the reactions. Then the equation for determining σ will be

$$\alpha(t - t_0) = \frac{\sigma}{E},$$

from which

$$\sigma = E\alpha(t - t_0). \tag{12}$$

As a second example, let us consider the system represented in Fig. 18 and assume that the vertical bar is heated from the assembly temperature t_0 to a new temperature t. The corresponding thermal expansion will be partially prevented by the two other bars of the system, and certain compressive stresses will develop in the vertical bar and tensile stresses in the inclined bars. The magnitude of the compressive force in the vertical bar will be given by eq. (11), in which instead of the magnitude a of the inaccuracy in length we substitute the thermal expansion $\alpha l(t - t_0)$ of the vertical bar.

Problems

1. The rails of a tramway are welded together at 50° F. What stresses will be produced in these rails when heated by the sun to 100° if the coefficient of thermal expansion of steel is $70 \cdot 10^{-7}$?

Answer. $\sigma = 10,500$ lb per sq in.

2. What change of stresses will be produced in the case represented in Fig. 24 by increasing the temperature from $t_0°$ to $t°$ if the coefficient of expansion of steel is α_s and that of copper α_c?

Solution. Since $\alpha_c > \alpha_s$ the increase of temperature produces compression in the copper and tension in the steel. The unit elongations of the copper and the steel must be equal. Denoting by X the increase in the tensile force in the bolt due to the change of temperature, we obtain

$$\alpha_s(t - t_0) + \frac{X}{A_s E_s} = \alpha_c(t - t_0) - \frac{X}{A_c E_c},$$

from which

$$X = \frac{(\alpha_c - \alpha_s)(t - t_0) A_s E_s}{1 + \dfrac{A_s E_s}{A_c E_c}}.$$

The change in the stresses in the bolt and in the tube may now be calculated in the usual way.

3. A strip of copper is soldered between two strips of steel (Fig. 27). What stresses will be produced in the steel and in the copper by a rise in the temperature of the strips from t_0 to t degrees?

Suggestion. The same method as in the previous problem should be used.

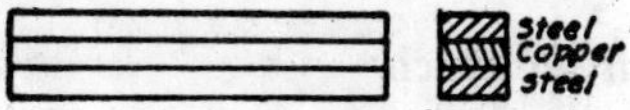

Fig. 27.

4. What stresses will be produced in the bars of the system represented in Fig. 18 if the temperature of all the bars is raised from t_0 to t?

Solution. Let X denote the tensile force produced in the steel bar by an increase in temperature. Then from the condition of equilibrium of the hinge O it can be seen that in the copper bars compressive forces act, equal to $X/2 \cos \alpha$; consequently the elongation of the steel bar becomes

$$\delta = \alpha_s(t - t_0)l + \frac{Xl}{A_s E_s}$$

and the elongation of the copper bars is

$$\delta_1 = \alpha_c(t - t_0)\frac{l}{\cos \alpha} - \frac{Xl}{2 A_c E_c \cos^2 \alpha}.$$

Furthermore from previous considerations (see p. 21),

$$\delta_1 = \delta \cos \alpha.$$

Therefore

$$\alpha_s(t - t_0)l + \frac{Xl}{A_s E_s} = \alpha_c(t - t_0)\frac{l}{\cos^2 \alpha} - \frac{Xl}{2 A_c E_c \cos^3 \alpha},$$

from which

$$X = \frac{(t - t_0)\left(\dfrac{\alpha_c}{\cos^2 \alpha} - \alpha_s\right) A_s E_s}{1 + \dfrac{1}{2 \cos^3 \alpha}\dfrac{A_s E_s}{A_c E_c}}.$$

The stresses in the steel and in the copper will now be obtained from the following equations:

$$\sigma_s = \frac{X}{A_s}, \qquad \sigma_c = \frac{X}{2 A_c \cos \alpha}.$$

5. Assuming that in the case shown in Fig. 20 a constant load $P = 100{,}000$ lb is applied at an initial temperature t_0, determine at

what increase in temperature the load will be completely transmitted to the copper if $\alpha_s = 70 \times 10^{-7}$ and $\alpha_c = 92 \times 10^{-7}$.

Solution.

$$(\alpha_c - \alpha_s)(t - t_0) = \frac{4P}{\pi(D^2 - d^2)E_c},$$

from which

$$t - t_0 = 75.4° \, F.$$

6. A steel bar consisting of two portions of lengths l_1 and l_2 and cross-sectional areas A_1 and A_2 is fixed at the ends. Find the thermal stresses if the temperature rises by $100° F$. Assume $l_1 = l_2$, $A_1 = 2A_2$ and $\alpha_s = 70 \times 10^{-7}$.

Answer. $\sigma_1 = 14,000$ lb per sq in., $\sigma_2 = 28,000$ lb per sq in.

7. Find the thermal stresses in the system shown in Fig. 27 if the temperature of all three strips rises by $100° F$. The thickness of each of the three strips is the same and the coefficients of thermal expansion are $\alpha_s = 70 \times 10^{-7}$ and $\alpha_c = 92 \times 10^{-7}$. Assume $E_c : E_s = 8 : 15$.

Answer. $\sigma_c = 2,780$ lb per sq in. compression, $\sigma_s = 1,390$ lb per sq in. tension.

8. The temperature of the system shown in Fig. 18 rises by $100° F$. Find the thermal stresses if all three bars are of steel and have equal cross-sectional areas. Take $\alpha_s = 70 \times 10^{-7}$ and $E_s = 30 \times 10^6$ lb per sq in.

Answer. Vertical bar, $\sigma = \dfrac{42,000 \cos \alpha \sin^2 \alpha}{2 \cos^3 \alpha + 1}$ lb per sq in. tension; inclined bars, $\sigma = \dfrac{21,000 \sin^2 \alpha}{2 \cos^3 \alpha + 1}$ lb per sq in. compression.

9. Find the stresses in the wires of the system shown in Fig. 21 if the cross-sectional area of the wires is 0.1 sq in., the load $Q = 4,000$ lb, and the temperature of the system rises after assembly by $10° F$. Assume $\alpha_c = 92 \times 10^{-7}$, $\alpha_s = 70 \times 10^{-7}$, $E_c = 16 \times 10^6$ lb per sq in., $E_s = 30 \times 10^6$ lb per sq in.

Answer. $\sigma_s = 19,700$ lb per sq in., $\sigma_c = 10,200$ lb per sq in.

10. Determine the stresses which will be built up in the system represented in Fig. 23 if the temperature of the upper horizontal bar rises from t_0 to t degrees.

Answer. The compressive force X in the upper horizontal bar is given by the equation:

$$\alpha(t - t_0)a = X\left(\frac{2h \tan^2 \alpha}{AE} + \frac{2h}{A_1 E \cos^2 \alpha \sin \alpha} + \frac{2a}{A_1 E}\right).$$

8. Extension of a Circular Ring.—If uniformly distributed radial forces act along the circumference of a thin circular ring (Fig. 28), uniform enlargement of the ring will be produced.

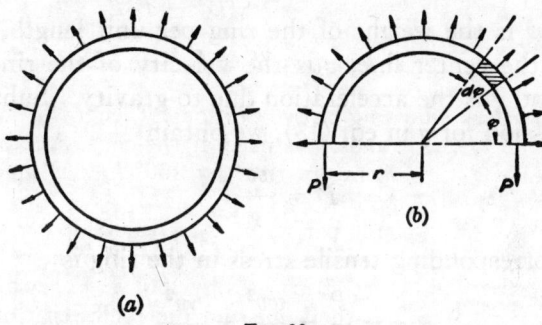

(a)

(b)

FIG. 28.

In order to determine the tensile force P in the ring let us imagine that the ring is cut at the horizontal diametral section (Fig. 28b) and consider the upper portion as a free body. If q denotes the uniform load per unit length of the center line of the ring and r is the radius of the center line, the force acting on an element of the ring cut out by two adjacent cross sections will be $qr\,d\varphi$, where $d\varphi$ is the central angle, corresponding to the element. Taking the sum of the vertical components of all the forces acting on the half ring, the following equation of equilibrium is obtained:

$$2P = 2\int_0^{\pi/2} qr \sin \varphi d\varphi = 2qr,$$

from which

$$P = qr. \tag{13}$$

The tensile stress in the ring may now be obtained by dividing the force P by the cross-sectional area of the ring.[8]

In practical applications the determination of tensile stresses in a rotating ring is frequently necessary. Then q represents

---

[8] It will be shown later (see Part II) that in the case of thin rings it is justifiable to assume that the stresses are uniformly distributed over the cross section of the ring.

the centrifugal force per unit length of the ring and is given by
the equation:

$$q = \frac{w\,v^2}{g\,r},\tag{14}$$

in which w is the weight of the ring per unit length, r is the
radius of the center line, v is the velocity of the ring at the
radius r and g is the acceleration due to gravity. Substituting
this expression for q in eq. (13), we obtain

$$P = \frac{wv^2}{g},$$

and the corresponding tensile stress in the ring is

$$\sigma = \frac{P}{A} = \frac{wv^2}{Ag} = \frac{\gamma v^2}{g}.\tag{15}$$

It is seen that the stress is proportional to the density γ/g of
the material and to the square of the peripheral velocity.[9] For
a steel ring and for the velocity $v = 100$ ft per sec this stress
becomes 1,060 lb per sq in. Then for the same material and
for any other velocity v_1 the stress will be $0.106v_1{}^2$ lb per sq
in., where v_1 is in ft per sec.

Problems

1. Determine the maximum tensile stress in the cylindrical wall
of the press shown in Fig. 5 if the inner diameter is 10 in. and the
thickness of the wall is 1 in.

Solution. The maximum hydrostatic pressure p in the cylinder
will be found from the equation:

$$p \cdot \frac{\pi 10^2}{4} = 100,000 \text{ lb,}$$

from which $p = 1,270$ lb per sq in. Cutting out from the cylinder
an elemental ring of width 1 in. in the direction of the axis of the
cylinder and using eq. (13) in which, for this case, $q = p = 1,270$ lb
per in. and $r = 5$ in., we obtain

$$\sigma = \frac{P}{A} = \frac{1,270 \times 5}{1 \times 1} = 6,350 \text{ lb per sq in.}$$

[9] For a thin ring the velocity at the center line may be taken equal to the
peripheral velocity.

2. A copper tube is fitted over a steel tube at a high temperature *t* (Fig. 29), the fit being such that no pressure exists between tubes at this temperature. Determine the stresses which will be produced in the copper and in the steel when cooled to room temperature t_0 if the outer diameter of the steel tube is *d*, the thickness of the steel tube is h_s and that of the copper tube is h_c.

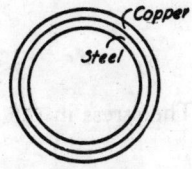

FIG. 29.

Solution. Due to the difference in the coefficients of expansion α_c and α_s there will be a pressure between the outer and the inner tubes after cooling. Let *x* denote the pressure per sq in.; then the tensile stress in the copper tube will be

$$\sigma_c = \frac{xd}{2h_c}$$

and the compressive stress in the steel will be

$$\sigma_s = \frac{xd}{2h_s}.$$

The pressure *x* will now be found from the condition that during cooling both tubes have the same circumferential contraction; hence

$$\alpha_c(t - t_0) - \frac{xd}{2E_c h_c} = \alpha_s(t - t_0) + \frac{xd}{2E_s h_s},$$

from which

$$\sigma_c = \frac{xd}{2h_c} = \frac{(\alpha_c - \alpha_s)(t - t_0)E_c}{1 + \dfrac{h_c}{h_s}\dfrac{E_c}{E_s}}.$$

In the same manner the stress in the steel may be calculated.

3. Referring to Fig. 29, what additional tensile stress in the tube will be produced by submitting it to an inner hydrostatic pressure $p = 100$ lb per sq in. if the inner diameter $d_1 = 4$ in., $h_s = 0.1$ in. and $h_c = \frac{1.5}{8} \times 0.1$ in.?

Solution. Cutting out of the tube an elemental ring of width 1 in., the complete tensile force in the ring will be

$$P = \frac{pd_1}{2} = 200 \text{ lb.}$$

Due to the fact that the unit circumferential elongations in copper and in steel are the same, the stresses will be in proportion to the moduli, i.e., the stress in the copper will be $\frac{8}{15}$ that in the steel. At the same time the cross-sectional area of the copper is $\frac{1.5}{8}$ that of the steel; hence the force P will be equally distributed between two metals and the tensile stress in the copper produced by the hydrostatic pressure will be

$$\sigma_c = \frac{P}{2 \times h_c} = \frac{200}{2 \times \frac{15}{8} \times 0.1} = 533 \text{ lb per sq in.}$$

The stress in the steel will be

$$\sigma_s = \tfrac{15}{8}\sigma_c = 1,000 \text{ lb per sq in.}$$

4. A built-up ring consists of an inner copper ring and an outer steel ring. The inner diameter of the steel ring is smaller than the outer diameter of the copper ring by the amount δ and the structure is assembled after preliminary heating of the steel ring. When cooled the steel ring produces pressure on the copper ring (shrink fit pressure). Determine the stresses in the steel and the copper after assembly if both rings have rectangular cross sections with the dimensions h_s and h_c in the radial direction and dimensions equal to unity in the direction perpendicular to the plane of the ring. The dimensions h_s and h_c may be considered small as compared with the diameter d of the surface of contact of the two rings.

Solution. Let x be the uniformly distributed pressure per sq in. of the surface of contact of the rings; then the compressive stress in the copper and the tensile stress in the steel will be found from the equations:

$$\sigma_c = \frac{xd}{2h_c}, \qquad \sigma_s = \frac{xd}{2h_s}. \tag{a}$$

The decrease in the outer diameter of the copper ring will be

$$\delta_1 = \frac{\sigma_c}{E_c} \cdot d = \frac{xd^2}{2h_cE_c}.$$

The increase of the inner diameter of the steel ring will be

$$\delta_2 = \frac{\sigma_s}{E_s} \cdot d = \frac{xd^2}{2h_sE_s}.$$

The unknown pressure x will be found from the equation:

$$\delta_1 + \delta_2 = \frac{xd^2}{2}\left(\frac{1}{h_cE_c} + \frac{1}{h_sE_s}\right) = \delta,$$

from which

$$x = \frac{2\delta h_sE_s}{d^2\left(1 + \dfrac{h_sE_s}{h_cE_c}\right)}.$$

Now the stresses σ_s and σ_c, from eqs. (a), will be

$$\sigma_c = \frac{\delta}{d} \cdot \frac{h_s}{h_c} \cdot \frac{E_s}{1 + \dfrac{h_s E_s}{h_c E_c}}, \qquad \sigma_s = \frac{\delta}{d} \cdot \frac{E_s}{1 + \dfrac{h_s E_s}{h_c E_c}}.$$

5. Determine the stresses which will be produced in the built-up ring of the preceding problem by rotation of the ring at constant speed n rpm.

Solution. Due to the fact that copper has a greater density and a smaller modulus of elasticity than steel, the copper ring will press on the steel ring during rotation. Let x denote the pressure per sq in. of contact surface between the two rings. Then the corresponding stresses will be given by eqs. (a) of the preceding problem. In addition to these stresses the stresses produced by centrifugal forces should be taken into consideration. Denoting by γ_s and γ_c the weights per unit volume of steel and copper and using eq. (15), we obtain

$$\sigma_s = \frac{\gamma_s}{g}\left(\frac{2\pi n}{60}\right)^2 \left(\frac{d + h_s}{2}\right)^2, \qquad \sigma_c = \frac{\gamma_c}{g}\left(\frac{2\pi n}{60}\right)^2 \left(\frac{d - h_c}{2}\right)^2.$$

Combining these stresses with the stresses due to pressure x and noting that the unit elongation for both rings should be the same, the following equation for determining x will be obtained:

$$\frac{1}{E_s}\left[\frac{\gamma_s}{g}\left(\frac{2\pi n}{60}\right)^2 \left(\frac{d + h_s}{2}\right)^2 + \frac{xd}{2h_s}\right]$$

$$= \frac{1}{E_c}\left[\frac{\gamma_c}{g}\left(\frac{2\pi n}{60}\right)^2 \left(\frac{d - h_c}{2}\right)^2 - \frac{xd}{2h_c}\right],$$

from which x may be calculated for each particular case. Knowing x, the complete stress in the copper and the steel may be found without difficulty.

6. Determine the limiting peripheral speed of a thin copper ring if the working stress is $\sigma_W = 3{,}000$ lb per sq in. and $\gamma_c = 550$ lb per cu ft.

Answer. $v = 159$ ft per sec.

7. Referring to Prob. 2 and Fig. 29, determine the stress in the copper at room temperature if $t - t_0 = 100°$ F, $\alpha_c - \alpha_s = 22 \times 10^{-7}$ and $h_s = h_c$.

Answer. $\sigma_c = 2{,}300$ lb per sq in.

8. Referring to Prob. 5, determine the number of revolutions n per minute at which the stress in the copper ring becomes equal

to zero if the initial assembly stress in the copper ring was a compression equal to σ_0, and $h_c = h_s$, $E_s = 2E_c$.

Answer. The number of revolutions n will be determined from
the equation:

$$3\sigma_0 = \left(\frac{2\pi n}{60}\right)^2 \left[\frac{\gamma_c}{g}\left(\frac{d - h_c}{2}\right)^2 + \frac{\gamma_s}{g}\left(\frac{d + h_s}{2}\right)^2\right].$$

9. Find the stresses in the built-up ring of Prob. 4, assuming
$\delta = 0.001$ in., $d = 4$ in., $h_s = h_c$ and $E_s/E_c = \frac{15}{8}$. Find the changes
of these stresses if the temperature of the rings increases after assem-
bly by 10° F. Take $\alpha_c = 92 \times 10^{-7}$ and $\alpha_s = 70 \times 10^{-7}$.

Answer. $\sigma_c = \sigma_s = 2,610$ lb per sq in. Change in stresses
$= 230$ lb per sq in.

10. Referring to Prob. 5, find the stresses in steel and in copper
if $n = 3,000$ rpm, $d = 2$ ft, $h_s = h_c = \frac{1}{2}$ in., $\gamma_s = 490$ lb per cu ft
and $\gamma_c = 550$ lb per cu ft.

CHAPTER II

ANALYSIS OF STRESS AND STRAIN

9. Stress on Inclined Planes for Simple Tension and Compression.—In discussing stresses in a prismatic bar submitted to an axial tension P we have previously considered (Art. 2) only the stress over cross sections perpendicular to the axis of the bar. We now take up the case in which the cross section pq (Fig. 30a), perpendicular to the plane of the figure, is inclined to the axis. Since all longitudinal fibers have the same elongation (see p. 4) the forces representing the action of the right portion of the bar on the left portion are uniformly distributed over the cross section pq. The left por-

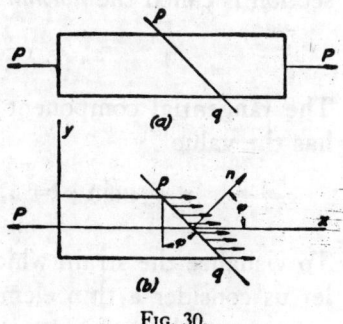

Fig. 30.

tion of the bar, isolated in Fig. 30b, is in equilibrium under the action of these forces and the external force P applied at the left end. Hence the resultant of the forces distributed over the cross section pq is equal to P. Denoting by A the area of the cross section normal to the axis of the bar and by φ the angle between the x axis and the normal n to the cross section pq, the cross-sectional area of pq will be $A/\cos \varphi$ and the stress s over this cross section is

$$s = \frac{P \cos \varphi}{A} = \sigma_x \cos \varphi, \tag{16}$$

where $\sigma_x = P/A$ denotes the stress on the cross section normal to the axis of the bar. It is seen that the stress s over any inclined cross section of the bar is smaller than the stress σ_x over the cross section normal to the axis of the bar and that it diminishes as the angle φ increases. For $\varphi = \pi/2$ the sec-

37

tion pq is parallel to the axis of the bar and the stress s becomes zero, which indicates that there is no pressure between the longitudinal fibers of the bar.

The stress s, defined by eq. (16), has the direction of the

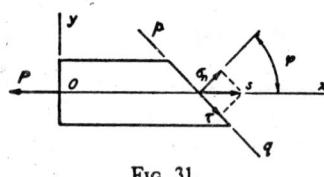

FIG. 31.

force P and is not perpendicular to the cross section pq. In such cases it is usual to resolve the total stress into two components, as is shown in Fig. 31. The stress component σ_n perpendicular to the cross section is called the *normal stress*. Its magnitude is

$$\sigma_n = s \cos \varphi = \sigma_x \cos^2 \varphi. \tag{17}$$

The tangential component τ is called the *shearing stress* and has the value

$$\tau = s \sin \varphi = \sigma_x \cos \varphi \sin \varphi = \frac{\sigma_x}{2} \sin 2\varphi. \tag{18}$$

To visualize the strain which each component stress produces, let us consider a thin element cut out of the bar by two adjacent parallel sections pq and p_1q_1, Fig. 32a. The stresses acting on this element are shown in Fig. 32a. Figs. 32b and 32c are obtained by resolving these stresses into normal and tangential components as explained above and showing separately the action of each of these components. It is seen that the *normal stresses σ_n*

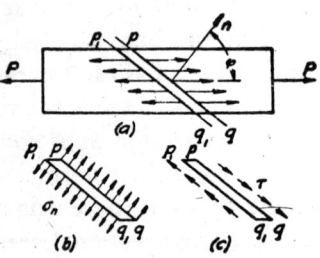

FIG. 32.

produce extension of the element in the direction of the normal n to the cross section pq and the shearing stresses produce sliding of section pq with respect to p_1q_1.

From eq. (17) it is seen that the *maximum normal stress* acts on cross sections normal to the axis of the bar and we have

$$(\sigma_n)_{\max} = \sigma_x.$$

The *maximum shearing stress*, as seen from eq. (18), acts on cross sections inclined at $45°$ to the axis of the bar, where $\sin 2\varphi = 1$, and has the magnitude

$$\tau_{\max} = \tfrac{1}{2}\sigma_x. \tag{19}$$

Although the maximum shearing stress is one-half the maximum normal stress, this stress is sometimes the controlling factor when considering the ultimate strength of materials which are much weaker in shear than in tension. For example, in a tensile test of a bar of mild steel with a polished surface, visible yielding of the metal occurs along inclined lines, called *Lueders' lines*, Fig. 33. Yielding occurs along the inclined

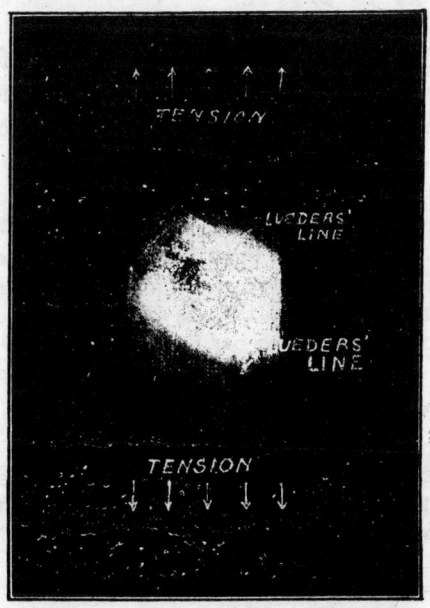

FIG. 33.

planes on which the shearing stress is a maximum and at the value of the force P which corresponds to the point B in Fig. 4a. This indicates that in the case of mild steel failure is produced by the maximum shearing stress although this stress is equal to only one-half of the maximum normal stress.

Formulas (17) and (18), derived for a bar in tension, can be used also in the case of compression. Tensile stress is assumed positive and compressive negative. Hence for a bar under axial compression we have only to take σ_x with a negative sign in formulas (17) and (18). The negative sign of σ_n indicates that in Fig. 32b we obtain, instead of tension, a compressive action on the thin element between the adjacent cross sections pq and p_1q_1. The negative sign for τ in formula (18) indicates that for

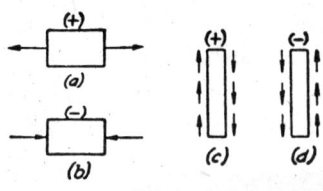

FIG. 34.

compression of the bar the shearing action on the element has the direction opposite to that shown in Fig. 32c. Fig. 34 illustrates the sign convention for normal and shearing stresses which will be used. Positive sign for shear is taken when the shear stresses form a couple in the clockwise direction and negative sign for the opposite direction.

Problems

1. Show that the couples in Figs. 32b and 32c balance each other.

2. A prismatic bar of cross-sectional area A is subjected to axial compression by a force $P = 10,000$ lb. Find σ_n and τ for a plane inclined at 45° to the axis of the bar.

Answer. $\sigma_n = \pm\tau = -\dfrac{5,000}{A}$ lb per sq in.

3. Find the change in distance between the planes pq and p_1q_1 in Fig. 32a produced by forces $P = 30,000$ lb if the initial distance between those planes is 0.5 in., the cross-sectional area $A = 1$ sq in. and $\varphi = 45°$.

Answer. $\delta = 0.00025$ in.

4. Find the angle φ (Fig. 32a) defining the plane pq for which (1) normal stress σ_n is one-half of the maximum stress σ_x, (2) shearing stress τ is one-third of σ_n.

Answer.   (1) $\varphi = \pm45°$; (2) $\varphi = \arctan \frac{1}{3}$.

10. Mohr's Circle.

10. Mohr's Circle.—Formulas (17) and (18) can be represented graphically.[1] We take an orthogonal system of coordi-

---

[1] This graphical representation is due to O. Mohr, *Civilingenieur*, p. 113, 1882. See also his *Abhandlungen aus dem Gebiete der technischen Mechanik*, Berlin, p. 219, 1906. In this book, references to other publications on the same subject are given.

nates with the origin at O and with the positive direction of the axes as shown in Fig. 35. Beginning with the cross section pq perpendicular to the axis of the bar we have for this case $\varphi = 0$, in Fig. 31, and we find, from formulas (17) and (18) $\sigma_n = \sigma_x, \tau = 0$. Selecting a scale for stresses and measuring normal components along the horizontal axis and shearing components along the vertical axis, the stress

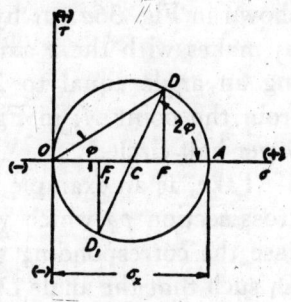

Fig. 35.

acting on the plane with $\varphi = 0$ is represented in Fig. 35 by a point A having the abscissa equal to σ_x and the ordinate equal to zero. Taking now a plane parallel to the axis of the bar we have $\varphi = \pi/2$, and observing that both stress components vanish for such a plane we conclude that the origin O, in Fig. 35, corresponds to this plane. Now constructing a circle on $\overline{OA}$ as diameter it can readily be proved that the stress components for any cross section pq with an arbitrarily chosen angle φ, Fig. 31, will be represented by the coordinates of a point on that circle. To obtain the point on the circle corresponding to a definite angle φ, it is only necessary to measure in the counter-clockwise direction from the point A the arc subtending an angle equal to 2φ. Let D be the point obtained in this manner; then, from the figure,

$$\overline{OF} = \overline{OC} + \overline{CF} = \frac{\sigma_x}{2} + \frac{\sigma_x}{2} \cos 2\varphi = \sigma_x \cos^2 \varphi,$$

$$\overline{DF} = \overline{CD} \sin 2\varphi = \frac{\sigma_x}{2} \sin 2\varphi.$$

Comparing these expressions for the coordinates of point D with expressions (17) and (18) it is seen that this point defines the stresses acting on the plane pq, Fig. 31. As the section pq rotates in the counter-clockwise direction about an axis perpendicular to the plane of Fig. 31, φ varying from 0 to $\pi/2$, the point D moves from A to O, so that the upper half-circle determines the stresses for all values of φ within these limits. If the angle φ is larger than $\pi/2$ we obtain a cross section as shown in Fig. 36a cut by a plane mm whose external normal [2] n_1 makes with the x axis an angle larger than $\pi/2$. Measuring an angle equal to 2φ in the counter-clockwise direction from the point A, in Fig. 35, we obtain now a point on the lower half-circle.

Take, as an example, the case when mm is perpendicular to cross section pq which was previously considered. In such a case the corresponding point on the circle in Fig. 35 is point D_1 such that the angle DOD_1 is equal to π; thus DD_1 is a diameter of the circle. Using the coordinates of point D_1, we find the stress components σ_{n_1} and τ_1 for the plane mm:

$$\sigma_{n_1} = \overline{OF_1} = \overline{OC} - \overline{F_1C} = \frac{\sigma_x}{2} - \frac{\sigma_x}{2}\cos 2\varphi = \sigma_x \sin^2 \varphi, \quad (20)$$

$$\tau_1 = -\overline{F_1D_1} = -\overline{CD_1}\sin 2\varphi = -\frac{\sigma_x}{2}\sin 2\varphi. \quad (21)$$

Comparing these results with expressions (17) and (18), we find

$$\sigma_n + \sigma_{n_1} = \sigma_x \cos^2 \varphi + \sigma_x \sin^2 \varphi = \sigma_x, \quad (22)$$

$$\tau_1 = -\tau. \quad (23)$$

This indicates that the sum of the normal stresses acting on two perpendicular planes remains constant and equal to σ_x. The shear stresses acting on two perpendicular planes are numerically equal but of opposite sign.

By taking the adjacent cross sections m_1m_1 and p_1q_1 parallel to mm and pq, Fig. 36a, an element is isolated as shown in Fig. 36b, and the directions of the stresses acting on this element are indicated. It is seen that the shearing stresses acting on the sides of the element parallel to the pq plane produce a couple in the clockwise direction, which, according to the accepted rule defined in Fig. 34c, must be considered positive. The shearing stresses acting on the other two sides of the ele-

---

[2] The portion of the bar on which the stresses act is indicated by shading. The external normal n_1 is directed outward from that portion.

[2] The minus sign is taken since point D_1 is on the side of negative ordinates.

ment produce a couple in the counter-clockwise direction which, according to the rule defined in Fig. 34*d*, is negative.

The circle in Fig. 35, called the *circle of stress* or *Mohr's circle*, is used to determine the stress components σ_n and τ for a cross section *pq* whose normal makes any angle φ with the *x* axis, Fig. 31. A similar construction can be used to solve the inverse problem, i.e., when the components σ_n and τ are given

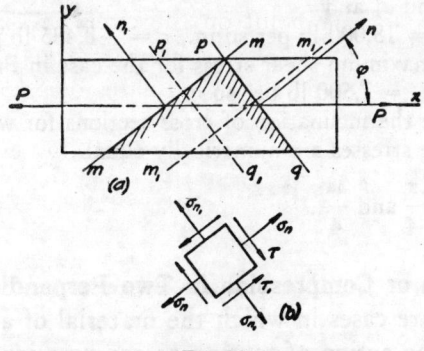

Fig. 36.

and it is required to find the tensile stress σ_x in the axial direction and the angle φ. We observe that the angle between the chord *OD* and the *x* axis is equal to φ, Fig. 35. Hence, after constructing the point *D* with coordinates σ_n and τ, we obtain φ by drawing the line *OD*. Knowing the angle φ, the radius *DC* making the angle 2φ with the axis *OC* can be drawn and the center *C* of the circle of stress obtained. The diameter of this circle gives the required stress σ_x.

Problems

1. Determine σ_n and τ analytically and graphically if $\sigma_x = 15,000$ lb per sq in. and $\varphi = 30°$ or $\varphi = 120°$. By using the angles $30°$ and $120°$ isolate an element as shown in Fig. 36*b* and show by arrows the directions of the stresses acting on the element.

2. Solve the preceding problem assuming that instead of tensile stress σ_x there acts compressive stress of the same amount. Observe that in this case the diameter of the circle, Fig. 35, must lie on the negative side of the abscissa.

3. On a plane pq, Fig. 31, there acts a normal stress $\sigma_n = 12,000$ lb per sq in. and a shearing stress $\tau = 4,000$ lb per sq in. Find the angle φ and the stress σ_x.

Answer. $\tan \varphi = \frac{1}{3}$, $\sigma_x = \dfrac{\sigma_n}{\cos^2 \varphi} = 13,330$ lb per sq in.

4. Acting on the two perpendicular sides of the element in Fig. 36b are the normal stresses $\sigma_n = 12,000$ lb per sq in. and $\sigma_{n_1} = 6,000$ lb per sq in. Find σ_x and τ.

Answer. $\sigma_x = 18,000$ lb per sq in., $\tau = \pm 8,485$ lb per sq in.

5. Find the maximum shear stress for the case in Prob. 1.

Answer. $\tau_{max} = 7,500$ lb per sq in.

6. Determine the inclination of cross sections for which the normal and shearing stresses are numerically equal.

Answer. $\varphi = \dfrac{\pi}{4}$ and $\dfrac{3\pi}{4}$.

11. Tension or Compression in Two Perpendicular Directions.—There are cases in which the material of a structure is submitted to the action of tension or compression in two perpendicular directions. As an example of such a stress condition let us consider stresses in the cylindrical wall of a boiler submitted to internal pressure p lb per sq in.[4] Let us cut out a small element from the cylindrical wall of the boiler by two adjacent axial sections and by two circumferential sections, Fig. 37a. Because of the internal pressure the cylinder will

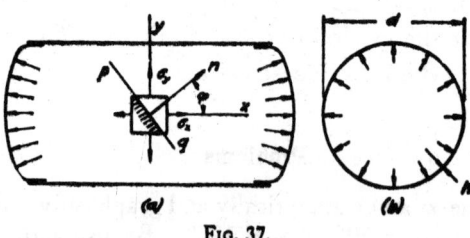

Fig. 37.

expand both in the circumferential and in the axial directions. The tensile stress σ_y in the circumferential direction will be determined in the same manner as in the case of a circular ring

---

[4] More accurately p denotes the difference between the internal pressure and the external atmospheric pressure.

(Art. 8). Denoting the inner diameter of the boiler by d and its wall thickness by h, this stress is

$$\sigma_y = \frac{pd}{2h}. \tag{24}$$

In calculating the tensile stress σ_x in the axial direction we imagine the boiler cut by a plane perpendicular to the x axis. Considering the equilibrium of one portion of the boiler it is seen that the tensile force producing longitudinal extension of the boiler is equal to the resultant of the pressure on the ends of the boiler, i.e., equal to

$$P = p\left(\frac{\pi d^2}{4}\right).$$

The cross-sectional area of the wall of the boiler is [5]

$$A = \pi dh.$$

Hence

$$\sigma_x = \frac{P}{A} = \frac{pd}{4h}. \tag{25}$$

It is seen that the element of the wall undergoes tensile stresses σ_x and σ_y in two perpendicular directions.[6] In this case the tensile stress σ_y in the circumferential direction is twice as large as the stress σ_x in the axial direction.

For the general case, we consider now the stress over any cross section pq, Fig. 37a, perpendicular to the xy plane and whose normal n makes an angle φ with the x axis. By using formulas (17) and (18) of the previous article, we conclude that the tensile stress σ_x acting in the axial direction produces on the plane pq normal and shearing stresses of magnitude

$$\sigma_n{}' = \sigma_x \cos^2 \varphi, \qquad \tau' = \tfrac{1}{2}\sigma_x \sin 2\varphi. \tag{a}$$

To calculate the stress components produced on the same plane pq by the tensile stress σ_y, we observe that the angle

---

[5] The thickness of the wall is assumed small in comparison with the diameter and the approximate formula for the cross-sectional area is used.

[6] There is also a pressure on the inner cylindrical surface of the element but this pressure is small in comparison with σ_x and σ_y and is neglected in further discussion.

between σ_y and the normal n, Fig. 37a, is $\frac{\pi}{2} - \varphi$ and is measured clockwise from the y axis, while φ is measured counter-
clockwise from the x axis. From this we conclude that in using
eqs. (17) and (18) we must substitute in this case σ_y for σ_x and
$- \left(\frac{\pi}{2} - \varphi \right)$ for φ. This gives

$$\sigma_n{}'' = \sigma_y \sin^2 \varphi, \qquad \tau'' = -\tfrac{1}{2}\sigma_y \sin 2\varphi. \qquad (b)$$

Summing up the stress components (a) and (b) produced by
stresses σ_x and σ_y, respectively, the resultant normal and shearing stresses on an inclined plane for the case of tension in two
perpendicular directions are obtained:

$$\sigma_n = \sigma_x \cos^2 \varphi + \sigma_y \sin^2 \varphi, \qquad (26)$$

$$\tau = \tfrac{1}{2}(\sigma_x - \sigma_y) \sin 2\varphi. \qquad (27)$$

12. Mohr's Circle for Combined Stresses.—Proceeding as
in Art. 10, a graphical representation of formulas (26) and (27)
can be readily obtained using Mohr's circle or the circle of
stress. Assuming again that the abscissas and the ordinates
represent to a certain scale the normal and the shearing stress
components, we conclude that the points A and B, in Fig. 38,

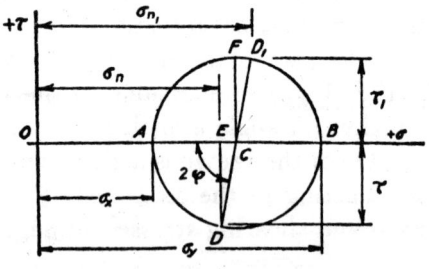

Fig. 38.

with abscissas equal to σ_x and σ_y, represent the stresses acting
on the sides of the element in Fig. 37a, perpendicular to the x
and y axes, respectively. To obtain the stress components on
any inclined plane, defined by an angle φ in Fig. 37a, we have
only to construct a circle on AB as a diameter and draw the

radius CD making the angle ACD, measured in the counter-clockwise direction from point A, equal to 2φ. From the figure we conclude that

$$\overline{OE} = \overline{OC} - \overline{CE} = \tfrac{1}{2}(\overline{OA} + \overline{OB}) - \tfrac{1}{2}(\overline{OB} - \overline{OA}) \cos 2\varphi$$

$$= \frac{\sigma_x + \sigma_y}{2} - \frac{\sigma_y - \sigma_x}{2} \cos 2\varphi = \sigma_x \cos^2 \varphi + \sigma_y \sin^2 \varphi.$$

This indicates that the abscissa $\overline{OE}$ of the point D on the circle, if measured to the assumed scale, gives the normal stress component σ_n, eq. (26).

The ordinate of the point D is

$$\overline{DE} = \overline{CD} \sin 2\varphi = \frac{\sigma_y - \sigma_x}{2} \sin 2\varphi.$$

Observing that this ordinate must be taken with negative sign, we conclude that the ordinate of the point D, taken with the proper sign, gives the shearing stress component τ, eq. (27).

When the plane pq rotates counter-clockwise with respect to an axis perpendicular to the xy plane, Fig. 37a, the corresponding point D moves in the counter-clockwise direction along the circle of stress in Fig. 38 so that for each value of φ the corresponding values of the components σ_n and τ are obtained as the coordinates of the point D.

From this graphical representation of formulas (26) and (27) it follows at once that the maximum normal stress component in the present case [7] is equal to σ_y and the maximum shearing stress represented by the radius $\overline{CF}$ of the circle in Fig. 38 is

$$\tau_{\max} = \frac{\sigma_y - \sigma_x}{2} \tag{28}$$

and occurs when $\sin 2\varphi = -1$ and $\varphi = 3\pi/4$. The same magnitude of shearing stress but with negative sign acts on the plane for which $\varphi = \pi/4$.

Taking two perpendicular planes defined by the angles φ and $\pi/2 + \varphi$, which the normals n and n_1 make with the x

---

[7] We consider only planes perpendicular to the xy plane. For a more general case, see Art. 18.

axis, the corresponding stress components are given by the co-ordinates of points D and D_1 in Fig. 38, and we conclude

$$\sigma_n + \sigma_{n_1} = \sigma_x + \sigma_y, \qquad (29)$$

$$\tau_1 = -\tau. \qquad (30)$$

This indicates that the sum of the normal stresses acting on two perpendicular planes remains constant as the angle φ varies. Shearing stresses acting on two perpendicular planes are numerically equal but of opposite sign.

The circle of stress, similar to that in Fig. 38, can be constructed also if one or both stresses σ_x and σ_y are compressive. It is only necessary to measure the compressive stresses on the negative side of the abscissa axis. Assuming, for example, that the stresses acting on an element are as shown in Fig. 39a, the corresponding circle is shown in Fig. 39b. The stress compo-

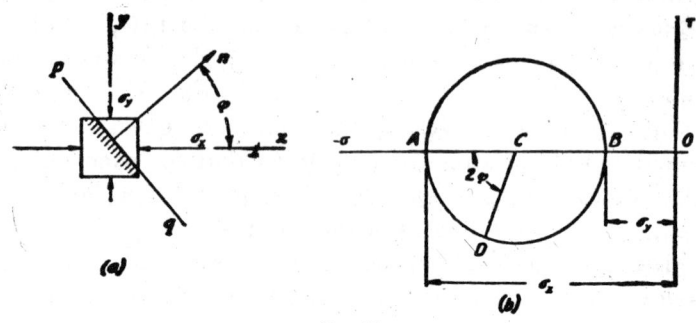

Fig. 39.

nents acting on a plane pq with normal n are given by the co-ordinates of the point D in the diagram.

Problems

1. The boiler shown in Fig. 37 has $d = 100$ in., $h = \frac{1}{2}$ in. Determine σ_x and σ_y if $p = 100$ lb per sq in. Isolate a small element by the planes for which $\varphi = 30°$ and $120°$ and show the magnitudes and the directions of the stress components acting on the lateral sides of that element.

2. Determine the stresses σ_n, σ_{n_1}, τ and τ_1 if, in Fig. 39a, $\sigma_x = 10,000$ lb per sq in., $\sigma_y = -5,000$ lb per sq in. and $\varphi = 30°$, $\varphi_1 = 120°$.

Answer. $\sigma_n = 6,250$ lb per sq in., $\sigma_{n_1} = -1,250$ lb per sq in., $\tau = -\tau_1 = 6,500$ lb per sq in.

3. Determine σ_n, σ_{n_1}, τ and τ_1 in the preceding problem, if the angle φ is chosen so that τ is a maximum.

Answer. $\sigma_n = \sigma_{n_1} = 2,500$ lb per sq in., $\tau = -\tau_1 = 7,500$ lb per sq in.

13. Principal Stresses.—It was shown in the preceding article that for tension or compression in two perpendicular directions x and y one of the two stresses σ_x or σ_y is the maximum and the other, the minimum normal stress. For all inclined planes, such as planes pq in Figs. 37a and 39a, the value of the normal stress σ_n lies between these limiting values. At the same time not only normal stresses σ_n, but also shearing stresses τ, act on all inclined planes. Stresses such as σ_x and σ_y, one of which is the maximum and the other the minimum normal stress, are called the *principal stresses* and the two perpendicular planes on which they act are called the *principal planes*. There are no shearing stresses acting on the principal planes.

In the example of the previous article, Fig. 37, the principal stresses σ_x and σ_y were found from very simple considerations and it was required to find the expressions for the normal and shearing stress components acting on any inclined plane, such as plane pq in Fig. 37a. In our further discussion (see p. 126) there will be cases of an inverse problem. It will be possible to determine the shear and normal stresses acting on two perpendicular planes, and it will be required to find the magnitudes and the directions of the principal stresses. The simplest way of solving this problem is by using the circle of stress as considered in Fig. 38. Assume that the stresses acting on an elementary rectangular parallelepiped *abcd* are as shown in Fig. 40a. The stresses σ_x and σ_y are not principal stresses, since not only normal but also shearing stresses act on the planes perpendicular to the x and y axes. To construct the circle of stress in this case, we first use the stress components σ_x, σ_y and τ and construct the points D and D_1 as shown in Fig. 40b. Since these two points represent the stresses acting on two perpendicular planes, the length DD_1 represents a diameter of the circle of stress. The intersection of this diameter with the x axis gives the center C of the circle, so that the circle can be readily constructed. Points A and B where the circle intersects the x axis define the magnitudes of the maximum and the minimum normal stresses, which are the principal stresses and

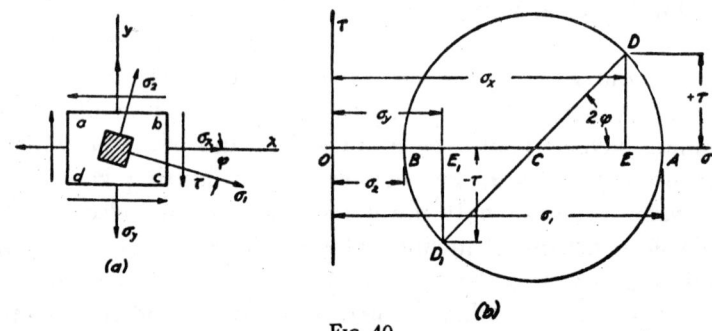

FIG. 40.

are denoted by σ_1 and σ_2. Using the circle, the formulas for calculating σ_1 and σ_2 can be easily obtained. From the figure we have

$$\sigma_1 = \overline{OA} = \overline{OC} + \overline{CD} = \frac{\sigma_x + \sigma_y}{2} + \sqrt{\left(\frac{\sigma_x - \sigma_y}{2}\right)^2 + \tau^2}, \quad (31)$$

$$\sigma_2 = \overline{OB} = \overline{OC} - \overline{CD} = \frac{\sigma_x + \sigma_y}{2} - \sqrt{\left(\frac{\sigma_x - \sigma_y}{2}\right)^2 + \tau^2}. \quad (32)$$

The directions of the principal stresses can also be obtained from the figure. We know that the angle DCA is the double angle between the stress σ_1 and the x axis and since 2φ is measured from D to A in the clockwise direction, the direction of σ_1 must be as indicated in Fig. 40a. If we isolate the element shaded in the figure with the sides normal and parallel to σ_1 there will be only normal stresses σ_1 and σ_2 acting on its sides. For the calculation of the numerical value of the angle φ we have, from the figure,

$$|\tan 2\varphi| = \frac{\overline{DE}}{\overline{CE}}.$$

Regarding the sign of the angle φ, it must be taken negative in this case since it is measured from the x axis in the clockwise direction, Fig. 40a. Hence

$$\tan 2\varphi = -\frac{\overline{DE}}{\overline{CE}} = -\frac{2\tau}{\sigma_x - \sigma_y}. \tag{33}$$

The maximum shearing stress is given by the magnitude of the radius of the circle of stress and we have

$$\tau_{max} = \frac{\sigma_1 - \sigma_2}{2} = \sqrt{\left(\frac{\sigma_x - \sigma_y}{2}\right)^2 + \tau^2} \tag{34}$$

Eqs. (31)–(34) completely solve the problem of the determination of the maximum normal and the maximum shearing stresses when the normal and shearing stresses acting on any two perpendicular planes are given.

Problems

1. An element, Fig. 40a, is submitted to the action of stresses $\sigma_x = 5,000$ lb per sq in., $\sigma_y = 3,000$ lb per sq in., $\tau = 1,000$ lb per sq in. Determine the magnitudes and the directions of principal stresses σ_1 and σ_2.

Solution. By using formulas (31) and (32) we obtain

$$\sigma_1 = \frac{5,000 + 3,000}{2} + \sqrt{\left(\frac{5,000 - 3,000}{2}\right)^2 + 1,000^2}$$

$$= 4,000 + 1,414 = 5,414 \text{ lb per sq in.,}$$

$$\sigma_2 = 4,000 - 1,414 = 2,586 \text{ lb per sq in.}$$

From formula (33) we have

$$\tan 2\varphi = -1, \quad 2\varphi = -45°, \quad \varphi = -22\tfrac{1}{2}°.$$

The minus sign indicates that φ is measured from the x axis in the clockwise direction as shown in Fig. 40a.

2. Determine the direction of the principal stresses in the preceding problem if $\sigma_x = -5,000$ lb per sq in.

Solution. The corresponding circle of stress is shown in Fig. 41

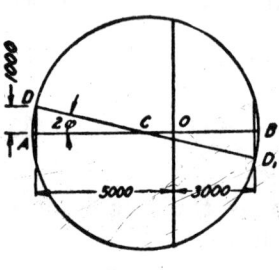

FIG. 41.

tan $2\varphi = \frac{1}{4}$, $2\varphi = 14°2'$. Hence the angle which the maximum compressive stress makes with the x axis is equal to $7°1'$ and is measured counter-clockwise from the x axis.

3. Find the circle of stress for the case of two equal tensions $\sigma_x = \sigma_y = \sigma$ and for two equal compressions $\sigma_x = \sigma_y = -\sigma$, $\tau = 0$ in both cases.

Answer. Circles become points on the horizontal axis with the abscissas σ and $-\sigma$, respectively.

4. On the sides of the element shown in Fig. 42a are acting the stresses $\sigma_x = -500$ lb per sq in., $\sigma_y = 1,500$ lb per sq in., $\tau = 1,000$ lb per sq in. Find, by using the circle of stress, the magnitudes of the normal and shearing stresses on (1) the principal planes, (2) the planes of maximum shearing stress.

Solution. The corresponding circle of stress is shown in Fig. 42b. The points D and D_1 represent stresses acting on the sides of the element in Fig. 42a perpendicular to the x and y axes. $\overline{OB}$ and $\overline{OA}$

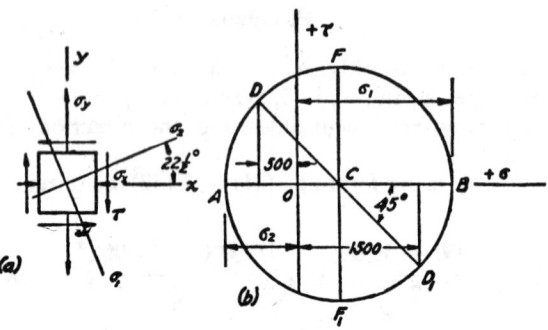

FIG. 42.

represent the principal stresses. Their magnitudes are $\sigma_1 = 1,914$ lb per sq in. and $\sigma_2 = -914$ lb per sq in., respectively. The direction of the maximum compressive stress σ_2 makes an angle of $22\frac{1}{2}°$ with the x axis, this angle being measured from the x axis in the counter-clockwise direction as shown in Fig. 42a. The points F and F_1 represent stresses acting on the planes subject to maximum shear. The magnitude of this shear is 1,414 lb per sq in. $\overline{OC}$ represents the normal stresses equal to 500 lb per sq in. acting on the same plane.

5. Solve the previous problem if $\sigma_x = -5,000$ lb per sq in., $\sigma_y = 3,000$ lb per sq in., $\tau = 1,000$ lb per sq in.

14. Lateral Contraction.—In Art. 2, the axial elongation of a bar in tension was discussed. Experiments show that such axial elongation is always accompanied by lateral contraction of the bar, and that within the elastic limit the ratio

$$\frac{\text{unit lateral contraction}}{\text{unit axial elongation}}$$

is constant for a given material. This constant will be called μ and is known as *Poisson's ratio*, after the name of the French mathematician who determined this ratio analytically by using the molecular theory of structure of the material. For materials which have the same elastic properties in all directions, called *isotropic materials*, Poisson found $\mu = \frac{1}{4}$. Experimental investigations of lateral contraction in structural metals [8] show that μ is usually close to the value calculated by Poisson. For instance, in the case of structural steel the value can be taken as $\mu = 0.30$. Knowing Poisson's ratio for a material, the change in volume of a bar in tension can be calculated. The length of the bar will increase in the ratio $(1 + \epsilon):1$. The lateral dimensions diminish in the ratio $(1 - \mu\epsilon):1$. Hence the cross-sectional area diminishes in the ratio $(1 - \mu\epsilon)^2:1$. Then the volume of the bar changes in the ratio $(1 + \epsilon)(1 - \mu\epsilon)^2:1$, which becomes $(1 + \epsilon - 2\mu\epsilon):1$ if we recall that ϵ is a small quantity and neglect its powers. Then the *unit volume expansion* is $\epsilon(1 - 2\mu)$. It is unlikely that any material diminishes its volume when in tension, hence μ must be less than 0.50. For such materials as rubber and paraffin μ approaches the above limit and the volume of these materials during extension remains approximately constant. On the other hand concrete has a small magnitude of μ ($\mu = \frac{1}{8}$ to $\frac{1}{12}$) and for cork μ can be taken equal to zero.

The above discussion of lateral contraction during tension can be applied with suitable changes to the case of compression. Longitudinal compression is accompanied by lateral expansion and for calculating this expansion the same value for μ is used as in the case of extension.

---

[8] These materials can be considered as isotropic (see Part II).

Problems

1. Determine the increase in unit volume of a bar in tension if $\sigma_W = 5{,}600$ lb per sq in., $\mu = 0.30$, $E = 30 \cdot 10^6$ lb per sq in.

Solution. Increase in unit volume is

$$\epsilon(1 - 2\mu) = \frac{\sigma_W}{E}(1 - 2\mu) = \frac{5{,}600}{30 \times 10^6}(1 - 0.6) = 74.7 \times 10^{-6}.$$

2. Determine the increase in volume of a bar due to a force P at the end and the weight of the bar (see Art. 5).

Answer. The increase in volume is equal to

$$\frac{Al(1 - 2\mu)}{E}\left(\frac{P}{A} + \frac{\gamma l}{2}\right).$$

3. A circular steel bar is subjected to an axial tensile force P of such a magnitude that the initial diameter of 5 in. is diminished by 0.001 in. Find P.

Answer. $P = 393{,}000$ lb.

4. The steel bar of the preceding problem is stretched by a force $P = 100{,}000$ lb. Find the decrease of the cross-sectional area.

Answer. 0.002 sq in.

15. Strain in the Case of Tension or Compression in Two Perpendicular Directions.—If a bar in the form of a rectangular parallelepiped is submitted to tensile forces acting in two perpendicular directions x and y (Fig. 37), the elongation in one of these directions will depend not only upon the tensile stress in this direction but also upon the stress in the perpendicular direction. The unit elongation in the direction of the x axis due to the tensile stress σ_x will be σ_x/E. The tensile stress σ_y will produce lateral contraction in the x direction equal to $\mu\sigma_y/E$. Then, if both stresses σ_x and σ_y act simultaneously, the unit elongation in the x direction will be

$$\epsilon_x = \frac{\sigma_x}{E} - \mu\frac{\sigma_y}{E}. \tag{35}$$

Similarly, for the y direction, we obtain

$$\epsilon_y = \frac{\sigma_y}{E} - \mu\frac{\sigma_x}{E}. \tag{36}$$

The contraction of the parallelepiped in the z direction will be

$$\epsilon_z = -\mu \frac{\sigma_x}{E} - \mu \frac{\sigma_y}{E} = -\frac{\mu}{E}(\sigma_x + \sigma_y).$$

In the particular case when the two tensions are equal, $\sigma_x = \sigma_y = \sigma$, we obtain

$$\epsilon_x = \epsilon_y = \frac{\sigma}{E}(1 - \mu). \tag{37}$$

From eqs. (35) and (36) the stresses σ_x and σ_y can be obtained as functions of the unit strains ϵ_x and ϵ_y as follows:

$$\sigma_x = \frac{(\epsilon_x + \mu\epsilon_y)E}{1 - \mu^2}, \qquad \sigma_y = \frac{(\epsilon_y + \mu\epsilon_x)E}{1 - \mu^2}. \tag{38}$$

If in the case shown in Fig. 37a the elongation ϵ_x in the axial direction and the elongation ϵ_y in the circumferential direction are measured by an extensometer, the corresponding tensile stresses σ_x and σ_y may be found from eqs. (38).

Problems

1. Determine the increase in the volume of the cylindrical steel boiler under internal pressure (Fig. 37), neglecting the deformation of the ends and taking $\sigma_y = 6,000$ lb per sq in.

Solution. By using eqs. (35) and (36),

$$\epsilon_x = \frac{3,000}{30 \times 10^6} - 0.3\frac{6,000}{30 \times 10^6} = \frac{1,200}{30 \times 10^6} = 4 \times 10^{-5}$$

$$\epsilon_y = \frac{6,000}{30 \times 10^6} - 0.3\frac{3,000}{30 \times 10^6} = \frac{5,100}{30 \times 10^6} = 17 \times 10^{-5}.$$

The volume of the boiler will increase in the ratio

$$(1 + \epsilon_x)(1 + \epsilon_y)^2:1 = (1 + \epsilon_x + 2\epsilon_y):1 = 1.00038:1.$$

2. A cube of concrete is compressed in two perpendicular directions by the arrangement shown in Fig. 43. Determine the decrease

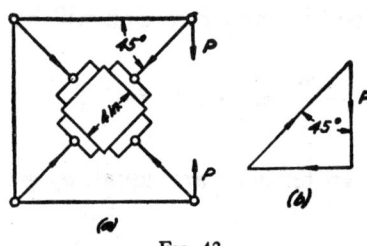

FIG. 43.

in the volume of the cube if it is 4 in. on a side, the compressive stress is uniformly distributed over the faces, $\mu = 0.1$ and $P = 20,000$ lb.

Solution. Neglecting friction in the hinges and considering the equilibrium of each hinge (Fig. 43b), it can be shown that the block is submitted to equal compression in two perpendicular directions and that the compressive force is equal to $P\sqrt{2} = 28,300$ lb. The corresponding strain, from eq. (37), is

$$\epsilon_x = \epsilon_y = -\frac{28,300}{16 \times 4 \times 10^6}(1 - 0.1) = -0.000398.$$

In the direction perpendicular to the plane of the figure a lateral expansion of the block takes place which is

$$\epsilon_z = -\mu\frac{\sigma_x}{E} - \mu\frac{\sigma_y}{E} = 0.2 \times \frac{28,300}{16 \times 4 \times 10^6} = 0.0000885.$$

The change per unit volume of the block will be

$$\epsilon_x + \epsilon_y + \epsilon_z = -2 \times 0.000398 + 0.0000885 = -0.000707.$$

3. Determine the increase in the cylindrical lateral surface of the boiler considered in Prob. 1 above.

Solution. Increase per unit area of lateral surface $= \epsilon_x + \epsilon_y = 21 \times 10^{-5}$.

4. Determine the unit elongation in the σ_1 direction of a bar of steel if the stress conditions are such as indicated in Prob. 1, p. 51.

Solution.

$$\epsilon_1 = \frac{1}{30 \times 10^6}(5,414 - 0.3 \times 2,586) = 154.6 \times 10^{-6}.$$

5. Under an axial tensile stress $\sigma_x = 30,000$ lb per sq in. a bar has unit elongation $\epsilon_x = 0.001$ and the ratio of the unit volume change to the unit change in cross-sectional area is $\frac{3}{4}$. Find E and μ.

Answer. $E = 30 \times 10^6$ lb per sq in., $\mu = \frac{2}{7}$.

6. A rectangular parallelepiped is subjected to tension in two perpendicular directions as shown in Fig. 44. Find the unit elongation ϵ in $\overline{OC}$ direction.

Solution. The coordinates of point C after deformation will be $a(1 + \epsilon_x)$ and $b(1 + \epsilon_y)$, and the length of $\overline{OC}$ after deformation will be

$$\sqrt{a^2(1 + 2\epsilon_x) + b^2(1 + 2\epsilon_y)}$$
$$\approx \sqrt{a^2 + b^2}\left(1 + \frac{a^2\epsilon_x}{a^2 + b^2} + \frac{b^2\epsilon_y}{a^2 + b^2}\right).$$

Subtracting from this the initial length $\sqrt{a^2 + b^2}$ and dividing by the initial length, we obtain

$$\epsilon = \epsilon_x \cos^2 \alpha + \epsilon_y \sin^2 \alpha.[9]$$

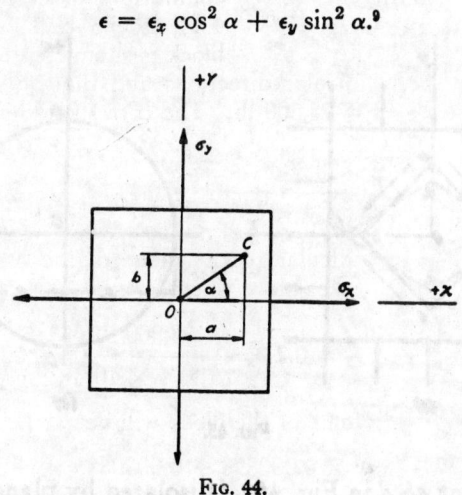

FIG. 44.

16. **Pure Shear.**—*Modulus in Shear.*—Let us consider the particular case of normal stresses acting in two perpendicular directions in which the tensile stress σ_x in the horizontal direction is numerically equal to the compressive stress σ_y in the vertical direction, Fig. 45a. The corresponding circle of stress is shown in Fig. 45b. Point D on this circle represents the stresses acting on the planes ab and cd perpendicular to the xy plane and inclined at 45° to the x axis. Point D_1 represents stresses acting on the planes ad and bc perpendicular to ab and cd. It is seen from the circle of stress that the normal stress on each of these planes is zero and that the shearing stress over these planes, represented by the radius of the circle, is numerically equal to the normal stress σ_x, so that

$$\tau = \sigma_x = -\sigma_y. \qquad (a)$$

---

[9] This equation is similar to eq. (26). Thus, a graphical representation of strain (strain circle), similar to Mohr's circle for stress, can be used.

If we imagine the element *abcd* to be isolated, it will be in equilibrium under the shearing stresses only, as shown in Fig. 45*a*. Such a state of stress is called *pure shear*. It may be concluded that pure shear is equivalent to the state of stress produced by tension in one direction and an equal compression in the perpendicular direction. If a square element, similar

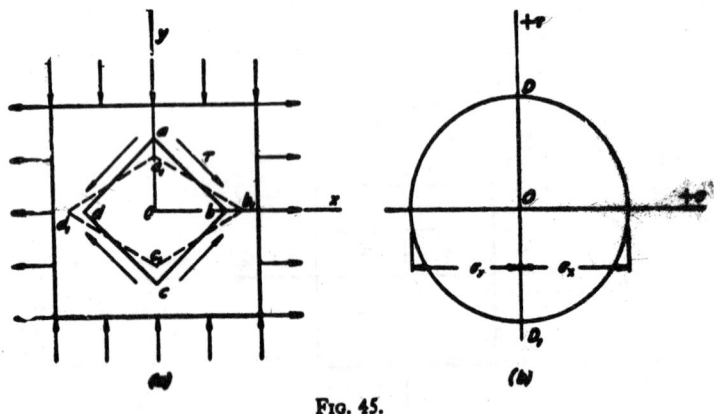

Fig. 45.

to the element *abcd* in Fig. 45*a*, is isolated by planes which are no longer at 45° to the *x* axis, normal stress as well as shearing stress will act on the sides of such an element. The magnitude of these stresses may be obtained from the circle of stress, Fig. 45*b*, in the usual way.

Let us consider now the deformation of the element *abcd*. Since there are no normal stresses acting on the sides of this element the lengths *ab*, *ad*, *bc* and *cd* will not change due to the deformation, but the horizontal diagonal *bd* will be stretched and the vertical diagonal *ac* will be shortened, changing the square *abcd* into a rhombus as indicated in the figure by dotted lines. The angle at *b*, which was $\pi/2$ before deformation, now becomes less than $\pi/2$, say $(\pi/2) - \gamma$, and at the same time the angle at *a* increases and becomes equal to $(\pi/2) + \gamma$. The small angle γ determines the distortion of the element *abcd*, and is called the *shearing strain*.

The shearing strain may also be visualized as follows: The element *abcd* of Fig. 45*a* is turned counter-clockwise through

45° and put into the position shown in Fig. 46. After the distortion produced by the shearing stresses τ, the same element takes the position indicated by the dotted lines. The shearing strain, represented by the magnitude of the small angle γ, may be taken equal to the ratio $\overline{aa_1}/\overline{ad}$, equal to the horizontal sliding aa_1 of the side ab with respect to the side dc divided by the distance between these two sides. If the material obeys Hooke's law, this sliding is proportional to the

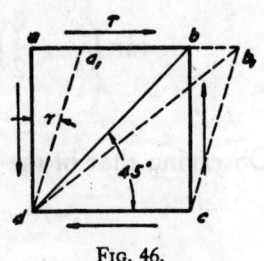

Fig. 46.

stress τ and we can express the relation between the shearing stress and the shearing strain by the equation

$$\gamma = \frac{\tau}{G}, \tag{39}$$

in which G is a constant depending on the mechanical properties of the material. Eq. (39) is analogous to eq. (4) which was established for simple tension, and the constant G is called the *modulus of elasticity in shear*, or *modulus of rigidity*.

Since the distortion of the element $abcd$, Fig. 46, is entirely defined by the elongation of the diagonal bd and the contraction of the diagonal ac, and since these deformations can be calculated by using the equations of the preceding article, it may be concluded that the modulus G can be expressed in terms of the modulus in tension E and Poisson's ratio μ. To establish this relationship we consider the triangle Oab, Fig. 45a. The elongation of the side Ob and the shortening of the side Oa of this triangle during deformation will be found by using eqs. (35) and (36). In terms of ϵ_x and ϵ_y we have

$$Ob_1 = Ob(1 + \epsilon_x), \qquad Oa_1 = Oa(1 + \epsilon_y),$$

and, from the triangle Oa_1b_1,

$$\tan(Ob_1a_1) = \tan\left(\frac{\pi}{4} - \frac{\gamma}{2}\right) = \frac{Oa_1}{Ob_1} = \frac{1 + \epsilon_y}{1 + \epsilon_x}. \tag{b}$$

For a small angle γ we have also

$$\tan\left(\frac{\pi}{4} - \frac{\gamma}{2}\right) = \frac{\tan\dfrac{\pi}{4} - \tan\dfrac{\gamma}{2}}{1 + \tan\dfrac{\pi}{4}\tan\dfrac{\gamma}{2}} \approx \frac{1 - \dfrac{\gamma}{2}}{1 + \dfrac{\gamma}{2}}. \qquad (c)$$

Observing that in the case of pure shear

$$\sigma_x = -\sigma_y = \tau,$$

$$\epsilon_x = -\epsilon_y = \frac{\sigma_x(1 + \mu)}{E} = \frac{\tau(1 + \mu)}{E},$$

and equating expressions (b) and (c), we obtain

$$\frac{1 - \dfrac{\tau(1 + \mu)}{E}}{1 + \dfrac{\tau(1 + \mu)}{E}} = \frac{1 - \dfrac{\gamma}{2}}{1 + \dfrac{\gamma}{2}},$$

from which

$$\frac{\gamma}{2} = \frac{\tau(1 + \mu)}{E}$$

or

$$\gamma = \frac{2\tau(1 + \mu)}{E}.$$

Comparing this result with formula (39), we conclude that

$$G = \frac{E}{2(1 + \mu)}. \qquad (40)$$

We see that the modulus of elasticity in shear can be easily calculated if the modulus in tension E and Poisson's ratio μ are known. In the case of steel, for example,

$$G = \frac{30 \cdot 10^6}{2(1 + 0.30)} = 11.5 \cdot 10^6 \text{ lb per sq in.}$$

It should be noted that the application of a uniform shearing stress to the sides of a block as assumed in Fig. 46 is very difficult to realize so the condition of pure shear is usually produced by the torsion of a circular tube, Fig. 47. Due to a small rotation of one end of the tube with respect to the other, the generators traced on the cylindrical surface become inclined to the axis of the cylinder and an element *abcd* formed by two generators and two adjacent circular cross sections undergoes a shearing strain similar to that shown in Fig. 46.

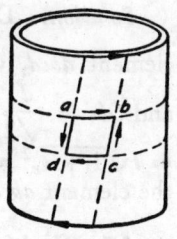

Fig. 47.

The problem of twist will be discussed later (see Chap. 10) where it will be shown how the shearing stress τ and the shearing strain γ of the element *abcd* can be calculated if the torque and the corresponding angle of twist of the shaft are measured. If τ and γ are found from such a torsion test, the value of the modulus G can be calculated from eq. (39). With this value of G, and knowing E from a tensile test, Poisson's ratio μ can be calculated from eq. (40). The direct determination of μ by measuring lateral contraction during a tensile test is more complicated since this contraction is very small and an extremely sensitive instrument is required to measure it with sufficient accuracy.

Problems

1. The block *abcd*, Fig. 46, is made of a material for which $E = 10 \cdot 10^6$ lb per sq in. and $\mu = 0.25$. Find γ and the unit elongation of the diagonal *bd* if $\tau = 10,000$ lb per sq in.
Answer. $\gamma = 0.0025$, $\epsilon = 0.00125$.

2. Find for the previous problem the sliding aa_1 of the side *ab* with respect to the side *cd* if the diagonal $bd = 2$ in.
Answer. $aa_1 = \gamma\sqrt{2}$.

3. Prove that the change in volume of the block *abcd* in Fig. 46 is zero if only the first powers of the strain components ϵ_x and ϵ_y are considered.

4. Prove that in the case of pure shear, Fig. 46, the unit elongation of the diagonal *bd* is equal to one-half of the shearing strain γ.

5. Find the unit elongation of the sides of the element *abcd* in Fig. 45a, taking into consideration small quantities of the second degree.

Solution. Denoting by l the initial length of the sides of the element *abcd*, we find that after deformation $Oa_1 = \dfrac{l}{\sqrt{2}}(1 - \epsilon_x)$, and $Ob_1 = \dfrac{l}{\sqrt{2}}(1 + \epsilon_x)$. Then $a_1b_1 = \dfrac{l}{\sqrt{2}}\sqrt{(1 - \epsilon_x)^2 + (1 + \epsilon_x)^2}$ $= l\sqrt{1 + \epsilon_x^2} \approx l(1 + \tfrac{1}{2}\epsilon_x^2)$. The unit elongation of the sides of the element *abcd* is $\tfrac{1}{2}\epsilon_x^2$.

17. Working Stress in Shear.—By submitting a material to pure shear (see Fig. 47) the relation between shearing stress and shearing strain can be established experimentally. Such a relationship is usually shown by a diagram, Fig. 48, in which the abscissa represents shearing strain and the ordinate represents shearing stress. The diagram is similar to that of a tensile test and we can mark on it the proportional limit A and the yield point B. Experiments show that for a

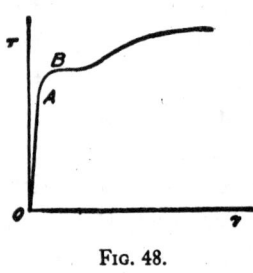

FIG. 48.

material such as structural steel the yield point in shear τ_{YP} is only about $0.55 - 0.60$ of σ_{YP}. Since at the yield point considerable distortion occurs without an appreciable change in stress, it is logical to take as the working stress in shear only a portion of the yield point stress, so that

$$\tau_W = \frac{\tau_{YP}}{n}, \tag{41}$$

where n is the factor of safety. Taking this factor of the same magnitude as in tension or compression, we obtain

$$\tau_W = 0.55 \text{ to } 0.60 \text{ of } \sigma_W,$$

which indicates that the working stress in shear should be taken much smaller than the working stress in tension.

It was already indicated that in practical applications we do not usually encounter a uniform distribution of shearing stress over the sides of a block as was assumed in Fig. 46 and that pure shear is realized in the case of torsion. We will see later that pure shear also occurs in the bending of beams. There are many practical problems in which a solution is obtained on the assumption that we are dealing with pure shear even though this assumption is only a rough approximation. Take, for example, the case of the joint in Fig. 49. It is evident

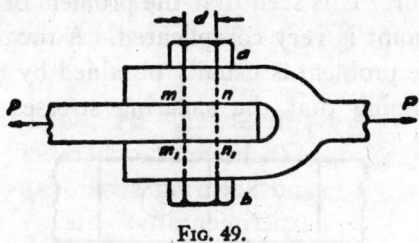

FIG. 49.

that if the diameter of the bolt ab is not large enough the joint may fail due to shear along the cross sections mn and m_1n_1. Although a more rigorous study of the problem indicates that the shearing stresses are not uniformly distributed over these cross sections and that the bolt undergoes not only shear but also bending under the action of the tensile forces P, a rough approximation for the required diameter of the bolt is obtained by assuming that along the planes mn and m_1n_1 we have a uniformly distributed shear stress τ which is obtained by dividing the force P by the sum of the cross-sectional areas mn and m_1n_1. Hence

$$\tau = \frac{2P}{\pi d^2},$$

and the required diameter of the bolt is obtained from the equation

$$\tau_W = \frac{2P}{\pi d^2}. \tag{42}$$

We have another example of such a simplified treatment of shear problems in the case of riveted joints, Fig. 50. Since the heads of the rivets are formed at high temperature the rivets produce after cooling considerable compression of the plates.[10] If tensile forces P are applied, relative motion between the plates is prevented by friction due to the pressure

---

[10] Experiments show that the tensile stress in rivets usually approaches the yield point of the material of which the rivets are made. See C. Bach. *Z. Ver. deut. Ing.*, 1912.

between the plates. Only after friction is overcome do the rivets begin to work in shear and if the diameter of the rivets is not sufficient failure due to shear along the planes mn and m_1n_1 may occur. It is seen that the problem of stress analysis for a riveted joint is very complicated. A rough approximate solution of the problem is usually obtained by neglecting friction and assuming that the shearing stresses are uniformly

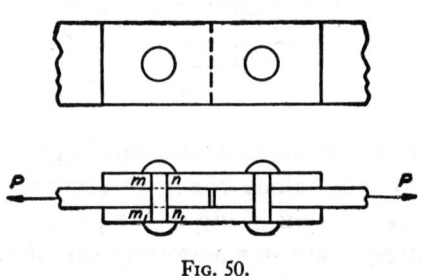

FIG. 50.

distributed along the cross section mn and m_1n_1. Then the correct diameter of the rivets is obtained by using eq. (42) as in the previous example.

Problems

1. Determine the diameter of the bolt in the joint shown in Fig. 49 if $F = 10,000$ lb and $\tau_w = 6,000$ lb per sq in.
Answer. $d = 1.03$ in.

2. Find the safe length $2l$ of the joint of two rectangular wooden bars, Fig. 51, submitted to tension, if $P = 10,000$ lb, $\tau_w = 100$ lb

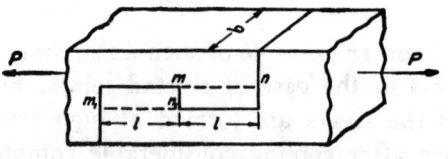

FIG. 51.

per sq in. for shear parallel to the fibers and $b = 10$ in. Determine the proper depth mn_1, if the safe limit for the local compressive stress along the fibers of the wood is 800 lb per sq in.
Answer. $2l = 20$ in., $mn_1 = 1.25$ in.

3. Find the diameter of the rivets in Fig. 50, if $\tau_W = 8,000$ lb per sq in. and $P = 8,000$ lb.

Answer. $d = 0.80$ in.

4. Determine the dimensions l and δ in the joint of two rectangular bars by steel plates, Fig. 52, if the forces, the dimensions and the working stresses are the same as in Prob. 2.

Answer. $l = 5$ in., $\delta = \frac{5}{8}$ in.

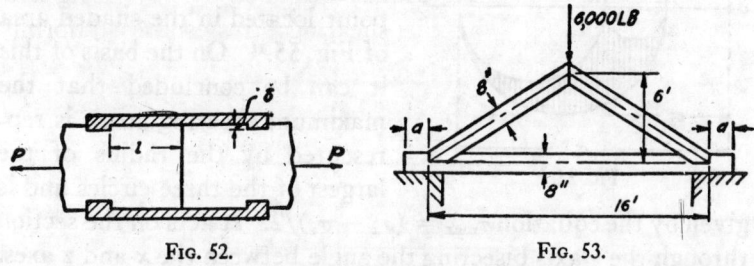

FIG. 52.   FIG. 53.

5. Determine the distance a which is required in the structure shown in Fig. 53, if the allowable shearing stress is the same as in Prob. 2 and the cross-sectional dimensions of all bars are 4 by 8 in. Neglect the effect of friction.

Answer. $a = 10$ in.

18. Tension or Compression in Three Perpendicular Directions.—If a bar in the form of a rectangular parallelepiped is submitted to the action of uniformly distributed forces P_x, P_y and P_z (Fig. 54), the normal stresses over cross sections perpendicular to the x, y and z axes are, respectively,

$$\sigma_x = \frac{P_x}{A_x}, \qquad \sigma_y = \frac{P_y}{A_y}, \qquad \sigma_z = \frac{P_z}{A_z}.$$

FIG. 54.

It is assumed below that $\sigma_x > \sigma_y > \sigma_z$.

Combining the effects of the forces P_x, P_y and P_z, it can be concluded that on a section through the z axis only the forces P_x and P_y produce stresses and therefore these stresses may be calculated from eqs. (26) and (27) and represented graphically by using Mohr's circle. In Fig. 55 the stress circle with diameter AB represents these stresses. In the same manner the stresses on any section through the x axis can be represented by a circle having BC as a diameter. The circle with the diameter AC represents stresses on any section through the y axis. The three Mohr's circles represent stresses on three series

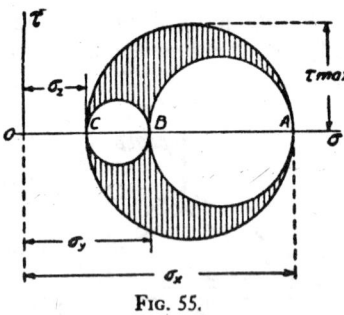

FIG. 55.

of sections through the x, y and z axes. For any section inclined to x, y and z axes the stress components are the coordinates of a point located in the shaded area of Fig. 55.[11] On the basis of this it can be concluded that the maximum shearing stress is represented by the radius of the largest of the three circles and is given by the equation $\tau_{\max} = (\sigma_x - \sigma_z)/2$. It acts on the section through the y axis bisecting the angle between the x and z axes.

The equations for calculating the unit elongations in the directions of the x, y and z axes may be obtained by combining the effects of P_x, P_y and P_z in the same manner as in considering tension or compression in two perpendicular directions (see Art. 15). In this manner we obtain

$$\epsilon_x = \frac{\sigma_x}{E} - \frac{\mu}{E}(\sigma_y + \sigma_z),$$

$$\epsilon_y = \frac{\sigma_y}{E} - \frac{\mu}{E}(\sigma_x + \sigma_z), \tag{43}$$

$$\epsilon_z = \frac{\sigma_z}{E} - \frac{\mu}{E}(\sigma_x + \sigma_y).$$

The volume of the bar increases in the ratio

$$(1 + \epsilon_x)(1 + \epsilon_y)(1 + \epsilon_z):1,$$

or, neglecting small quantities of higher order,

$$(1 + \epsilon_x + \epsilon_y + \epsilon_z):1.$$

It is seen that the unit volume expansion is

$$\Delta = \epsilon_x + \epsilon_y + \epsilon_z. \tag{44}$$

---

[11] The proof of this statement can be found in the book by A. Föppl, *Technische Mechanik*, Vol. 5, p. 18, 1918. See also H. M. Westergaard, *Z. angew. Math. u. Mech.*, Vol. 4, p. 520, 1924.

The relation between the unit volume expansion and the stresses acting on the sides of the bar will be obtained by adding together eqs. (43). In this manner we obtain

$$\Delta = \epsilon_x + \epsilon_y + \epsilon_z = \frac{(1 - 2\mu)}{E} (\sigma_x + \sigma_y + \sigma_z). \qquad (45)$$

In the particular case of uniform hydrostatic pressure we have

$$\sigma_x = \sigma_y = \sigma_z = -p.$$

Then from eqs. (43)

$$\epsilon_x = \epsilon_y = \epsilon_z = -\frac{p}{E}(1 - 2\mu), \qquad (46)$$

and from eqs. (45)

$$\Delta = -\frac{3(1 - 2\mu)}{E} p, \qquad (47)$$

or, using the notation

$$\frac{E}{3(1 - 2\mu)} = K, \qquad (48)$$

we obtain

$$\Delta = -\frac{p}{K}. \qquad (49)$$

The unit volume contraction is proportional to the compressive stress p and inversely proportional to the quantity K, which is called the *bulk modulus of elasticity*.

Problems

1. Determine the decrease in the volume of a solid steel sphere of 10 in. diameter submitted to a uniform hydrostatic pressure $p = 10,000$ lb per sq in.

Solution. From eq. (49),

$$\Delta = -\frac{p}{K} = -\frac{10,000 \times 3(1 - 2 \times 0.3)}{30 \times 10^6} = -\frac{4}{10^4}.$$

The decrease in the volume is, therefore.

$$\frac{4}{10^4} \times \frac{\pi d^3}{6} = 0.209 \text{ cu in.}$$

2. Referring to Fig. 56, a rubber cylinder A is compressed in a
steel cylinder B by a force P. Determine the pressure between the

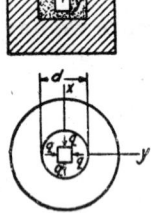

rubber and the steel if $P = 1,000$ lb, $d = 2$ in.,
Poisson's ratio for rubber $\mu = 0.45$. Friction between rubber and steel is neglected.

Solution. Let p denote the compressive stresses
over any cross section perpendicular to the axis of
the cylinder and q the pressure between the rubber
and the inner surface of the steel cylinder. Compressive stress of the same magnitude will act between the lateral surfaces of the longitudinal fibers
of the rubber cylinder, from which we isolate an element in the form of a rectangular parallelepiped,
with sides parallel to the axis of the cylinder (see
Fig. 56). This element is in equilibrium under the

FIG. 56.   compressive stresses q on the lateral faces of the element and the axial compressive stress p. Assuming
that the steel cylinder is absolutely rigid, the lateral expansion of
the rubber in the x and y directions must be equal to zero and from
eqs. (43) we obtain

$$0 = \frac{q}{E} - \frac{\mu}{E} (p + q),$$

from which

$$q = \frac{\mu p}{1 - \mu} = \frac{0.45}{1 - 0.45} \cdot \frac{1,000 \times 4}{\pi \times 2^2} = 260 \text{ lb per sq in.}$$

3. A concrete column is enclosed in a steel tube (Fig. 57). Determine the pressure between the steel and concrete and the circumferential tensile stress in the tube, assuming that there is no friction
between concrete and steel and that all the dimensions and the longitudinal compressive stress in the column are known.

Solution. Let p denote the longitudinal and q
the lateral compressive stress, d the inner diameter
and h the thickness of the tube, E_s the modulus of
elasticity for steel, E_c, μ_c the modulus of elasticity
and Poisson's ratio for concrete. The expansion of
the concrete in the lateral direction will be, from
eqs. (43),

$$\epsilon_x = -\frac{q}{E_c} + \frac{\mu_c}{E_c} (p + q). \qquad (a)$$

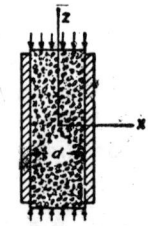

FIG. 57.

This expansion equals the circumferential expansion of the tube
(see eq. 13, p. 31),

$$\epsilon = \frac{qd}{2hE_s}. \qquad (b)$$

From eqs. (*a*) and (*b*) we obtain

$$\frac{qa}{2hE_s} = -\frac{q}{E_c} + \frac{\mu_c}{E_c}(p+q),$$

from which

$$q = p\,\frac{\mu_c}{\dfrac{d}{2h}\dfrac{E_c}{E_s} + 1 - \mu_c}.$$

The circumferential tensile stress in the tube will now be calculated from the equation

$$= \frac{qd}{2h}.$$

4. Determine the maximum shearing stress in the concrete column of the previous problem, assuming that $p = 1,000$ lb per sq in., $\mu_c = 0.10$, $d/2h = 7.5$.

Solution.

$$\tau_{max} = \frac{p-q}{2} = \frac{p}{2}\left(1 - \frac{0.1}{1.9}\right) = 474 \text{ lb per sq in.}$$

5. A steel spherical shell (Fig. 58) is subjected to a uniform inner and outer pressure p. Find the reduction δ of the inner diameter. Construct Mohr's circles as in Fig. 55.

Answer. $\delta = \dfrac{pd(1 - 2\mu)}{E}$; in this case all three circles are reduced to a point with abscissa $\sigma = -p$.

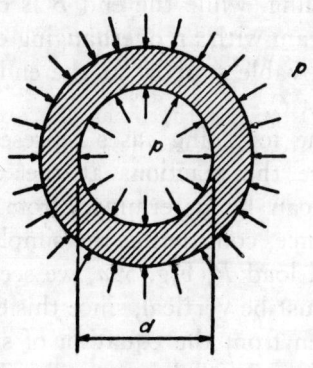

Fig. 58.

CHAPTER III

BENDING MOMENT AND SHEARING FORCE

19. Types of Beams.—In this chapter we will discuss the simplest types of beams having a vertical plane of symmetry through the longitudinal axis, and supported as shown in Fig. 59. It is assumed that all the applied forces are vertical and act in the plane of symmetry so that bending occurs in

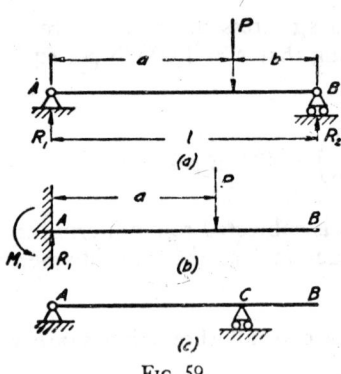

FIG. 59.

the same plane. Fig. 59a represents a *beam with simply supported ends.* Points of support *A* and *B* are hinged so that the ends of the beam can rotate freely during bending. It is also assumed that one of the supports is mounted on rollers and can move freely in the horizontal direction. Fig. 59b represents a *cantilever beam.* The end *A* of this beam is built into the wall and cannot rotate during bending, while the end *B* is entirely free. Fig. 59c represents a beam with an overhanging end. This beam is hinged to an immovable support at the end *A* and rests on a movable support at *C*.

All three of the foregoing cases represent *statically determinate beams* since the reactions at the supports produced by a given load can be determined from the equations of statics. For instance, considering the simply supported beam carrying a vertical load *P*, Fig. 59a, we see that the reaction R_2 at the end *B* must be vertical, since this end is free to move horizontally. Then from the equation of statics, $\Sigma X = 0$, it follows that reaction R_1 is also vertical. The magnitudes of R_1 and R_2 are then determined from the equations of moments. Equating to zero the sum of the moments of all forces

70

with respect to point B, we obtain

$$R_1 l - Pb = 0,$$

from which

$$R_1 = \frac{Pb}{l}.$$

In a similar way, by considering the moments with respect to point A, we obtain

$$R_2 = \frac{Pa}{l}.$$

The reactions for the beam with an overhanging end, Fig. 59c, can be calculated in the same manner.

In the case of the cantilever beam, Fig. 59b, the load P is balanced by the reactive forces acting on the built-in end. From the equations of statics, $\Sigma X = 0$ and $\Sigma Y = 0$, we conclude at once that the resultant of the reactive forces R_1 must be vertical and equal to P. From the equation of moments, $\Sigma M = 0$, it follows that the moment M_1 of the reactive forces with respect to point A is equal to Pa and acts in the counter-clockwise direction as shown in the figure.

The reactions produced by any other kind of loading on the above types of beams can be calculated by a similar procedure.

It should be noted that the special provisions permitting free rotation of the ends and free motion of the support are used in practice only in beams of large spans, such as those found in bridges. In beams of shorter span, the conditions at the support are usually as illustrated in Fig. 60. During bending of such a beam, friction forces between the supporting surfaces and the beam will be

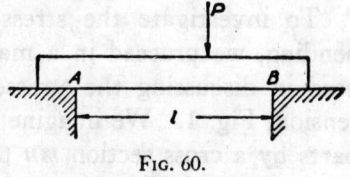

FIG. 60.

produced such as to oppose horizontal movement of the ends of the beam. These forces can be of some importance in the case of flexible bars and thin metallic strips (see p. 179), but in the case of a stiff beam for which the deflection is very small in comparison with the span length l these forces can be neglected, and the reactions can be calculated as though the beam were simply supported, Fig. 59a.

20. Bending Moment and Shearing Force.—Let us now consider a simply supported beam acted on by vertical forces P_1, P_2 and P_3, Fig. 61a. We assume that the beam has an

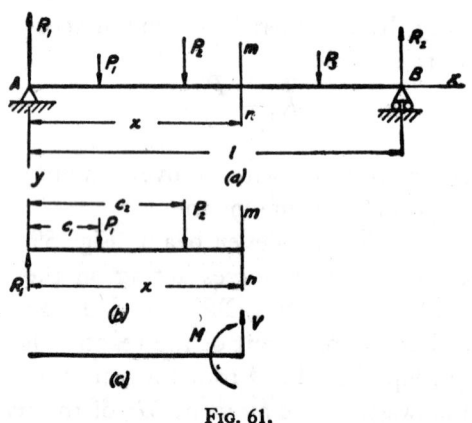

Fig. 61.

axial plane of symmetry and that the loads act in this plane. Then, from considerations of symmetry, we conclude that bending must also occur in this same plane. In most practical cases this condition of symmetry is fulfilled since the usual cross-sectional shapes of beams, such as a circle, rectangle, I-section or T-section, are symmetrical. The more general case of a non-symmetrical cross section will be discussed later (see Chap. 8).

To investigate the stresses produced in a beam during bending, we proceed in a manner similar to the one already used in discussing the stresses produced in a bar by simple tension, Fig. 1. We imagine that the beam AB is cut in two parts by a cross section mn taken at any distance x from the left support A, Fig. 61a, and that the portion of the beam to the right is removed. In discussing the equilibrium of the remaining left-hand portion of the beam, Fig. 61b, we must consider not only the external forces such as loads P_1, P_2, and reaction R_1, but also the internal forces which are distributed over the cross section mn and which represent the action of the right portion of the beam on the left portion. These internal forces must be of such a magnitude as to equilibrate the above-mentioned external forces P_1, P_2 and R_1.

In the ensuing discussion it will be advantageous to reduce the actual system of external forces to a simplified equivalent system. From statics we know that a system of parallel forces can be replaced by one force equal to the algebraic sum of the given forces together with a couple. In our particular case we can replace the forces P_1, P_2 and R_1 by the vertical force V acting in the plane of the cross section mn and by the couple M. The magnitude of the force is

$$V = R_1 - P_1 - P_2, \qquad (a)$$

and the magnitude of the couple is

$$M = R_1 x - P_1(x - c_1) - P_2(x - c_2). \qquad (b)$$

The force V, which is equal to the algebraic sum of the external forces to the left of the cross section mn, is called the *shearing force* at the cross section mn. The couple M, which is equal to the algebraic sum of the moments of the external forces to the left of the cross section mn with respect to the centroid of this cross section, is called the *bending moment* at the cross section mn. Thus the system of external forces to the left of the cross section mn can be replaced by the statically equivalent system consisting of the shearing force V acting in the plane of the cross section and the couple M, Fig. 61c. The stresses which are distributed over the cross section mn and which represent the action of the right portion of the beam on its left portion must then be such as to balance the bending moment M and the shearing force V.

If a distributed load rather than a number of concentrated forces acts on a beam, the same reasoning can be used as in the previous case. Take, as an example, the uniformly loaded beam shown in Fig. 62a. Denoting the load per unit length by q, the reactions in this case are

$$R_1 = R_2 = \frac{ql}{2}.$$

To investigate the stresses distributed over a cross section mn we again consider the equilibrium of the left portion of the beam, Fig. 62b. The external forces acting on this portion of

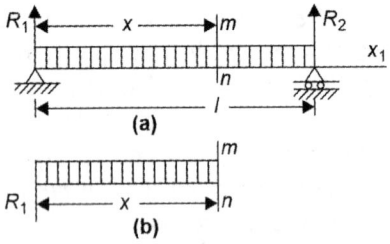

Fig. 62

the beam are the reaction R_1 and the load uniformly distributed along the length x. This latter load has, of course, a resultant equal to qx. The algebraic sum of all forces to the left of the cross section mn is thus R_1-qx. The algebraic sum of the moments of all forces to the left of the cross section mn with respect to the centroid of this cross section is obtained by subtracting the moment of the resultant of the distributed load from the moment R_1x of the reaction. The moment of the distributed load is evidently equal to

$$qx \times \frac{x}{2} = \frac{qx^2}{2} \qquad (a)$$

Thus, we obtain for the algebraic sum of the moments the expression

$$R_1x - \frac{qx^2}{2} \qquad (b)$$

All the forces acting on the left portion of the beam can now be replaced by one force acting in the plane of the cross section mn and equal to

$$V = R_1 - qx = q\left(\frac{l}{2} - x\right), \qquad (c)$$

together with a couple equal to

$$M = R_1x - \frac{qx^2}{2} = \frac{qx}{2}(l-x) \qquad (d)$$

Expressions *(c)* and *(a)* represent the shearing force and the bending moment, respectively, at the cross section *mn*.

In the above examples the equilibrium of the left portion of the beam has been discussed. If the right portion of the beam is considered, the algebraic sum of the forces to the right of the cross section and the algebraic sum of the moments of those forces will have the same magnitudes V and M as have already been found, but will be of opposite sense. This follows from the fact that the loads acting on a beam together with the reactions R_1 and R_2 represent a system of forces in equilibrium, and the moment of all these forces with respect to any point in their plane, as well as their algebraic sum, must be equal to zero. Hence, the moment of the forces acting on the left portion of the beam with respect to the centroid of a cross section *mn* must be equal and opposite to the moment with respect to the same point of the forces acting on the right portion of the beam. Also the algebraic sum of forces acting on the left portion of the beam must be equal and opposite to the algebraic sum of forces acting on rhe right portion.

In the following discussion the bending moment and the shearing force at a cross section *mn* are taken as positive if in considering the left portion of a beam the directions obtained are such as shown in Fig. 61c. To visualize this sign convention for bending moments, let us isolate an element of the beam by two adjacent cross sections *mn* and m_1n_1, Fig. 63. If the

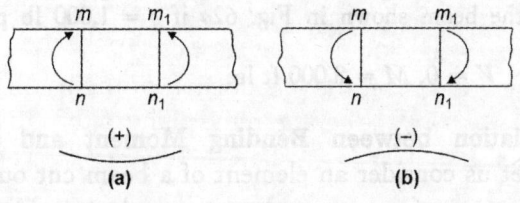

Fig. 63

bending moments in these cross sections are positive the forces to the left of the cross section *mn* give a moment in the clockwise direction and the forces to the right of the cross section m_1n_1, a moment in the counter-clockwise direction as shown in Fig. 63a. It is thus seen that the directions of the moments are such that a bending is produced which is convex down-wards. If the bending moments in the cross sections mn and m_1n_1 are negative, a bending convex upwards is produced as shown in Fig. 63b. Thus in portions of a beam where the bending moment is positive the deflection curve is convex downwards, while in portions where bending moment is negative the deflection curve is convex upwards.

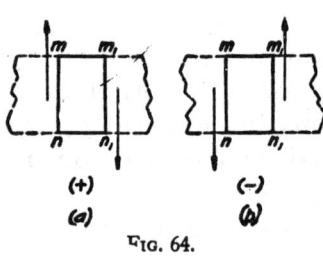

(+)   (−)
(a)   (b)

FIG. 64.

The sign convention for shearing force is represented in Fig. 64.

Problems

1. Find shearing force V and bending moment M at a cross section 4 ft from the left end of the beam shown in Fig. 59a, if $a = 6$ ft, $l = 10$ ft and $P = 10,000$ lb.

Answer. $V = 4,000$ lb, $M = 16,000$ ft lb.

2. Find shearing force V and bending moment M at a cross section 2 ft from the left end of the beam shown in Fig. 59b, if $a = 8$ ft, $P = 12,000$ lb.

Answer. $V = 12,000$ lb, $M = -72,000$ ft lb.

3. Find shearing force V and bending moment M for the cross section mn of the beam shown in Fig. 61a, if $P_1 = P_2 = 12,000$ lb, $P_3 = 0$, $c_1 = 2$ ft, $c_2 = 4$ ft, $x = 6$ ft, $l = 12$ ft.

Answer. $V = -6,000$ lb, $M = 36,000$ ft lb.

4. Find the shearing force V and the bending moment M at the middle of the beam shown in Fig. 62a if $q = 1,000$ lb per ft and $l = 8$ ft.

Answer. $V = 0$, $M = 8,000$ ft lb.

21. Relation between Bending Moment and Shearing Force.

—Let us consider an element of a beam cut out by two adjacent cross sections mn and m_1n_1 which are a distance dx apart, Fig. 65. Assuming that there is a positive bending moment and a positive shearing force at the cross section mn, the action of the left portion of the beam on the element is represented by the force V and the couple M, as indicated in Fig. 65a. In the same manner, assuming that at section m_1n_1 the bending moment and the shearing force are positive, the action of the right portion of the beam on the element is represented by the couple and the force shown. If no forces act on the beam between cross sections mn and m_1n_1, Fig. 65a, the shearing forces at these two cross sections are equal.[1] Regarding the bending moments, it can be seen from the equilibrium of the element that they are not equal at two adjacent cross sections

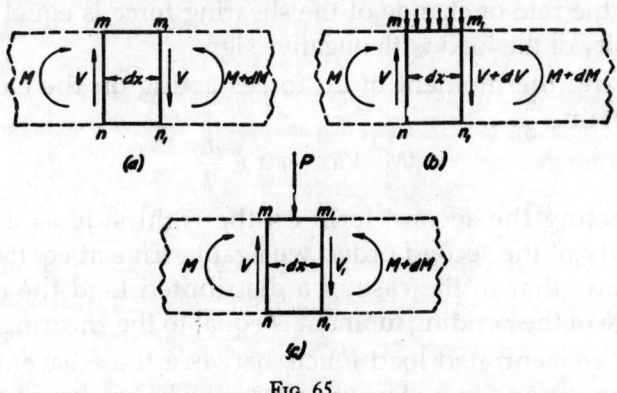

FIG. 65.

and that the increase dM in the bending moment equals the moment of the couple represented by the two equal and opposite forces V, i.e.,

$$dM = V\,dx$$

and

$$\frac{dM}{dx} = V. \tag{50}$$

Thus, on all portions of a beam between loads the shearing force is the rate of change of the bending moment with respect to x.

Let us now consider the case in which a distributed load of intensity q acts between the cross sections mn and m_1n_1, Fig. 65b. Then the total load acting on the element is $q\,dx$. If q is considered positive when the load acts downward, it may be concluded from the equilibrium of the element that the

---

[1] The weight of the element of the beam is neglected in this discussion.

shearing force at the cross section m_1n_1 is different from that at mn by an amount

$$dV = - qax,$$

from which it follows that

$$\frac{dV}{dx} = -q \qquad\qquad (51)$$

Thus, the rate of change of the shearing force is equal to the intensity of the load with negative sign.

Taking the moment of all forces acting on the element we obtain

$$dM = Vdx - qdx \times \frac{dx}{2}$$

Neglecting the second term on the right side as a small quantity of the second order, we again arrive at eq. (50) and conclude that in the case of a distributed load the rate of change of the bending moment is equal to the shearing force.

If a concentrated load P acts between the adjacent cross sections mn and m_1n_1, Fig. 65c, there will be an abrupt change in the magnitude of the shearing force. Let V denote the shearing force at the cross section mn and V_1 that at the cross section m_1n_1. Then from the equilibrium of the element mm_1n_1n, we find

$$V_1 = V - P$$

Thus, the magnitude of the shearing force changes by the amount P as we pass the point of application of the load. From eq. (50) it can then be concluded that at the point of application of a concentrated force there is an abrupt change in the magnitude of the derivative dM/dx.

22. Bending Moment and Shearing Force Diagrams: It was shown in the preceding discussion that the stresses acting on a cross section mn of a beam are such as to balance the bending moment M and shearing force V at that cross section. Thus, the magnitudes of M and V at any cross section determine the magnitude of the stresses acting on that cross section. To simplify the investigation of stresses in a beam it is advisable to use a graphical representation of the variation of the bending moment and the shearing force along the axis of the beam. In such a representation the abscissa indicates the position of the cross section and the ordinate represents the value of the bending moment or shearing force which acts at this cross section, positive values being plotted above the horizontal axis and negative values below. Such graphical representations are called *bending moment* and *shearing force diagrams*, respectively.

Let us consider, as an example, a simply supported beam with a single concentrated load P, Fig. 66.[2] The reactions in this case are

$$R_1 = \frac{Pb}{l} \quad \text{and} \quad R_2 = \frac{Pa}{l}$$

Taking a cross section mn to the left of P, it can be concluded that at such a cross section

$$V = \frac{Pb}{l} \quad \text{and} \quad M = \frac{Pb}{l} x \qquad (a)$$

The shearing force and the bending moment have the same sense as those in Figs 63a and 64a and are therefore positive. It is seen that the shearing force remains constant along the portion of the beam to the left of the load and that the bending moment varies directly as x. For $x = 0$ the moment is zero and for $x = a$, i.e. at the cross section where the load is applied, the moment is equal to Pab/l. The corresponding portions of the shearing force and bending moment diagrams are shown in Fig. 66b and 66c, respectively, by the straight lines ac and a_1c_1. For a cross section to the right of the load, we obtain

$$V = \frac{Pb}{l} - P \quad \text{and} \quad M = \frac{Pb}{l} x - P(x - a), \qquad (b)$$

x always being the distance from the left end of the beam. The shearing force for this portion of the beam remains constant and negative. In Fig. 66b this force is represented by of

---

[2]For simplicity the rollers under the movable supports will usually be omitted in subsequent figures.

the line $c'b$ parallel to the x axis. The bending moment is a linear function of x which for $x = a$ is equal to Pab/l and for $x = l$ is equal to zero. It is always positive and its variation along the right portion of the beam is represented in Fig. 66c by the straight line c_1b_1. The broken lines $acc'b$ and $a_1c_1b_1$ in Figs. 66b and 66c represent, respectively, the shearing force and bending moment diagrams for the whole length of the beam.

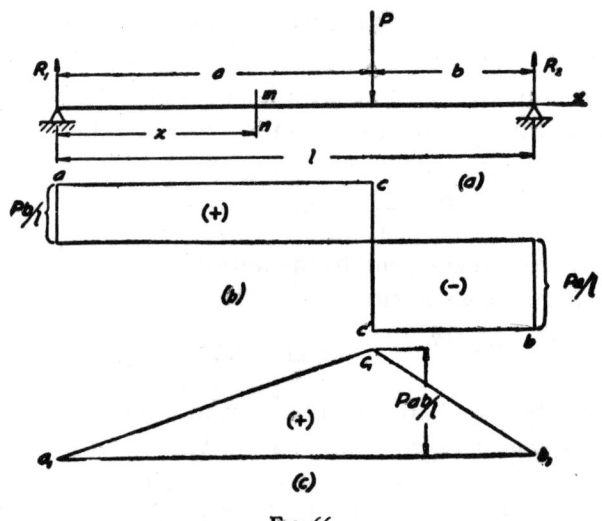

FIG. 66.

At the load P there is an abrupt change in the magnitude of the shearing force from the positive value Pb/l to the negative value $-Pa/l$ and a sharp change in the slope of the bending moment diagram.

In deriving expressions (b) for the shearing force and bending moment, we considered the left portion of the beam, a portion which is acted upon by the two forces R_1 and P. It would have been simpler in this case to consider the right portion of the beam where only the reaction Pa/l acts. Following this procedure and using the rule of signs indicated in Figs. 63 and 64, we obtain

$$V = -\frac{Pa}{l} \quad \text{and} \quad M = \frac{Pa}{l}(l - x). \qquad (c)$$

Expressions (b) previously obtained can also be brought to this simpler form if we observe that $a = l - b$.

It is interesting to note that the shearing force diagram consists of two rectangles with equal areas. Taking into consideration the opposite signs of these areas, we conclude that the total area of the shearing force diagram is zero. This result is not accidental. By integrating eq. (50), we have

$$\int_A^B dM = \int_A^B V dx, \qquad (d)$$

where the limits A and B indicate that the integration is taken over the entire length of the beam from the end A to the end B. The right side of eq. (d) then represents the total area of the shearing force diagram. The left side of the same equation, after integration, gives the difference $M_B - M_A$ of the bending moments at the ends B and A. In the case of a simply supported beam the moments at the ends vanish. Hence the total area of the shearing force diagram is zero.

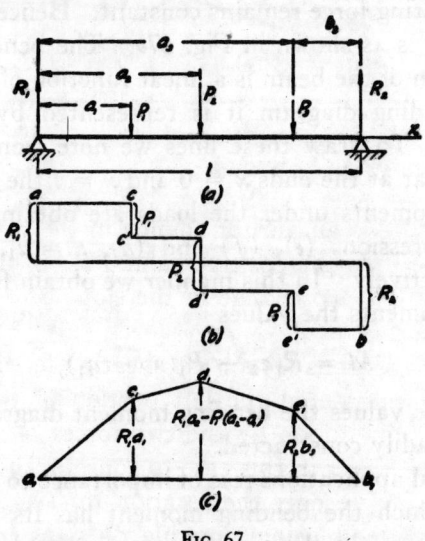

Fig. 67.

If several loads act on a beam, Fig. 67, the beam is divided into several portions and expressions for V and M must be established for each portion. Measuring x from the left end

of the beam and taking $x < a_1$, we obtain for the first portion
of the beam

$$V = R_1 \quad \text{and} \quad M = R_1 x. \tag{e}$$

For the second portion of the beam, i.e., for $a_1 < x < a_2$, we
obtain

$$V = R_1 - P_1 \quad \text{and} \quad M = R_1 x - P_1(x - a_1). \tag{f}$$

For the third portion of the beam, i.e., for $a_2 < x < a_3$, it is
advantageous to consider the right portion of the beam rather
than the left. In this way we obtain

$$V = -(R_2 - P_3)$$

and

$$M = R_2(l - x) - P_3(l - x - b_3). \tag{g}$$

Finally for the last portion of the beam we obtain

$$V = -R_2, \quad M = R_2(l - x). \tag{h}$$

From expressions (e)–(h) we see that in each portion of the
beam the shearing force remains constant. Hence the shearing
force diagram is as shown in Fig. 67b. The bending moment
in each portion of the beam is a linear function of x. Hence in
the corresponding diagram it is represented by an inclined
straight line. To draw these lines we note from expressions
(e) and (h) that at the ends $x = 0$ and $x = l$ the moments are
zero. The moments under the loads are obtained by substituting in expressions (e), (f) and (h), $x = a_1$, $x = a_2$ and
$x = a_3$, respectively. In this manner we obtain for the above-
mentioned moments the values

$$M = R_1 a_1, \quad M = R_1 a_2 - P_1(a_2 - a_1), \quad M = R_2 b_3.$$

By using these values the bending moment diagram, shown in
Fig. 67c, is readily constructed.

In practical applications it is of importance to find the cross
sections at which the bending moment has its maximum or
minimum values. In the case of concentrated loads just considered, Fig. 67, the maximum bending moment occurs under
the load P_2. This load corresponds in the bending moment
diagram to point d_1, at which point the slope of the diagram

changes sign. Further, from eq. (50), we know that the slope
of the bending moment diagram at any point is equal to the
shearing force. Hence the bending moment has its maximum
or minimum values at the cross sections in which the shearing
force changes its sign. If as we proceed along the x axis the
shearing force changes from a positive to a negative value, as
under the load P_2 in Fig. 67, the slope in the bending moment
diagram also changes from positive to negative. Hence we
have the maximum bending moment at this cross section. A
change in V from a negative to a positive value indicates a
minimum bending moment. In the general case a shearing
force diagram may intersect the horizontal axis in several
places. To each such intersection point there then corresponds
a maximum or a minimum in the bending moment diagram.
The numerical values of all these maximums and minimums
must be investigated to find the numerically largest bending
moment.

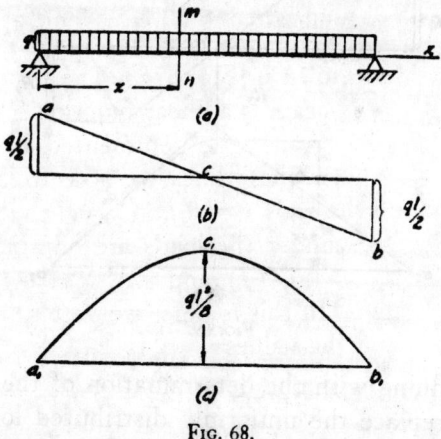

Fig. 68.

Let us next consider the case of a uniformly distributed
load, Fig. 68. From our previous discussion (p. 74), we have
for a cross section at distance x from the left support,

$$V = q\left(\frac{l}{2} - x\right) \quad \text{and} \quad M = \frac{qx}{2}(l - x). \qquad (i)$$

We see that the shearing force diagram consists in this case of an inclined straight line for which the ordinates at $x = 0$ and $x = l$ are equal to $ql/2$ and $-ql/2$, respectively, as shown in Fig. 68b. As can be seen from expression (i) the bending moment in this case is a parabolic curve with its vertical axis at the middle of the span of the beam, Fig. 68c. The moments at the ends $x = 0$ and $x = l$ are zero; and the maximum value of the moment occurs at the middle of the span where the shearing force changes sign. This maximum is obtained by substituting $x = l/2$ in expression (i), which gives $M_{max} = ql^2/8$.

If a uniform load q covers only a part of the span, Fig. 69, we must consider separately the three portions of length a, b

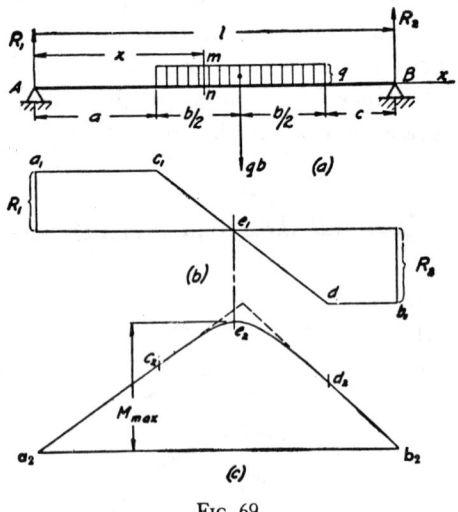

Fig. 69.

and c. Beginning with the determination of the reactions R_1 and R_2, we replace the uniformly distributed load by its resultant qb. From the equations of statics for the moments with respect to B and A, we then obtain

$$R_1 = \frac{qb}{l}\left(c + \frac{b}{2}\right) \quad \text{and} \quad R_2 = \frac{qb}{l}\left(a + \frac{b}{2}\right).$$

The shearing force and the bending moment for the unloaded portion at the left of the beam ($0 < x < a$) are

$$V = R_1 \quad \text{and} \quad M = R_1 x. \tag{j}$$

For a cross section mn taken in the loaded portion of the beam the shearing force is obtained by subtracting the load $q(x - a)$ to the left of the cross section from the reaction R_1. The bending moment in the same cross section is obtained by subtracting the moment of the load to the left of the cross section from the moment of the reaction R_1. In this manner we find

$$V = R_1 - q(x - a)$$

and

$$M = R_1 x - q(x - a) \times \frac{x - a}{2}. \qquad (k)$$

For the unloaded portion at the right of the beam, considering the forces to the right of any cross section, we find

$$V = -R_2 \qquad \text{and} \qquad M = R_2(l - x). \qquad (l)$$

By using expressions (j), (k) and (l) the shearing force and bending moment diagrams are readily constructed. The shearing force diagram, Fig. 69b, consists of the horizontal portions $a_1 c_1$ and $d_1 b_1$ corresponding to the unloaded portions of the beam and the inclined line $c_1 d_1$ corresponding to the uniformly loaded portion. The bending moment diagram, Fig. 69c, consists of the two inclined lines $a_2 c_2$ and $b_2 d_2$ corresponding to the unloaded portions and of the parabolic curve $c_2 e_2 d_2$ with vertical axis corresponding to the loaded portion of the beam. The maximum bending moment is at the point e_2, which corresponds to the point e_1 where the shearing force changes sign. At points c_2 and d_2 the parabola is tangent to the inclined lines $a_2 c_2$ and $d_2 b_2$, respectively. This follows from the fact that at points c_1 and d_1 of the shearing force diagram there is no abrupt change in the magnitude of the shearing force. Hence, by virtue of eq. (50), there cannot occur an abrupt change in slope of the bending moment diagram at the corresponding points c_2 and d_2.

In the case of a cantilever beam, Fig. 70, the same method as before is used to construct the shearing force and bending moment diagrams. Measuring x from the left end of the beam and considering the portion to the left of the load P_2 $(0 < x < a)$, we obtain

$$V = -P_1 \qquad \text{and} \qquad M = -P_1 x.$$

The minus sign in these expressions follows from the rule of signs indicated in Figs. 63b and 64b. For the right portion of the beam ($a < x < l$) we obtain

$$V = -P_1 - P_2 \quad \text{and} \quad M = -P_1x - P_2(x - a).$$

The corresponding diagrams of shearing force and bending moment are shown in Figs. 70b and 70c. The total area of the

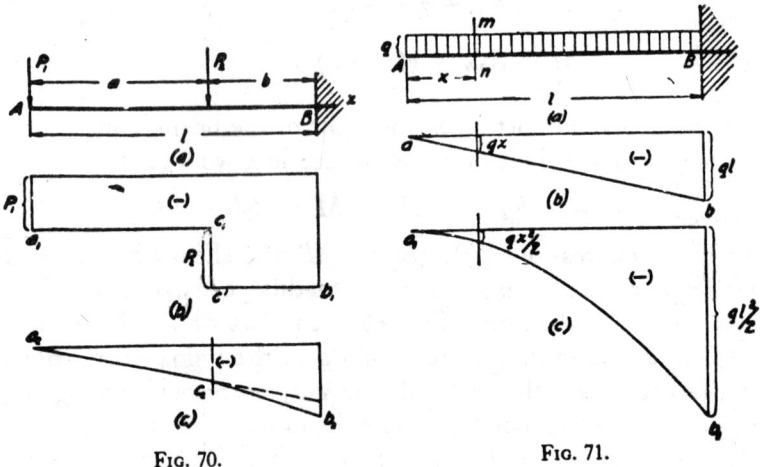

FIG. 70.   FIG. 71.

shearing force diagram does not vanish in this case and is equal to $-P_1l - P_2b$, which is the bending moment M_B at the end B of the beam. The bending moment diagram consists of the two inclined lines a_2c_2 and c_2b_2 the slopes of which are equal to the values of the shearing force at the corresponding portions of the cantilever. The numerical maximum of the bending moment is at the built-in end B of the beam.

If a cantilever carries a uniform load, Fig. 71, the shearing force and bending moment at a distance x from the left end are

$$V = -qx \quad \text{and} \quad M = -qx \times \frac{x}{2} = -\frac{qx^2}{2}.$$

The shearing force is represented in Fig. 71b by the inclined line ab and the bending moment in Fig. 71c by the parabola a_1b_1 which has a vertical axis and is tangent to the horizontal axis at a_1, where the shearing force vanishes. The numerical maximums of the bending moment and shearing force occur at the end B of the beam.

If concentrated loads and distributed loads act on the beam simultaneously, it is advantageous to draw the diagrams separately for each kind of loading and obtain the total values of V or M at any cross section by summing up the corresponding ordinates of the two partial diagrams. If, for example, we have concentrated loads P_1, P_2 and P_3, Fig. 67, acting simultaneously with a uniform load, Fig. 68, the bending moment at any cross section is obtained by summing up the corresponding ordinates of the diagrams in Figs. 67c and 68c.

Problems

1. Draw approximately to scale the shearing force and bending moment diagrams and label the values of the largest positive and negative shearing forces and bending moments for the beams shown in Fig. 72.

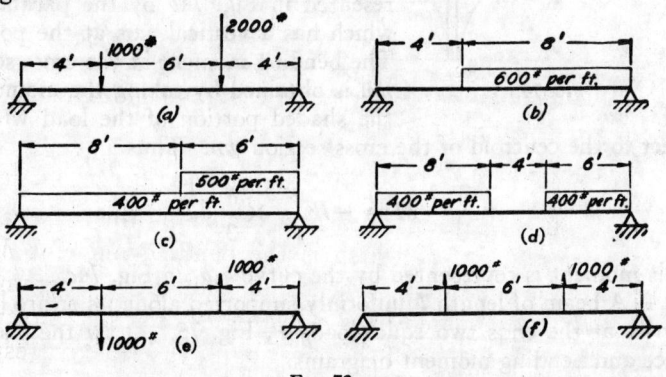

Fig. 72.

2. Draw approximately to scale the shearing force and bending moment diagrams and label the values of the largest positive and negative shearing forces and bending moments for the cantilever beams shown in Fig. 73.

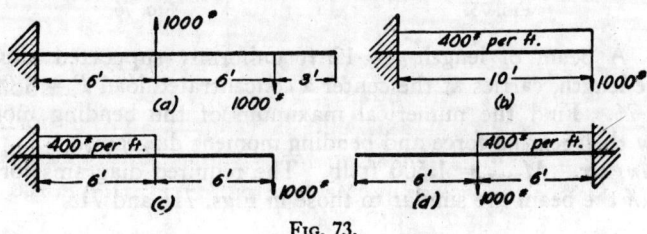

Fig. 73.

3. A cantilever carrying a total load W which increases uniformly in intensity from zero at the left end as shown by the inclined line AC, Fig. 74a, is built in at the right-hand end. Draw the shearing force and bending moment diagrams.

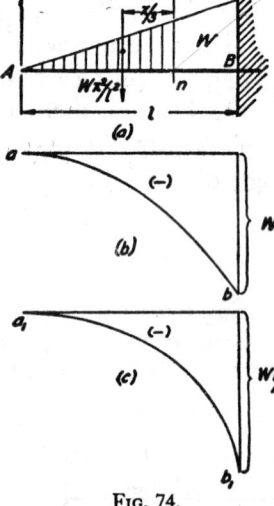

Solution. The shearing force at a cross section mn at a distance x from the left end of the cantilever is numerically equal to the shaded portion of the load. Since the total load W is represented by the triangle ACB the shaded portion is Wx^2/l^2. By using the sign convention previously adopted, Fig. 64, we obtain

$$V = -W\frac{x^2}{l^2}.$$

The shearing force diagram is thus represented in Fig. 74b by the parabola ab which has a vertical axis at the point a. The bending moment at the cross section mn is obtained by taking the moment of the shaded portion of the load with respect to the centroid of the cross section mn. Thus

Fig. 74.

$$M = -W\frac{x^2}{l^2} \times \frac{x}{3}.$$

This moment is represented by the curve a_1b_1 in Fig. 74c.

4. A beam of length l uniformly supported along its entire length carries at the ends two equal loads P, Fig. 75. Draw the shearing force and bending moment diagrams.

Answer. The diagrams are obtained from Figs. 68b and 68c by substituting $-2P$ for ql.

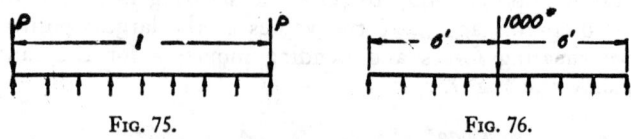

Fig. 75.   Fig. 76.

5. A beam of length $l = 12$ ft uniformly supported along its entire length, carries at the center a concentrated load $P = 1,000$ lb, Fig. 76. Find the numerical maximum of the bending moment. Draw the shearing force and bending moment diagrams.

Answer. $M_{max} = 1,500$ ft lb. The required diagrams for each half of the beam are similar to those in Figs. 71b and 71c.

6. A simply supported beam of length l carries a total distributed load W which increases in intensity uniformly from zero at the left end, as shown in Fig. 77a. Draw approximately to scale the shearing force and bending moment diagrams if $W = 12,000$ lb and $l = 24$ ft.

Solution. The reactions at the supports in this case are $R_1 = \frac{1}{3}W = 4,000$ lb and $R_2 = 8,000$ lb. The shearing force at a cross section mn is obtained by subtracting the shaded portion of the load from the reaction R_1. Hence

$$V = R_1 - W\frac{x^2}{l^2} = W\left(\frac{1}{3} - \frac{x^2}{l^2}\right).$$

The shearing force diagram is represented by the parabolic curve acb in Fig. 77b. The bending moment at a cross section mn is

$$M = R_1 x - W\frac{x^2}{l^2} \times \frac{x}{3}$$

$$= \frac{1}{3}Wx\left(1 - \frac{x^2}{l^2}\right).$$

This moment is represented by the curve $a_1 c_1 b_1$ in Fig. 77c. The maximum bending moment is at c_1 where the shearing force changes its sign, i.e., where $x = l/\sqrt{3}$.

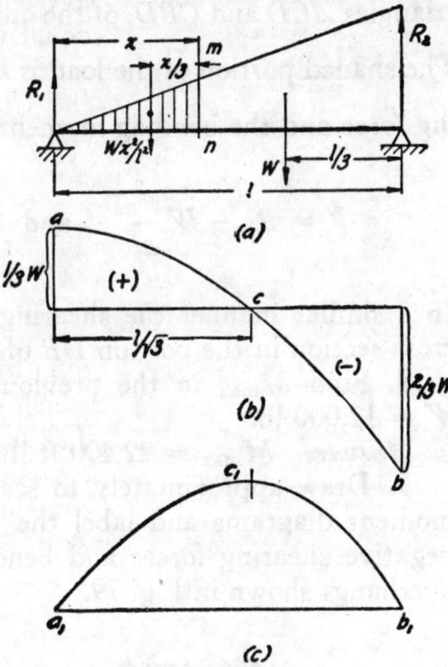

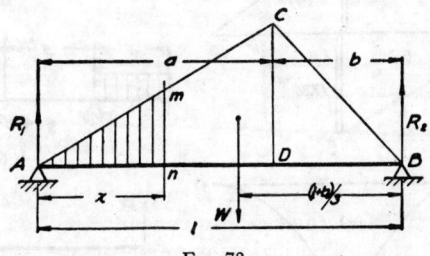

Fig. 77.

7. A simply supported beam AB carries a distributed load the intensity of which is represented by the line ACB, Fig. 78. Find

Fig. 78.

the expressions for the shearing force and the bending moment at a cross section mn.

Solution. Assuming the total load W to be applied at the centroid of the triangle ACB, the reactions at the supports are

$$R_1 = W \frac{l+b}{3l} \quad \text{and} \quad R_2 = W \frac{l+a}{3l}.$$

The total load is then divided into two parts, represented by the triangles ACD and CBD, of the amount Wa/l and Wb/l, respectively. The shaded portion of the load is $W \frac{a}{l} \times \frac{x^2}{a^2} = W \frac{x^2}{al}$. For the shearing force and the bending moment at mn we then obtain

$$V = R_1 - W \frac{x^2}{al} \quad \text{and} \quad M = R_1 x - W \frac{x^2}{al} \times \frac{x}{3}.$$

In a similar manner the shearing force and bending moment for a cross section in the portion DB of the beam can be obtained.

8. Find M_{max} in the previous problem if $l = 12$ ft, $b = 3$ ft, $W = 12,000$ lb.

Answer. $M_{max} = 22,400$ ft lb.

9. Draw approximately to scale the shearing force and bending moment diagrams and label the values of the largest positive and negative shearing forces and bending moments for the beams with overhangs shown in Fig. 79.

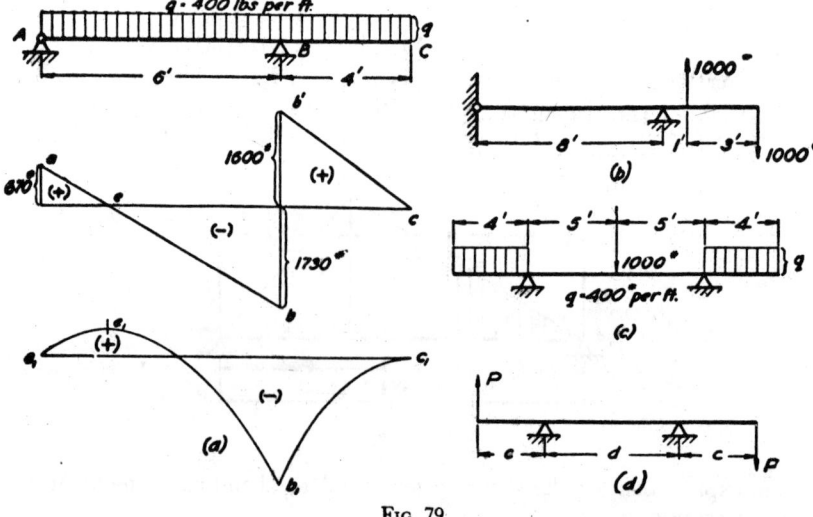

Fig. 79.

Solution. In the case shown in Fig. 79a the reactions are 670 lb and 3,330 lb. The shearing force for the left portion of the beam is $V = 670 - 400x$. It is represented in the figure by the inclined line ab. The shearing force for the right portion of the beam is found as for a cantilever beam and is shown by the inclined line $b'c$. The bending moment for the left portion of the beam is $M = 670x - 400x^2/2$. It is represented by the parabola $a_1e_1b_1$. The maximum of the moment at e_1 corresponds to the point e, at which the shearing force changes its sign. The bending moment diagram for the right portion is the same as for a cantilever and is represented by the parabola b_1c_1 tangent at c_1.

10. A beam with two equal overhangs, Fig. 80, loaded by a uniformly distributed load, has a length l. Find the distance d between

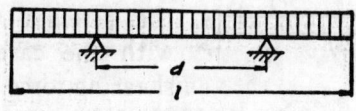

FIG. 80.

the supports such that the bending moment at the middle of the beam is numerically equal to the moments at the supports. Draw the shearing force and bending moment diagrams for this case.

Answer. $d = 0.586l$.

CHAPTER IV

STRESSES IN LATERALLY LOADED SYMMETRICAL BEAMS

23. Pure Bending.—It was mentioned in the preceding chapter that the magnitude of the stresses at any cross section of a beam is defined by the magnitude of the shearing force and bending moment at that cross section. To calculate the stresses we shall begin with the case in which the shearing force is zero and only bending moment is present. This case is called *pure bending*. An example of such bending is shown in Fig. 81. From symmetry we conclude that the reactions in this case are equal to P. Considering the equilibrium of the portion of the beam to the left of cross section mn, it can be concluded that the internal forces which are distributed over the cross section mn and which represent the action of the removed right portion of the beam on the left portion must be statically equivalent to a couple equal and opposite to the bending moment Pa. To find the distribution of these internal forces over the cross section, the deformation of the beam must be considered. For the simple case of a beam having a longitudinal plane of symmetry with the external bending couples acting in this plane, bending will take place in this same plane. If the beam is of rectangular cross section and two adjacent vertical lines mm and pp are drawn on its sides, direct experiment shows that these lines remain straight during bending and rotate so as to remain perpendicular to the longitudinal fibers of the beam (Fig. 82). The following theory of bending is based on the assumption that not only such lines as mm remain straight, but that the entire transverse section of the beam, originally

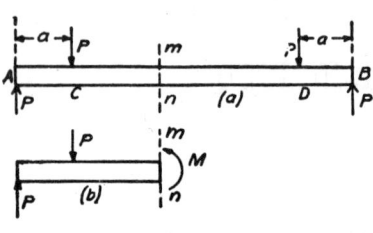

Fig. 81.

92

plane, remains plane and normal to the longitudinal fibers of the beam after bending. Experiment shows that the theory based on this assumption gives very accurate results for the deflection of beams and the strain of longitudinal fibers. From the above assumption it follows that during bending the cross sections mm and pp rotate with respect to each other about axes perpendicular to the plane of bending, so that longitudinal fibers on the convex side suffer extension and those on

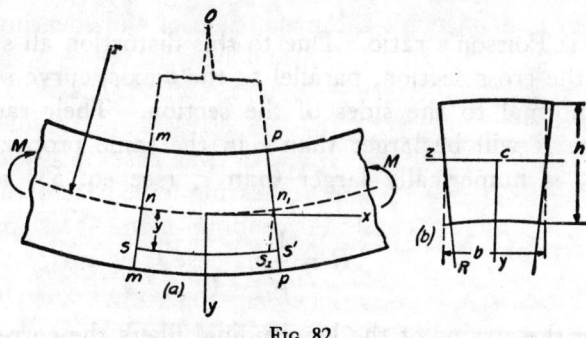

FIG. 82.

the concave side, compression. The line nn_1 is the trace of the surface in which the fibers do not undergo strain during bending. This surface is called the *neutral surface* and its intersection with any cross section is called the *neutral axis*. The elongation $s's_1$ of any fiber at distance y from the neutral surface is obtained by drawing the line n_1s_1 parallel to mm (Fig. 82a). Denoting by r the radius of curvature of the deflected axis [1] of the beam and using the similarity of the triangles non_1 and s_1n_1s', the unit elongation of the fiber ss' is

$$\epsilon_x = \frac{s's_1}{nn_1} = \frac{y}{r}. \tag{52}$$

It can be seen that the strains of the longitudinal fibers are proportional to the distance y from the neutral surface and inversely proportional to the radius of curvature.

Experiments show that longitudinal extension in the fibers on the convex side of the beam is accompanied by *lateral con-*

[1] The axis of the beam is the line through the centroids of its cross sections. O denotes the center of curvature.

traction and longitudinal compression on the concave side is accompanied by lateral expansion of the same magnitude, as in the case of simple tension or compression (see Art. 14). As a result of this the shape of the cross section changes, the vertical sides of the rectangular section becoming inclined to each other as in Fig. 82*b*. The unit strain in the lateral direction is

$$\epsilon_z = -\mu\epsilon_x = -\mu\frac{y}{r}, \tag{53}$$

where μ is Poisson's ratio. Due to this distortion all straight lines in the cross section, parallel to the z axis, curve so as to remain normal to the sides of the section. Their radius of curvature R will be larger than r in the same proportion in which ϵ_x is numerically larger than ϵ_z (see eq. 53) and we obtain

$$R = \frac{1}{\mu}r. \tag{54}$$

From the strains of the longitudinal fibers the corresponding stresses follow from Hooke's law (eq. 4, p. 4):

$$\sigma_x = \frac{Ey}{r}. \tag{55}$$

The distribution of these stresses is shown in Fig. 83. The stress in any fiber is proportional to its distance from the neutral axis *nn*. The position of the neutral axis and the radius of curvature r, the two unknowns in eq. (55), can now be determined from the condition that the forces distributed over any cross section of the beam must give rise to a *resisting couple* which balances the external couple M (Fig. 81).

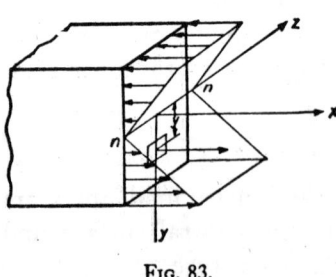

Fig. 83.

Let dA denote an elemental area of cross section at distance y from the neutral axis (Fig. 83). The force acting on this elemental area is the product of the stress (eq. 55) and the area dA, i.e., $(Ey/r)dA$. Due to the fact that all such forces distributed over the cross section represent a system equivalent to a couple, the resultant of these forces in the x direction must be equal to zero and we obtain

$$\int \frac{Ey}{r}\, dA = \frac{E}{r} \int y dA = 0,$$

i.e., the moment of the area of the cross section with respect ·to the neutral axis is equal to zero. Hence the *neutral axis passes through the centroid of the section.*

The moment of the force acting on the element dA with respect to the neutral axis is $(Ey/r)\cdot dA\cdot y$. Adding all such moments over the cross section and putting the resultant equal to the moment M of the external forces, the following equation for determining the radius of curvature r is obtained.

$$\int \frac{E}{r} y^2 dA = \frac{EI_z}{r} = M \quad \text{or} \quad \frac{1}{r} = \frac{M}{EI_z}, \qquad (56)$$

in which

$$I_z = \int y^2 dA$$

is the *moment of inertia* of the cross section with respect to the neutral axis z (see Appendix, p. 417). From eq. (56) it is seen that the curvature varies directly as the bending moment and inversely as the quantity EI_z, which is called the *flexural rigidity* of the beam. Elimination of r from eqs. (55) and (56) gives the following equation for the stresses:

$$\sigma_x = \frac{My}{I_z}. \qquad (57)$$

In this equation, M is positive when it produces a deflection of the bar convex down, as in Fig. 82; y is positive in the downward direction.

The preceding discussion was for the case of a rectangular cross section. It will also hold for a bar of any type of cross section which has a longitudinal plane of symmetry and which is bent by end couples acting in this plane. For such cases bending takes place in the plane of the couples and cross-sectional planes remain plane and normal to the longitudinal fibers after bending.

The maximum tensile and compressive stresses occur in the outermost fibers, and for a rectangular cross section, or any other cross section which has its centroid at the middle of the depth h, this will be for $y = \pm h/2$. Then for positive M we obtain

$$(\sigma_x)_{\max} = \frac{Mh}{2I_z} \quad \text{and} \quad (\sigma_x)_{\min} = -\frac{Mh}{2I_z}. \tag{58}$$

For simplicity the following notation is used:

$$Z = \frac{2I_z}{h}. \tag{59}$$

Then

$$(\sigma_x)_{\max} = \frac{M}{Z}, \quad (\sigma_x)_{\min} = -\frac{M}{Z}. \tag{60}$$

The quantity Z is called the *section modulus*. In the case of a rectangular cross section (Fig. 82b) we have

$$I_z = \frac{bh^3}{12}, \quad Z = \frac{bh^2}{6}.$$

For a circular cross section of diameter d,

$$I_z = \frac{\pi d^4}{64}, \quad Z = \frac{\pi d^3}{32}.$$

For the various profile sections in commercial use, such as WF beams, I beams, channels and so on, the magnitudes of I_z and Z for the sizes manufactured are tabulated in handbooks. An abridged listing of such sections is given in the Appendix.

When the centroid of the cross section is not at the middle of the depth, as, for instance, in the case of a T beam, let h_1 and h_2 denote the distances from the neutral axis to the outermost fibers in the downward and upward directions, respectively. Then for a positive bending moment we obtain

$$(\sigma_x)_{\max} = \frac{Mh_1}{I_z}, \quad (\sigma_x)_{\min} = -\frac{Mh_2}{I_z}. \tag{61}$$

For a negative bending moment we obtain

$$(\sigma_x)_{max} = -\frac{Mh_2}{I_z}, \qquad (\sigma_x)_{min} = \frac{Mh_1}{I_z}. \qquad (62)$$

Problems

1. Determine the maximum stress in a locomotive axle (Fig. 84) if $c = 13.5$ in., diameter d of the axle is 10 in. and the spring-borne load P per journal is 26,000 lb.

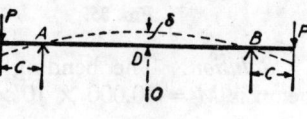

Solution. The bending moment acting in the middle portion of the axle is $M = P \times c = 26,000 \times 13.5$ in. lb. The maximum stress, from eq. (60), is

FIG. 84.

$$\sigma_{max} = \frac{M}{Z} = \frac{32 \cdot M}{\pi d^3} = \frac{32 \times 26,000 \times 13.5}{\pi \times 10^3} = 3,580 \text{ lb per sq in.}$$

2. Determine the radius of curvature r and the deflection of the axle of the preceding problem, if the material is steel and the distance AB is 59 in.

Solution. The radius of curvature r is determined from eq. (55) by substituting $y = d/2 = 5$ in., $(\sigma_x)_{max} = 3,580$ lb per sq in. Then

$$r = \frac{E}{\sigma} \cdot \frac{d}{2} = \frac{30 \times 10^6 \times 5}{3,580} = 41,900 \text{ in.}$$

For calculating δ (Fig. 84), observe that the deflection curve is an arc of a circle of radius r and $\overline{DB}$ is one leg of the right triangle DOB, where O is the center of curvature. Therefore

$$\overline{DB}^2 = r^2 - (r - \delta)^2 = 2r\delta - \delta^2.$$

δ is very small in comparison with the radius r and the quantity δ^2 can be neglected in the above equation. Then

$$\delta = \frac{\overline{DB}^2}{2r} = \frac{59^2}{8 \times 41,900} = 0.0104 \text{ in.}$$

3. A wooden beam of square cross section 10×10 in. is supported at A and B, Fig. 84, and the loads P are applied at the ends. Determine the magnitude of P and the deflection δ at the middle if $AB = 6'$, $c = 1'$, $(\sigma_x)_{max} = 1,000$ lb per sq in. and $E = 1.5 \times 10^6$

lb per sq in. The weight of the beam is to be neglected. Construct bending moment and shearing force diagrams.

Answer. $P = 13,900$ lb, $\delta = 0.0864$ in.

4. A standard 30" WF beam is supported as shown in Fig. 85

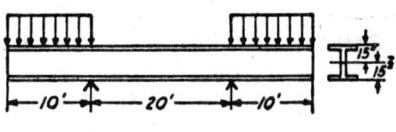

FIG. 85.

and loaded on the overhangs by a uniformly distributed load of 10,000 lb per ft. Determine the maximum stress in the middle portion of the beam and the deflection at the middle of the beam if $I_z = 7,892$ in.[4]

Solution. The bending moment for the middle portion of the beam is $M = 10,000 \times 10 \times 60 = 6 \times 10^6$ in. lb.

$$(\sigma_x)_{max} = \frac{M}{Z} = \frac{6 \times 10^6 \times 15}{7,892} = 11,400 \text{ lb per sq in.,}$$

$$\delta = 0.182 \text{ in.}$$

5. Determine the maximum stress produced in a steel wire of diameter $d = \frac{1}{32}$ in. when coiled around a pulley of diameter $D = 20$ in.

Solution. The maximum elongation due to bending, from eq. (52), is

$$\epsilon = \frac{d}{D + d} \approx \frac{1}{32 \times 20}$$

and the corresponding tensile stress is

$$(\sigma_x)_{max} = \epsilon E = \frac{30 \times 10^6}{32 \times 20} = 46,900 \text{ lb per sq in.}$$

6. A steel rule having a cross section $\frac{1}{32} \times 1$ in. and a length $l = 10$ in. is bent by couples at the ends into a circular arc of 60°. Determine the maximum stress and deflection.

Solution. The radius of curvature r is determined from the equation $l = 2\pi r/6$, from which $r = 9.55$ in., and the maximum stress will be given by eq. (55),

$$(\sigma_x)_{max} = \frac{E \times \frac{1}{64}}{r} = \frac{30 \times 10^6}{64 \times 9.55} = 49,100 \text{ lb per sq in.}$$

The deflection, calculated as for a circular arc, will be

$$\delta = r(1 - \cos 30°) = 1.28 \text{ in.}$$

7. Determine the maximum stress and the magnitude of the couples applied at the ends of the rule in the preceding problem if the maximum deflection at the middle is 1 in.

Answer. $(\sigma_x)_{max}$ = 38,300 lb per sq in., M = 6.23 in. lb.

8. Determine the curvature produced in a freely supported steel beam of rectangular cross section by nonuniform heating over the depth h of the cross section. The temperature at any point at distance y from the middle plane xz of the beam (Fig. 82) is given by the equation:

$$t = \frac{t_1 + t_0}{2} + \frac{(t_1 - t_0)y}{h},$$

where t_1 is the temperature at the bottom of the beam, t_0 is the temperature at the top, $t_1 - t_0$ = 123° F, and the coefficient of expansion $\alpha_s = 70 \times 10^{-7}$. What stresses will be produced if the ends of the beam are clamped?

Solution. The temperature of the middle plane xz is the constant $(t_1 + t_0)/2$, and the change in temperature of the other fibers is proportional to y. The corresponding unit thermal expansions are also proportional to y, i.e., they follow the same law as the unit elongations given by eq. (52). As a result of this nonuniform thermal expansion of the fibers, curving of the beam will occur and the radius of curvature r is found from eq. (52), using $\alpha_s(t_1 - t_0)/2$ for ϵ_x and $h/2$ for y. Then

$$r = \frac{h}{\alpha_s(t_1 - t_0)} = 1{,}160h.$$

If the ends of the beam are clamped, reactive couples M at the ends will be produced of magnitude such as to eliminate the curvature due to nonuniform heating. Hence

$$M = \frac{E \cdot I_z}{r} = \frac{EI_z}{1{,}160h}.$$

Substituting this in eq. (57), we obtain

$$\sigma_x = \frac{Ey}{1{,}160h},$$

and the maximum stress is

$$(\sigma_x)_{max} = \frac{E}{2 \times 1{,}160} = 12{,}900 \text{ lb per sq in.}$$

9. Solve Probs. 6 and 7 if the rule is bent into a circular arc of 10° and the material is copper.

10. Solve Prob. 4, assuming that the beam is of wood, has a square cross section 12″ × 12″ and the intensity of distributed load is 1,000 lb per ft. Construct bending moment and shearing force diagrams.

24. Various Shapes of Cross Sections of Beams.[2]—From the discussion in the previous article it follows that the maximum tensile and compressive stresses in a beam in pure bending are proportional to the distances of the most remote fibers from the neutral axis of the cross section. Hence if the material has the same strength in tension and compression, it will be logical to choose those shapes of cross section in which the centroid is at the middle of the depth of the beam. In this manner the same factor of safety for fibers in tension and fibers in compression will be obtained. This is the underlying idea in the use of sections which are symmetrical with respect to the neutral axis for such materials as structural steel, which have the same yield point stress in tension and compression. If the section is not symmetrical with respect to the neutral axis, for example a rail section, the material is so distributed between the head and the base as to have the centroid at the middle of its height.

For a material of small strength in tension and high strength in compression, as in the case of cast iron or concrete, the advisable cross section for a beam will not be symmetrical with respect to the neutral axis but will be such that the distances h_1 and h_2 from the neutral axis to the most remote fibers in tension and compression are in the same ratio as the strengths of the material in tension and in compression. In this manner equal strength in tension and compression is obtained. For example, with a T section, the centroid of the section may be put in any prescribed position along the height of the section by properly proportioning its flange and web.

For a given bending moment the maximum stress depends upon the section modulus and it is interesting to note that there are cases in which increase in cross-sectional area does

---

[2] A very complete discussion of various shapes of cross sections of beams is given by Barré de Saint-Venant in his notes to the book by Navier, *Résistance des corps solides*, 3d Ed., pp. 128–62, 1864.

not give a decrease in this stress. For example, a bar of square cross section bent by couples acting in the vertical plane through ⸱⸱ diagonal of the cross section (Fig. 86) will have ⸱ lower maximum stress if the corners, shown shaded on the fig ure, are cut off. Letting a denote the length of the side of the square cross section, the moment of inertia of the square with respect to the z axis is (see Appendix) $I_z = a^4/12$ and the corresponding section modulus is

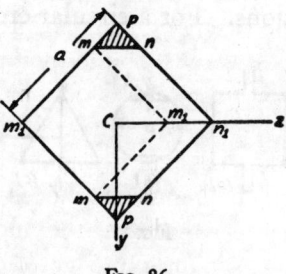

$$Z = \frac{I_z \sqrt{2}}{a} = \frac{\sqrt{2}}{12} a^3.$$

FIG. 86.

Let us now cut off the corners so that $\overline{mp} = \alpha a$, where α is a fraction to be determined later. The new cross section consists of a square mm_1mm_1 with the sides $a(1 - \alpha)$ and of two parallelograms mnn_1m_1. The moment of inertia of this new cross section with respect to the z axis is

$$I_z' = \frac{a^4(1 - \alpha)^4}{12} + 2 \cdot \frac{\alpha a \sqrt{2}}{3} \left[\frac{a(1 - \alpha)}{\sqrt{2}} \right]^3$$

$$= \frac{a^4(1 - \alpha)^3}{12} (1 + 3\alpha),$$

and the corresponding section modulus is

$$Z' = \frac{I_z' \sqrt{2}}{a(1 - \alpha)} = \frac{\sqrt{2}}{12} \cdot a^3 (1 - \alpha)^2 (1 + 3\alpha).$$

Now if we determine the value of α so as to make this section modulus a maximum, we find $\alpha = 1/9$. Using this value of α in Z', it is found that cutting off the corners diminishes the maximum bending stress by about 5 per cent.

This result is easily understood once we consider that the section modulus is the quotient of the moment of inertia and half the depth of the cross section. By cutting off the corners the moment of inertia of the cross section is diminished in a

smaller proportion than the depth is diminished, hence the
section modulus increases and $(\sigma_x)_{max}$ decreases. A similar
effect may be obtained in other cases. For a rectangle with
narrow outstanding portions (Fig. 87a) the section modulus is
increased, under certain conditions, by cutting off these portions. For a circular cross section (Fig. 87b) the section modulus is increased by 0.7 per cent by
cutting off the two shaded seg-
ments which have a depth $\delta =$
0.011d. In the case of a triangular
section (Fig. 87c) the section modulus can be increased by cutting
off the shaded corner.

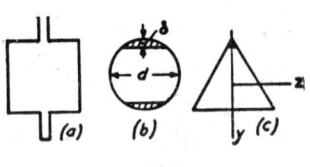

Fig. 87.

In designing a beam to undergo pure bending, not only th
conditions of strength should be satisfied but also the conditio
of economy in the weight of the beam itself. Of two cross sections having the same section modulus, i.e., satisfying the
condition of strength with the same factor of safety, the section with the smaller cross-sectional area is more economical.
In comparing various shapes of cross sections, we consider
first the rectangle of depth h and width b. The section modulus is

$$Z = \frac{bh^2}{6} = \frac{1}{6} Ah, \qquad (a)$$

where A denotes the cross-sectional area. It is seen that the
rectangular cross section becomes more and more economical
with increase in its depth h. However, there is a certain limit
to this increase, and the question of the stability of the beam
arises as the section becomes narrower. The collapse of a
beam of very narrow rectangular section may be due not to
overcoming the strength of the material but to sidewise buckling (see Part II).

In the case of a circular cross section we have

$$Z = \frac{\pi d^3}{32} = \frac{1}{8} A \cdot d. \qquad (b)$$

Comparing circular and square cross sections of the same

area, we find that the side h of the square will be $h = d\sqrt{\pi}/2$, for which eq. (a) gives

$$Z = 0.147A \cdot d.$$

Comparison of this with (b) shows a square cross section to be more economical than a circular one.

Consideration of the stress distribution along the depth of the cross section (Fig. 83) leads to the conclusion that for economical design most of the material of the beam should be put as far as possible from the neutral axis. The most favorable case for a given cross-sectional area A and depth h would be to distribute each half of the area at a distance $h/2$ from the neutral axis. Then

$$I_z = 2 \times \frac{A}{2} \times \left(\frac{h}{2}\right)^2 = \frac{Ah^2}{4}, \qquad Z = \tfrac{1}{2}Ah. \qquad (c)$$

This is a limit which may be somewhat approached in practice by the use of an I section or WF section with most of the material in the flanges. Due to the necessity of putting part of the material in the web of the beam, the limiting condition (c) can never be realized, and for standard WF profiles we have approximately

$$Z \approx 0.35Ah \qquad (d)$$

Comparison of (d) with (a) shows that a WF section is much more economical than a rectangular section of the same depth. In addition, due to its wide flanges, a WF beam will always be more stable with respect to sidewise buckling than a beam of rectangular section of the same depth and section modulus. From this brief discussion we see the reason for the wide application of WF beams in steel structures.

Problems

1. Determine the width x of the flange of a cast-iron beam having the section shown in Fig. 88, such that the maximum tensile stress is one-third of the maximum compressive stress. The depth of the beam $h = 4$ in., the thickness of the web and of the flange $t = 1$ in.

Solution. In order to satisfy the conditions of the problem it is necessary for the beam to have such dimensions that the distance c of the centroid from the extreme bottom edge will be equal to $\frac{1}{4}h$.

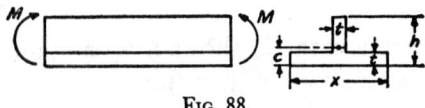

FIG. 88.

Now, referring to Fig. 88, we obtain the equation:

$$c = \frac{ht \cdot \dfrac{h}{2} + (x - t)\dfrac{t^2}{2}}{ht + (x - t)t} = \frac{h}{4},$$

from which

$$x = t + \frac{h^2}{h - 2t} = 1 + \frac{16}{4 - 2} = 9 \text{ in.}$$

2. Determine the ratio $(\sigma_x)_{\max} : (\sigma_x)_{\min}$ for the channel section shown in Fig. 89, if $t = 2$ in., $h = 10$ in., $b = 24$ in.
Answer. $(\sigma_x)_{\max} : (\sigma_x)_{\min} = 3 : -7$.

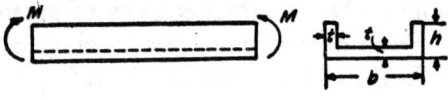

Fig. 89.

3. Determine the condition at which any further decrease of the depth h_1 of the section shown in Fig. 90 is accompanied by an increase in section modulus.
Solution.

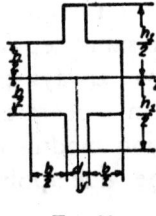

FIG. 90.

$$Z = \frac{bh^3}{6h_1} + \frac{dh_1^2}{6},$$

$$\frac{dZ}{dh_1} = -\frac{bh^3}{6h_1^2} + \frac{dh_1}{3}.$$

The condition for increase in Z with decrease of h_1 is

$$\frac{bh^3}{6h_1^2} > \frac{dh_1}{3} \quad \text{or} \quad \frac{b}{2d} > \frac{h_1^3}{h^3}.$$

4. Determine what amount should be cut from an equilateral triangular cross section (Fig. 87c) in order to obtain the maximum Z.

5. Determine the ratio of the weights of three beams of the same length under the same M and $(\sigma_x)_{\max}$ and having as cross sections, respectively, a circle, a square and a rectangle with proportions $h = 2b$.

Answer. 1.12:1:0.793.

6. Make a comparison of the section moduli for two beams of the same weight if the first beam is a solid circular beam of diameter d and the second is a circular tube of outer diameter D and inner diameter D_1.

Solution. The cross-sectional area of both beams is $A = \dfrac{\pi d^2}{4} = \dfrac{\pi(D^2 - D_1^2)}{4}$. For the solid beam $Z = Ad/8$, for the tubular beam $Z_1 = \dfrac{\pi(D^4 - D_1^4)}{32D} = \dfrac{AD}{8}\left(1 + \dfrac{D_1^2}{D^2}\right)$. Observing that $D_1^2 = D^2 - \dfrac{4A}{\pi}$, we find for the tubular beam $Z_1 = \dfrac{AD}{8}\left(2 - \dfrac{4A}{\pi D^2}\right)$, so that

$$\frac{Z_1}{Z} = \frac{D}{d}\left(2 - \frac{4A}{\pi D^2}\right).$$

Thus, for very thick tubes D approaches d and Z_1 approaches Z. For very thin tubes D is large in comparison with d and the ratio $Z_1:Z$ approaches the value $2D/d$.

25. General Case of Laterally Loaded Symmetrical Beams.

—In the general case of beams laterally loaded in a plane of symmetry, the stresses distributed over a cross section of a beam must balance the shearing force and the bending moment at that cross section. The calculation of the stresses is usually made in two steps by determining first the stresses produced by the bending moment, called the *bending stresses*, and afterwards the *shearing stresses* produced by the shearing force. In this article we shall limit ourself to the calculation of the bending stresses; the discussion of shearing stresses will be given in the next article. In calculating bending stresses we assume that these stresses are distributed in the same manner as in the case of pure bending and the formulas for the stresses derived in Art. 23 will be valid. (A more complete discussion of stresses near the points of application of concentrated forces is given in Part II.)

The calculation of bending stresses is usually made for the cross sections at which the bending moment has the largest positive or negative value. Having the numerical maximum of the bending moment and the magnitude of the allowable stress σ_W in bending, the required cross-sectional dimensions of a beam can be obtained from the equation

$$\sigma_W = \frac{M_{max}}{Z}. \qquad (63)$$

The application of this equation will now be shown by a number of examples.

Problems

1. Determine the necessary dimensions of a standard I beam to support a distributed load of 400 lb per ft, as shown in Fig. 91, if the working stress $\sigma_W = 16,000$ lb per sq in. Only the normal stresses σ_x are to be taken into consideration and the weight of the beam may be neglected.

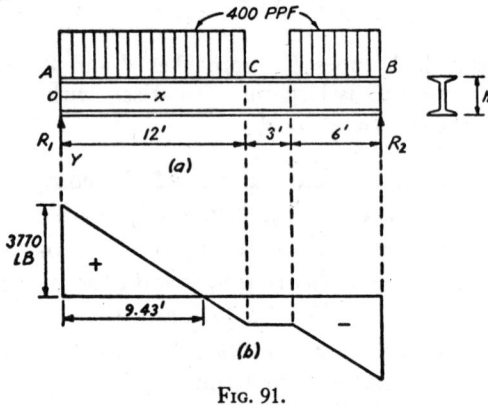

Fig. 91.

Solution. To obtain the section of maximum moment the shearing force diagram should be constructed, Fig. 91b. The reaction at the left support is

$$R_1 = \frac{12 \times 400 \times 15 + 6 \times 400 \times 3}{21} = 3,770 \text{ lb.}$$

The shearing force for any cross section of the portion AC of the beam is

$$V = R_1 - qx = 3,770 - 400 \times x.$$

This force is zero for $x = 3{,}770/400 = 9.43$ ft. For this section the bending moment is a maximum,

$$M_{max} = 3{,}770 \times 9.43 - 400 \times \tfrac{1}{2} \times 9.43^2 = 17{,}800 \times 12 \text{ in. lb.}$$

The required section modulus (eq. 63) is

$$Z \geqq \frac{17{,}800 \times 12}{16{,}000} = 13.4 \text{ in.}^3$$

This condition is satisfied by an 8 I 18.4 beam, with $Z = 14.2$ in.^{3} (See Appendix.)

2. A wooden dam (Fig. 92) is built up of vertical beams such as AB of rectangular cross section with dimension $h = 1$ ft and supported at the ends. Determine $(\sigma_x)_{max}$ if the length of the beams $l = 18$ ft and the weight of the beams is neglected.

Solution. If b is the width of one beam, the complete hydrostatic pressure on the beam, represented by the triangular prism ABC, is $W = \tfrac{1}{2}bl^2 \times 62.4$ lb. The reaction at A is $R_1 = \tfrac{1}{3}W = \tfrac{1}{6}bl^2 \times 62.4$ lb and the shearing force at any cross section mn is equal to the reaction R_1 minus the weight of the prism Amn of water, i.e.,

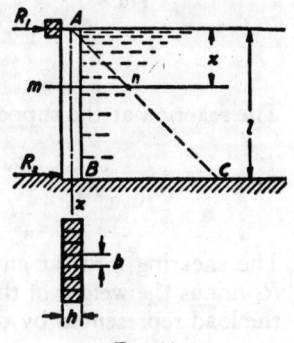

FIG. 92.

$$V = R_1 - W\frac{x^2}{l^2} = W\left(\frac{1}{3} - \frac{x^2}{l^2}\right).$$

The position of the cross section corresponding to M_{max} is found from the condition $V = 0$ or

$$\frac{1}{3} - \frac{x^2}{l^2} = 0,$$

from which

$$x = \frac{l}{\sqrt{3}} = 10.4 \text{ ft.}$$

The bending moment at any cross section mn is equal to the moment of the reaction R_1 minus the moment of the distributed load represented by the triangular prism Amn. Then

$$M = R_1 x - \frac{Wx^2}{l^2} \cdot \frac{x}{3} = \frac{Wx}{3}\left(1 - \frac{x^2}{l^2}\right).$$

Substituting, from the above, $x^2/l^2 = \frac{1}{3}$ and $x = 10.4$ ft, we obtain

$$M_{max} = \tfrac{1}{9}bl^2 \times 62.4 \times 10.4 \text{ ft lb},$$

$$(\sigma_x)_{max} = \frac{M_{max}}{Z} = \frac{6M_{max}}{bh^2} = \frac{2}{3}\left(\frac{l}{h}\right)^2 \frac{62.4 \times 10.4}{12^2} = 973 \text{ lb per sq in.}$$

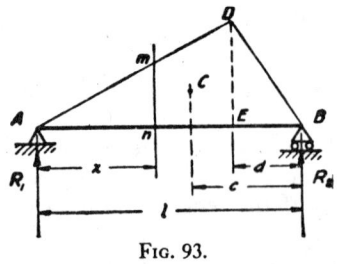

Fig. 93.

3. Determine the magnitude of M_{max} in a beam loaded by a triangular load ADB equal to $W = 12,000$ lb if $l = 12$ ft and $d = 3$ ft (Fig. 93).

Solution. The distance c to the vertical through the center of gravity C from the support B is, in the case of a triangle,

$$c = \tfrac{1}{3}(l + d) = 5 \text{ ft.}$$

The reaction at the support A is then

$$R_1 = \frac{W \cdot c}{l} = \frac{12,000 \times 5}{12} = 5,000 \text{ lb.}$$

The shearing force at any cross section mn is equal to the reaction R_1 minus the weight of the load represented by the area Amn. Since the load represented by the area ADE is

$$\frac{W(l-d)}{l} = \frac{3}{4}W,$$

we obtain

$$V = R_1 - \frac{3}{4}W\frac{x^2}{(l-d)^2}.$$

The position of the section with maximum moment is found from the condition

$$R_1 - \frac{3}{4}W\frac{x^2}{(l-d)^2} = 0$$

or

$$\frac{x^2}{(l-d)^2} = \frac{4R_1}{3W} = \frac{5}{9},$$

from which

$$x = 6.71 \text{ ft.}$$

The bending moment at any cross section mn is equal to the moment of the reaction R_1 minus the moment of the load Amn. Then

$$M = R_1 x - \frac{3}{4} W \frac{x^2}{(l-d)^2}$$

Substituting $x = 6.71$ ft, we obtain

$$M_{max} = 22,400 \text{ ft lb.}$$

4. Construct the bending moment and shearing force diagrams for the beam shown in Fig. 94a and determine the size of I beam required to carry the load if $a = c = l/4 = 6$ ft, $P = 2,000$ lb, $q = 400$ lb per ft, $\sigma_W = 15,000$ lb per sq in. The weight of the beam may be neglected.

Solution. In Figs. 94b and 94c the bending moment and shearing force diagrams produced by the distributed loads are shown. To these must be added the moment and shearing force produced by P. The maximum bending moment will be at the middle of the span and is

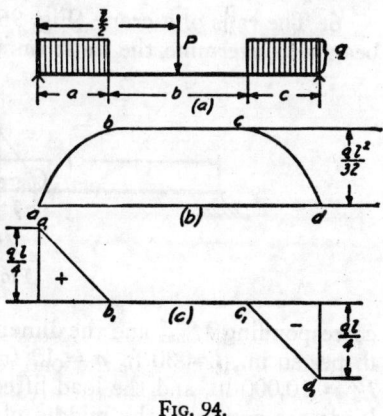

Fig. 94.

$$M_{max} = \frac{ql^2}{32} + \frac{Pl}{4} = 19,200 \text{ ft lb.}$$

The required section modulus is $Z = \dfrac{19,200 \times 12}{15,000} = 15.4 \text{ in.}^3$ The standard 8 I 23 beam of depth 8 in. and cross-sectional area 6.71 sq in., $Z = 16.0$ in.^{3}, is the nearest cross section satisfying the strength requirements (see Appendix).

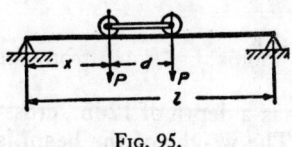

Fig. 95.

5. Determine the most unfavorable position of the hoisting carriage of a crane which rides on a beam as shown in Fig. 95. Find M_{max} if the load per wheel is $P = 10,000$ lb, $l = 24$ ft, $d = 6$ ft. The weight of the beam may be neglected.

Solution. It x is the distance of the left wheel from the left support of the beam, the bending moment under this wheel is

$$\frac{2P(l - x - \tfrac{1}{2}d)x}{l}.$$

This moment becomes a maximum when

$$x = \frac{l}{2} - \frac{d}{4}.$$

Hence in order to obtain the maximum bending moment under the left wheel the carriage must be displaced from the middle position by a distance $d/4$ towards the right support. The same magnitude of bending moment can also be obtained under the right wheel by displacing the carriage by $d/4$ from the middle position towards the left support.

$$M_{max} = \frac{2P(l/2 - d/4)^2}{l} = 91{,}900 \text{ ft lb.}$$

6. The rails of a crane (Fig. 96) are supported by two standard I beams. Determine the most unfavorable position of the crane, the

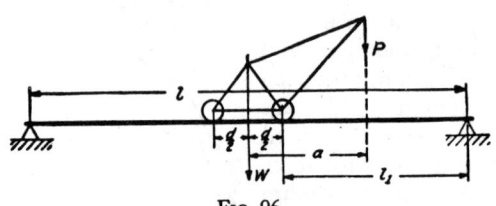

FIG. 96.

corresponding M_{max} and the dimensions of the I beams if $\sigma_W = 15{,}000$ lb per sq in., $l = 30$ ft, $a = 12$ ft, $d = 6$ ft, the weight of the crane $W = 10{,}000$ lb, and the load lifted by the crane $P = 2{,}000$ lb. The loads are acting in the middle plane between the two I beams and are equally distributed between them.

Solution. The maximum bending moment will be under the right wheel when the distance of this wheel from the right support is equal to $l_1 = \frac{1}{2}(l - \frac{1}{6}d)$; $M_{max} = 1{,}009{,}000$ in. lb. Dividing the moment equally between the two beams, we find the necessary section modulus

$$Z = \frac{M_{max}}{2\sigma_W} = 33.6 \text{ in.}^3$$

The necessary I beam is 12 I 31.8, which has a depth of 12 in., cross-sectional area 9.26 sq in., $Z = 36.0$ in.^{3} The weight of the beam is neglected.

7. A circular wooden beam supported at C and attached to the foundation at A (Fig. 97) carries a load $q = 300$ lb per ft uniformly distributed along the portion BC. Construct the bending moment diagram and determine the necessary diameter d if $\sigma_W = 1{,}200$ lb per sq in., $a = 3$ ft, $b = 6$ ft.

Solution. The bending moment diagram is shown in Fig. 97*b*. The numerically largest moment will be at C and is equal to 64,800 in. lb. Then from eq. (63),

$$d = \sqrt[3]{\frac{32}{\pi} \cdot \frac{M}{\sigma_W}} = 8.2 \text{ in.}$$

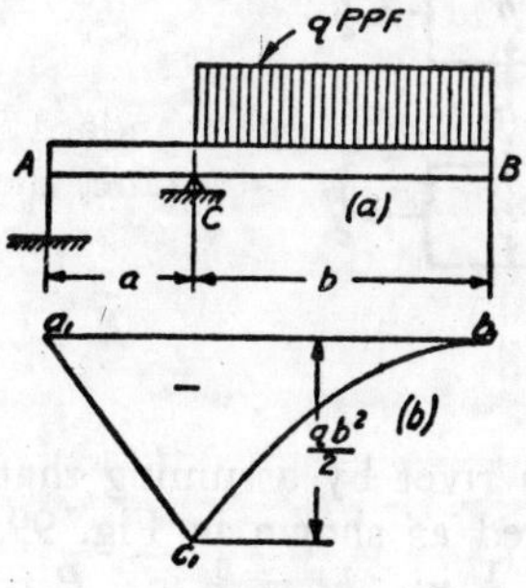

FIG. 97. 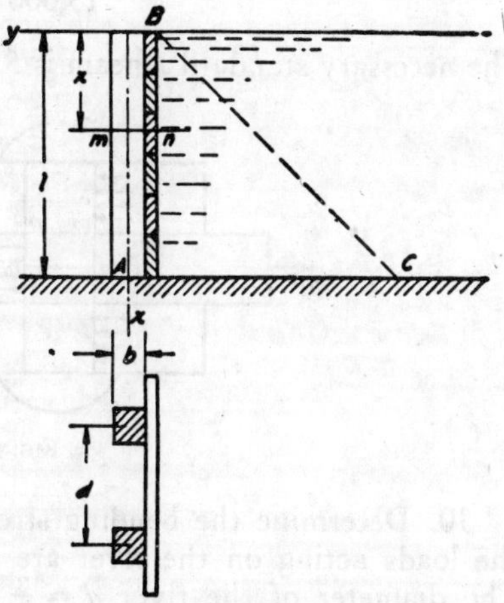   FIG. 98.

8. A wooden dam consists of horizontal boards backed by vertical pillars built in at the lower ends (Fig. 98). Determine the dimension of the square cross section of the pillars if $l = 6$ ft, $d = 3$ ft and $\sigma_W = 500$ lb per sq in. Construct the bending moment and shearing force diagrams.

Solution. The total lateral load on one pillar is represented by the weight W of the triangular prism ABC of water. At any cross section mn, the shearing force and the bending moment are

$$V = -\frac{W \cdot x^2}{l^2}, \qquad M = -\frac{W x^2}{l^2} \cdot \frac{x}{3}.$$

In determining the signs of V and M it is assumed that Fig. 98 is rotated 90° in the counter-clockwise direction so as to bring the axes x and y into coincidence with those of Fig. 61. The necessary dimension b is found from eq. (63),

$$Z = \frac{b^3}{6} = \frac{M_{max}}{\sigma_W} = \frac{3 \times 6^2 \times 62.4 \times 12}{500},$$

from which

$$b = 9.90 \text{ in.}$$

The construction of diagrams is left to the reader.

9. Determine the necessary dimensions of a cantilever beam of standard I section which carries a uniform load $q = 200$ lb per ft and a concentrated load $P = 500$ lb at the end if the length $l = 5$ ft and $\sigma_w = 15{,}000$ lb per sq in.

Answer. $Z = \dfrac{(500 \times 5 + 1{,}000 \times 2.5)12}{15{,}000} = 4 \text{ in.}^3$

The necessary standard i beam is 5 I 10 (see Appendix).

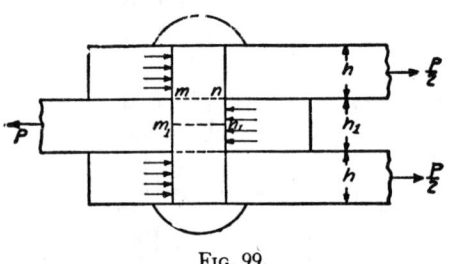

Fig. 99.

10. Determine the bending stresses in a rivet by assuming that the loads acting on the rivet are distributed as shown in Fig. 99. The diameter of the rivet $d = \frac{3}{4}$ in., $h = \frac{1}{4}$ in., $h_1 = \frac{3}{8}$ in., $P = 10{,}000$ lb.

Solution. The bending moment at the cross section mn is $P/2 \times h/2$. The bending moment at the middle cross section m_1n_1 is

$$\frac{P}{2}\left(\frac{h}{2} + \frac{h_1}{4}\right).$$

This latter moment is maximum and is the one to be taken into account in calculating the stresses. Then

$$(\sigma_x)_{\max} = \frac{P}{2}\left(\frac{h}{2} + \frac{h_1}{4}\right) \div \frac{\pi d^3}{32} = \frac{4P}{\pi d^2} \cdot \frac{2h + h_1}{d} = 26{,}400 \text{ lb per sq in.}$$

11. Determine the required section moduli Z_1, Z_2 and Z_3, and the necessary standard I beams for the cases shown in Figs. 72a, 72c and 73b, assuming a working stress of 16,000 lb per sq in.

Answer. $Z_1 = 5.14 \text{ in.}^3$, $Z_2 = 11.1 \text{ in.}^3$, $Z_3 = 22.5 \text{ in.}^3$

12. Determine the section modulus Z and the necessary dimensions of a simply supported beam of standard WF section such as to carry a uniform load of 400 lb per ft and a concentrated load $P = 4{,}000$ lb placed at the middle. The length of the beam is 15 ft and the working stress $\sigma_w = 16{,}000$ lb per sq in.

Answer. $Z = 19.7 \text{ in.}^3$

13. A channel with the cross section shown in Fig. 89 is simply supported at the ends and carries a concentrated load P at the middle. Calculate the maximum value of the load which the beam will carry if the working stress is 1,000 lb per sq in. for tension, 2,000 lb per sq in. for compression, $t = 2$ in., $h = 10$ in., $b = 24$ in. and the length $l = 10$ ft.

Answer. $P = 6,350$ lb.

26. Shearing Stresses in Bending.—It was shown in the preceding article that when a beam is bent by transverse loads not only normal stresses σ_x but also shearing stresses τ are produced in any cross section mn of the beam, Fig. 100.

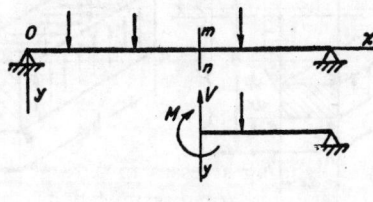

Fig. 100.

Considering the action on the right portion of the beam, Fig. 100, it can be concluded from the conditions of equilibrium that the magnitude of these shearing stresses is such that their sum gives the shearing force V. In investigating the law of distribution of these shearing stresses over the area of the cross section, we begin with the simple case of a rectangular cross section $mmnn$, Fig. 101. In such a case it is natural to assume that the shearing stress at each point of the cross section is parallel to the shearing force V, i.e., parallel to the sides mn of the cross section. We denote the stress in this case by τ_{xy}. The subscript y in τ_{xy} indicates that the shearing stress is parallel to the y axis and the subscript x that the stress acts in a plane perpendicular to the x axis. As a second assumption we take the distribution of the shearing stresses to be uniform across the width of the beam cc_1. These two assumptions will enable us to determine completely the distribution of the shearing stresses. A more elaborate investigation of the problem shows that the approximate solution thus obtained is usually sufficiently accurate and that for a narrow rectangle

(h large in comparison with b, Fig. 101) it practically coincides with the exact solution.[3]

If an element be cut out from the beam by two adjacent cross sections and by two adjacent planes parallel to the neutral plane, as element $acdea_1c_1d_1e_1$ in Fig. 101b, then in accordance with our assumption there is a uniform distribution of the

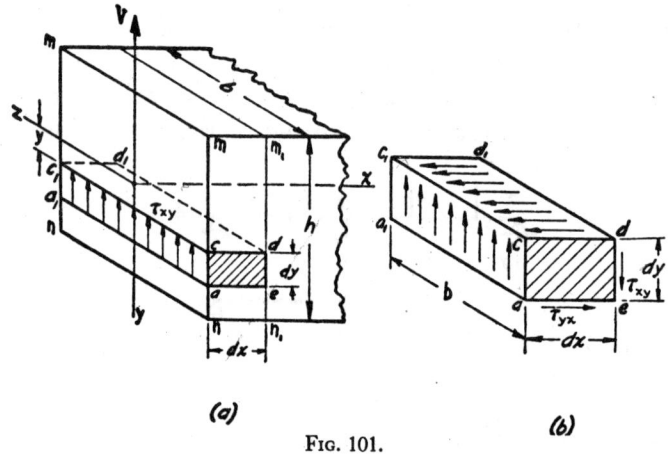

(a) *(b)*

Fig. 101.

shearing stresses τ_{xy} over the vertical face acc_1a_1. These stresses have a moment $(\tau_{xy}bdy)dx$ about the lower rear edge ee_1 of the element, which must be balanced by the moment $(\tau_{yx}bdx)dy$ due to shearing stresses distributed over the horizontal face of the element, cdd_1c_1. Then

$$\tau_{xy}bdydx = \tau_{yx}bdxdy \quad \text{and} \quad \tau_{yx} = \tau_{xy},$$

i.e., the shearing stresses acting on the two perpendicular faces of the element are equal.[4] The same conclusion was

---

[3] The exact solution of this problem is due to St.-Venant, *J. math.* (Liouville), 1856. An account of St.-Venant's famous work is given in Todhunter and Pearson's *History of the Theory of Elasticity*, Cambridge, 1886–93. The approximate solution given in this article is by Jourawski. For the French translation of his work, see *Ann. ponts et chaussées*, 1856. The exact theory shows that when the depth of the beam is small in comparison with the width, the discrepancy between the exact and the approximate theories becomes considerable.

[4] We consider here only the absolute value of these stresses.

met before in simple tension (see p. 42) and also in tension or compression in two perpendicular directions (see p. 48). The existence of shearing stresses in the planes parallel to the neutral plane can be demonstrated by simple experiments. Take two equal rectangular bars placed together on simple supports as shown in Fig. 102 and bent by a concentrated load P. If there is no friction between the bars, the bending of each bar will be independent of that of the other; each will have compression of the upper and tension of the lower longitudinal fibers and the condition will be that indicated in Fig. 102b. The lower

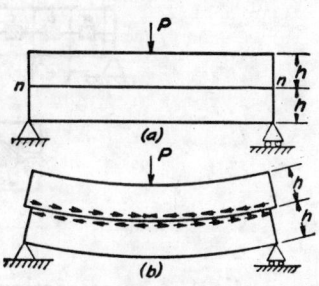

FIG. 102.

longitudinal fibers of the upper bar slide with respect to the upper fibers of the lower bar. In a solid bar of depth $2h$ (Fig. 102a) there will be shearing stresses along the neutral plane nn of such magnitude as to prevent this sliding of the upper portion of the bar with respect to the lower, shown [5] in Fig. 102b. Due to this prevention of sliding the single bar of depth $2h$ is much stiffer and stronger than two bars each of depth h. In practice, keys such as a, b, c, $\cdots$ are sometimes used with built-up wooden beams in order to prevent sliding (Fig. 103a). Observation of the clearances around a key, Fig. 103b, enables us to determine the direction of sliding in the case of a built-up beam and therefore the direction of the shearing stresses over the neutral plane in the case of a solid beam.[6]

The above discussion shows that the shearing stress τ_{xy} at any point of the vertical cross section of the beam is vertical in direction and numerically equal to the horizontal shearing stress τ_{yx} in the horizontal plane through the same point.

---

[5] The upper row of arrows indicates the action of the lower half of the beam on the upper half. The lower row of arrows shows the action of the upper half of the beam on the lower half.

[6] For an analysis of built-up wooden beams see the paper by F. Stüssi, "Schwere Notbrücke mit verdübelten Balken," Gesellschaft für militärische Bautechnik, Zürich.

This latter stress can easily be calculated from the condition of equilibrium of the element pp_1nn_1 cut out from the beam by

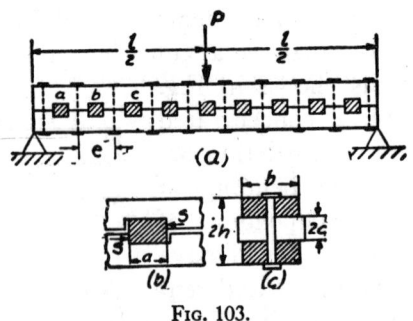

Fig. 103.

two adjacent cross sections mn and m_1n_1 and by the horizontal plane pp_1, Figs. 104a and 104b. The only forces on this element in the direction of the x axis are the shearing stresses τ_{yx} over the side pp_1 and the normal stress σ_x over the sides pn and p_1n_1. If the bending moments at cross sections mn and m_1n_1 are equal, i.e., in the case of pure bending, the normal

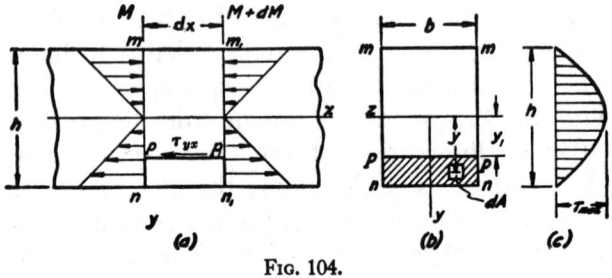

Fig. 104.

stresses σ_x over the sides np and n_1p_1 will also be equal and will be in balance between themselves. Then the shearing stress τ_{yx} must be equal to zero.

Let us consider now the more general case of a varying bending moment, denoting by M and $M + dM$ the moments in the cross sections mn and m_1n_1, respectively. Then the normal force acting on an elemental area dA of the side $nppn$ will be (eq. 57)

$$\sigma_x dA = \frac{My}{I_z} dA.$$

The sum of all these forces distributed over the side *nppn* of the element will be

$$\int_{y_1}^{h/2} \frac{My}{I_z} dA. \qquad (a)$$

In the same manner the sum of the normal forces acting on the side $n_1p_1p_1n_1$ is

$$\int_{y_1}^{h/2} \frac{(M + dM)y}{I_z} dA. \qquad (b)$$

The force due to the shearing stresses τ_{yx} acting on the top side pp_1 of the element is

$$\tau_{yx}bdx. \qquad (c)$$

The forces given in (*a*), (*b*) and (*c*) must satisfy $\Sigma X = 0$, hence

$$\tau_{yx}bdx = \int_{y_1}^{h/2} \frac{(M + dM)y}{I_z} dA - \int_{y_1}^{h/2} \frac{My}{I_z} dA,$$

from which

$$\tau_{yx} = \frac{dM}{dx} \frac{1}{b \cdot I_z} \int_{y_1}^{h/2} y dA,$$

or, by using eq. (50),

$$\tau_{xy} = \tau_{yx} = \frac{V}{bI_z} \int_{y_1}^{h/2} y dA. \qquad (64)$$

The integral in this equation represents the moment of the shaded portion of the cross section, Fig. 104*b*, with respect to the neutral axis *z*. For the rectangular section discussed,

$$dA = bdy$$

and the integral becomes

$$\int_{y_1}^{h/2} bydy = \left| \frac{by^2}{2} \right|_{y_1}^{h/2} = \frac{b}{2}\left(\frac{h^2}{4} - y_1{}^2 \right). \qquad (d)$$

The same result can be obtained by multiplying the area $b[(h/2) - y_1]$ of the shaded portion by the distance

$$\tfrac{1}{2}[(h/2) + y_1]$$

of its centroid from the neutral axis.

Substituting (d) in eq. (64), we obtain for the rectangular section

$$\tau_{xy} = \tau_{yx} = \frac{V}{2I_z}\left(\frac{h^2}{4} - y_1{}^2\right). \tag{65}$$

It is seen that the shearing stresses τ_{xy} are not uniformly distributed from top to bottom of the beam. The maximum value of τ_{xy} occurs for $y_1 = 0$, i.e., for points on the neutral axis, and is, from eq. (65),

$$(\tau_{xy})_{\max} = \frac{Vh^2}{8I_z}$$

or, since $I_z = bh^3/12$,

$$(\tau_{xy})_{\max} = \frac{3}{2} \cdot \frac{V}{bh}. \tag{66}$$

Thus the maximum shearing stress in the case of a rectangular cross section is 50 per cent greater than the average shearing stress, obtained by dividing the shearing force by the area of the cross section.

In the preceding derivation we took the element pnp_1n_1 from the lower portion of the beam. The same result is obtained by taking the element from the upper portion.

For points at the bottom and top of the cross section, $y_1 = \pm h/2$ and eq. (65) gives $\tau_{xy} = 0$. The graph of eq. (65) (Fig. 104c) shows that the distribution of the shearing stresses along the depth of the beam follows a parabolic law. The

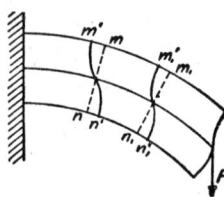

shaded area bounded by the parabola when multiplied by the width b of the beam gives $\frac{2}{3}(\tau_{xy})_{\max}hb = V$, as it should.

A natural consequence of the shearing stresses is shearing strain which causes cross sections, initially plane, to become warped. This warping can be easily demonstrated by bending with a force on the

end a rectangular piece of rubber (Fig. 105), on whose sides vertical lines have been drawn. The lines will not remain straight as indicated by the dotted lines, but become curved, so that the maximum shear strain occurs at the neutral surface. At the points m', m_1', n', n_1' the shearing strain is zero and the curves $m'n'$ and $m_1'n_1'$ are normal to the upper and lower surfaces of the bar after bending. At the neutral surface the angles between the tangents to the curves $m'n'$ and $m_1'n_1'$ and the normal sections mn and m_1n_1 are equal to $\gamma = (\tau_{xy})_{max}/G$. As long as the shearing force V remains constant along the beam the warping of all cross sections is the same, so that $mm' = m_1m_1'$, $nn' = n_1n_1'$ and the stretching or the shrinking produced by the bending moment in the longitudinal fibers is unaffected by shear. This fact explains the validity here of eq. (57), which was developed for pure bending and was based on the assumption that cross sections of a bar remain plane during bending.

A more elaborate investigation of the problem [7] shows also that the warping of cross sections does not substantially affect the strain in longitudinal fibers if a distributed load acts on the beam and the shearing force varies continuously along the beam. In the case of concentrated loads the stress distribution near the loads is more complicated, but this deviation from the straight line law is of a local type (see Part II).

Problems

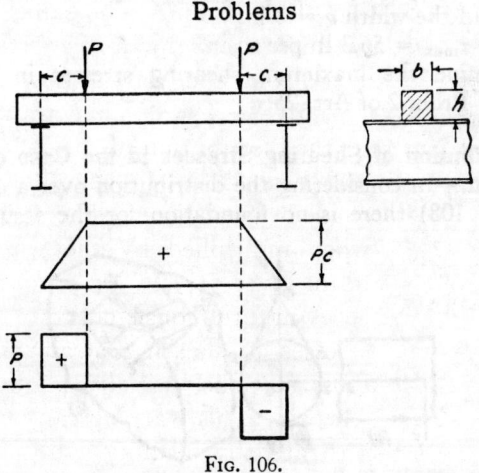

FIG. 106.

1. Determine the limiting values of the loads P acting on the wooden rectangular beam, Fig. 106, if $b = 8$ in., $h = 10$ in., $\sigma_W = 800$ lb per sq in., $\tau_W = 200$ lb per sq in., $c = 1.5$ ft.

[7] See W. Voigt, *Göttingen Abhandl.*, Vol. 34, 1887; J. H. Michell, *Quart. J. Math.*, Vol. 32, p. 63, 1901; L. N. G. Filon, *Trans. Roy. Soc. (London)*, *A*, Vol. 201, 1903, and *Proc. Roy. Soc. (London)*, Vol. 72, 1904. See also Th. Kármán, *Abhandl. Aerodyn. Inst.*, *Tech. Hochschule (Aachen)*, Vol. 7, 1927.

Solution. The bending moment and shearing force diagrams are given in Fig. 106.

$$V_{max} = P, \qquad M_{max} = P \cdot c.$$

From equations

$$\frac{Pc}{Z} = \sigma_W \qquad \text{and} \qquad \frac{3}{2}\frac{P}{bh} = \tau_W,$$

we obtain

$$P = 5,930 \text{ lb} \qquad \text{and} \qquad P = 10,700 \text{ lb}.$$

Therefore $P = 5,930$ lb is the limiting value of the load P.

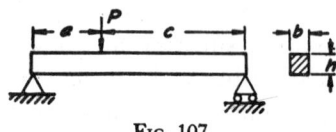

FIG. 107.

2. Determine the maximum normal stress σ_x and the maximum shearing stress τ_{yx} in the neutral plane of the beam represented in Fig. 107 if $a = 2$ ft, $c = 4$ ft, $b = 8$ in., $h = 10$ in. and $P = 6,000$ lb.

Answer. $(\sigma_x)_{max} = 720$ lb per sq in., $(\tau_{yx})_{max} = 75$ lb per sq in.

3. Determine the maximum shearing stress in the neutral plane of a uniformly loaded rectangular beam if the length of the beam $l = 6$ ft, the load $q = 1,000$ lb per ft, the depth of the cross section $h = 10$ in. and the width $b = 8$ in.

Answer. $\tau_{max} = 56.3$ lb per sq in.

4. Determine the maximum shearing stresses in the vertical beams AB of Prob. 2 of Art. 25.

27. Distribution of Shearing Stresses in the Case of a Circular Cross Section.—In considering the distribution over a circular cross section (Fig. 108) there is no foundation for the assumption that

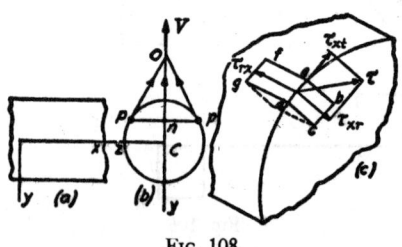

FIG. 108.

the shearing stresses are all parallel to the shearing force V. In fact we can readily show that at points p (Fig. 108b) of the cross section along the boundary, the shearing stress must be tangent to the boundary. Let us consider an infinitesimal element $abcd$ (Fig. 108c) in the form of a rectangular parallelepiped with the face $adfg$ in the surface of the beam and the face *abcd* in the plane *yz* of the cross section. If the shearing stress acting over the side *abcd* of the element has a direction such as τ, it can always be resolved into two components τ_{xr} in a radial direction and τ_{xt} in the direction of the tangent to the boundary. Now it has been proved before (see p. 114), by using the equation of equilibrium of an element, that if a shearing stress τ acts over an elemental area, a numerically equal shearing stress will act also over an elemental area perpendicular to τ. Applying this in our case, it must be concluded that if a shearing stress τ_{xr} is acting on the element *abcd* in a radial direction there must be an equal shearing stress τ_{rx} on the side *adfg* of the element lying in the surface of the beam. If the lateral surface of the beam is free from shearing stresses, the radial component τ_{xr} of the shearing stress τ must be equal to zero, i.e., τ must be in the direction of the tangent to the boundary of the cross section of the beam. At the midpoint *n* of the chord *pp*, symmetry requires that the shearing stress has the direction of the shearing force V. Then the directions of the shearing stresses at the points *p* and *n* will intersect at some point *O* on the *y* axis (Fig. 108*b*). Assuming now that the shearing stress at any other point of the line *pp* is also directed toward the point *O*, we define completely the directions of the shearing stresses. As a second assumption we take the vertical components of the shearing stresses equal for all points of the line *pp*.[8] As this assumption coincides completely with that made in the case of a rectangular cross section, we can use eq. (64) for calculating this component. In such a case, *b* will denote the length of the chord *pp*. Knowing the actual direction of the shearing stress and its vertical component, its magnitude may be easily calculated for any point of the cross section.

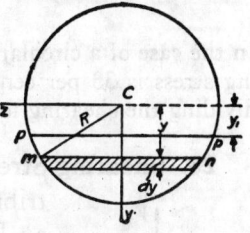

Fig. 109.

Let us calculate now the shearing stresses along the line *pp* of the cross section (Fig. 109). In applying eq. (64) to the calculation of the vertical component τ_{xy} of these stresses, we must find the moment of the segment of the circle below the line *pp* with respect to the *z* axis. The elemental area *mn* has the length $2\sqrt{R^2 - y^2}$ and the width *dy*. The area is $dA = 2\sqrt{R^2 - y^2}\, dy$. The moment

---

[8] The approximate theory based on the above two assumptions gives satisfactory accuracy, and comparison with the exact theory shows that the error in the magnitude of the maximum shearing stress is about 5 per cent which is not high for practical application. See St.-Venant, *loc. cit.*, p. 100. See also, A. E. H. Love, *Mathematical Theory of Elasticity*, 4th Ed., p. 346, 1927.

of this strip about Cz is ydA and the total moment for the entire segment is

$$\int_{y_1}^{R} 2\sqrt{R^2 - y^2} \cdot y\,dy = \tfrac{2}{3}(R^2 - y_1^2)^{3/2}.$$

Substituting this into eq. (64) and taking $2\sqrt{R^2 - y_1^2}$ for b, we obtain for the vertical shearing stress component

$$\tau_{xy} = \frac{V(R^2 - y_1^2)}{3I_z},\tag{67}$$

and the total shearing stress at points p (Fig. 109) is

$$\tau = \frac{\tau_{xy} \cdot R}{\sqrt{R^2 - y_1^2}} = \frac{VR\sqrt{R^2 - y_1^2}}{3I_z}.$$

It is seen that the maximum τ is obtained for $y_1 = 0$, i.e., for the neutral axis of the cross section. Then, substituting $I_z = \pi R^4/4$, we obtain

$$\tau_{max} = \frac{4}{3}\frac{V}{\pi R^2} = \frac{4}{3} \cdot \frac{V}{A}.\tag{68}$$

In the case of a circular cross section, therefore, the maximum shearing stress is 33 per cent larger than the average value obtained by dividing the shearing force by the cross-sectional area.

28. Shearing Stresses in I Beams.

—In considering the distribution of shearing stresses in the web of an I beam or WF beam (Fig. 110), the same assumptions are made as for a rectangular cross section; these were that the shearing stresses are parallel to the shearing force V and are uniformly distributed over the thickness b_1 of the web. Then eq. (64) may be used for calculating the stresses τ_{xy}. For points on the line pp at a distance y_1 from the neutral axis, where the width of the cross section is b_1, the moment of the shaded portion of the cross section with respect to the neutral axis z is

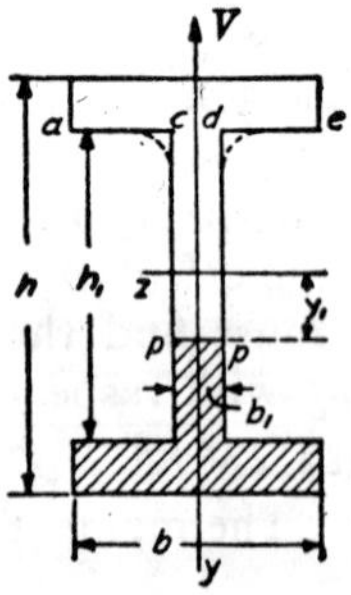

FIG. 110.

$$\int_{y_1}^{h/2} ydA = \frac{b}{2}\left(\frac{h^2}{4} - \frac{h_1^2}{4}\right) + \frac{b_1}{2}\left(\frac{h_1^2}{4} - y_1^2\right).$$

Substituting in eq. (64), we obtain

$$\tau_{xy} = \frac{V}{b_1 I_z}\left[\frac{b}{2}\left(\frac{h^2}{4} - \frac{h_1^2}{4}\right) + \frac{b_1}{2}\left(\frac{h_1^2}{4} - y_1^2\right)\right]. \quad (69)$$

It is seen that the stress τ_{xy} varies along the depth of the beam following a parabolic law. The maximum and minimum values of τ_{xy} in the web of the beam are obtained by putting $y_1 = 0$ and $y_1 = h_1/2$, which gives

$$(\tau_{xy})_{\max} = \frac{V}{b_1 I_z}\left[\frac{bh^2}{8} - \frac{h_1^2}{8}(b - b_1)\right], \quad (70)$$

$$(\tau_{xy})_{\min} = \frac{V}{b_1 I_z}\left(\frac{bh^2}{8} - \frac{bh_1^2}{8}\right). \quad (71)$$

When b_1 is very small in comparison with b there is no great difference between $(\tau_{xy})_{\max}$ and $(\tau_{xy})_{\min}$ and the distribution of the shearing stresses over the cross section of the web is practically uniform.

A good approximation for $(\tau_{xy})_{\max}$ is obtained by dividing the complete shearing force V by the cross-sectional area of the web alone. This follows from the fact that the shearing stresses distributed over the cross section of the web yield a force which is nearly equal to V, which means that the web takes nearly all the shearing force and the flanges have only a secondary part in its transmission. To prove this, we calculate the sum

$$V_1 = \int_{-h_1/2}^{h_1/2} \tau_{xy} b_1 dy.$$

Substituting expression (69) for τ_{xy}, we obtain:

$$V_1 = \frac{V}{b_1 I_z}\int_{-h_1/2}^{h_1/2}\left[\frac{b}{2}\left(\frac{h^2}{4} - \frac{h_1^2}{4}\right) + \frac{b_1}{2}\left(\frac{h_1^2}{4} - y_1^2\right)\right]b_1 dy,$$

and, after integration,

$$V_1 = \frac{V}{I_z}\left[\frac{b(h - h_1)}{2} \cdot \frac{h + h_1}{2} \cdot \frac{h_1}{2} + \frac{b_1 h_1^3}{12}\right]. \quad (a)$$

For small thickness of flanges, i.e., when h_1 approaches h, the moment of inertia I_z is represented with sufficient accuracy by the equation

$$I_z = \frac{b(h - h_1)}{2} \cdot \frac{(h + h_1)^2}{8} + \frac{b_1 h_1^3}{12}, \qquad (b)$$

in which the first term represents the cross-sectional area of the flanges multiplied by the square of the distance $(h + h_1)/4$ of their centers from the z axis, which is approximately the moment of inertia of the cross section of the flanges. The second term is the moment of inertia of the cross section of the web. Comparing (a) and (b), we see that as h_1 approaches h the force V_1 approaches V and the shearing force will be practically taken by the web alone.

In considering the distribution of the shearing stresses over the cross sections of the flanges, the assumption of no variation along the width of the section cannot be made. For example, at the level ae (Fig. 110), along the lower boundary of the flange, ac and de, the shearing stress τ_{xy} must be zero since the corresponding equal stress τ_{yx} in the free bottom surface of the flange is zero (see p. 114 and also Fig. 108c). In the part cd, however, the shearing stresses are not zero, but have the magnitudes calculated above for $(\tau_{xy})_{min}$ in the web. This indicates that at the junction cd of the web and the flange the distribution of shearing stresses follows a more complicated law than can be investigated by our elementary analysis. In order to reduce the stress concentration at the points c and d, the sharp corners are usually replaced by fillets, as indicated in the figure by dotted lines. A more detailed discussion of the distribution of shearing stresses in flanges will be given later (see Part II).

Problems

1. Determine $(\tau_{xy})_{max}$ and $(\tau_{xy})_{min}$ in the cross section of the web of a WF beam, Fig. 110, if $b = 5$ in., $b_1 = \frac{1}{2}$ in., $h = 12$ in., $h_1 = 10\frac{1}{2}$ in., $V = 30,000$ lb. Determine the shearing force transmitted by the web V_1.

Answer. $(\tau_{xy})_{max} = 5,870$ lb per sq in., $(\tau_{xy})_{min} = 4,430$ lb per sq in., $V_1 = 0.945V$.

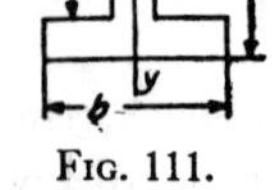

2. Determine the maximum shearing stress in the web of a T beam (Fig. 111) if $h = 8$ in., $h_1 = 7$ in., $b = 4$ in., $b_1 = 1$ in., $V = 1,000$ lb.

FIG. 111.

Answer. Using the same method as in the case of an I beam, we find $(\tau_{xy})_{max} = 176$ lb per sq in.

3. Determine the maximum shearing stress in Probs. 1 and 6 of Art. 25. Use the standard I beam section and assume that the total shearing force is uniformly distributed over the cross section of the web.

4. Determine the maximum shearing stress in the channel of Prob. 2, p. 104, if $V = 12,000$ lb.

Answer. $(\tau_{xy})_{max} = 441$ lb per sq in.

29. Principal Stresses in Bending.—By using eqs. (57) and (64) the normal stress σ_x and the shearing stress τ_{xy} can be calculated for any point of a cross section, provided the bending moment M and the shearing force V are known for this cross section. The maximum numerical value of σ_x will be in the fiber most remote from the neutral axis and the maximum value of τ_{xy} is usually at the neutral axis. In the majority of cases only the maximum values of σ_x and τ_{xy} obtained in this manner are used in design and the cross-sectional dimensions of beams are taken so as to satisfy the conditions

$$(\sigma_x)_{max} \leqq \sigma_W \quad \text{and} \quad (\tau_{xy})_{max} \leqq \tau_W.$$

It is assumed here that the material is equally strong in tension and compression and σ_W is the same for both. Otherwise the conditions of strength in tension and in compression must be satisfied separately and we obtain

$$(\sigma_x)_{max} \leqq \sigma_W \text{ in tension;} \mid (\sigma_x)_{min} \mid \leqq \sigma_W \text{ in compression.}$$

There are cases, however, which require a more detailed analysis of stresses and require a calculation of the principal stresses. Let us show such a calculation for a beam simply supported and loaded at the middle (Fig. 112). For a point A below the neutral axis in the cross section mn, the magnitudes of the stresses σ_x and $\tau_{yx} = \tau_{xy}$ are given by eqs. (57) and (64). In Fig. 112b those stresses are shown acting on an infinitesimal element cut out of the beam at the point A, their senses being easily determined from those of M and V. For an infinitesimal element the changes in stresses σ_x and τ_{xy} for various points of the element can be neglected and it can be assumed that the element is in a *homogeneous state of stress*, i.e., the quantities σ_x and τ_{xy} may be regarded as constant throughout the element. Such a state of stress is illustrated by the element of finite dimensions in Fig. 40a.

From our previous investigation (see p. 46) we know that the stresses on the sides of an element cut out from a stressed

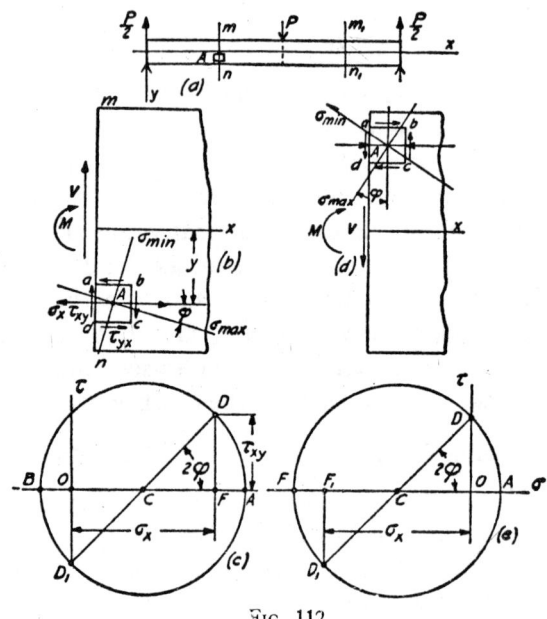

Fig. 112.

body vary with the directions of these sides and that it is possible to so rotate the element that only normal stresses are present (see p. 49). The directions of the sides are then called *principal directions* and the corresponding stresses, *principal stresses*. The magnitudes of these stresses for the present case can be found from eqs. (31) and (32) by substituting in these equations $\sigma_y = 0$. Then we obtain

$$\sigma_{\max} = \frac{\sigma_x}{2} + \sqrt{\left(\frac{\sigma_x}{2}\right)^2 + \tau_{xy}^2}, \tag{72}$$

$$\sigma_{\min} = \frac{\sigma_x}{2} - \sqrt{\left(\frac{\sigma_x}{2}\right)^2 + \tau_{xy}^2}. \tag{73}$$

It should be noted that σ_{max} is tension and σ_{min} is compression. Knowing principal stresses, the maximum shearing stress at any point will be obtained from eq. (34) (see p. 51):

$$\tau_{max} = \frac{\sigma_{max} - \sigma_{min}}{2} = \sqrt{\left(\frac{\sigma_x}{2}\right)^2 + \tau_{xy}^2}. \qquad (74)$$

For determining the directions of principal stresses Mohr's circle will be used. For an element such as at point A (Fig. 112b), the corresponding Mohr's circle is shown in Fig. 112c. By taking the distance $\overline{OF} = \sigma_x$ and $\overline{DF} = \tau_{xy}$, the point D, representing stresses over the sides bc and ad of the element, is obtained. The distance $\overline{OF}$ is taken in the direction of positive σ and $\overline{DF}$ in the upward direction because σ_x is tensile stress and shearing stresses τ_{xy} over sides bc and ad give a clockwise couple (see p. 76). Point D_1 represents the stresses over the sides ab and dc of the element on which the normal stresses are zero and the shearing stresses are negative. The circle constructed on the diameter DD_1 determines $\sigma_{max} = \overline{OA}$ and $\sigma_{min} = -\overline{OB}$. From the same construction the angle 2φ is determined and the direction of σ_{max} in Fig. 112b is obtained by measuring φ from the x axis in the clockwise direction. Of course σ_{min} is perpendicular to σ_{max}.

By taking a section m_1n_1 to the right of the load P (Fig. 112a) and considering a point A above the neutral axis, the direction of the stresses acting on an element $abcd$ at A will be that indicated in Fig. 112d. The corresponding Mohr's circle is shown in Fig. 112e. Point D represents the stresses for the sides ab and dc of the element $abcd$ and point D_1, the stresses over the sides ad and bc. The angle φ determining the direction σ_{max} must be measured in the clockwise direction from the outer normal to the side ab or cd as shown in Fig. 112d.

If we take a point at the neutral surface, then σ_x becomes zero. An element at this point will be in the condition of pure shear. The directions of the principal stresses will be at 45° to the x and y axes.

It is possible to construct two systems of orthogonal curves whose tangents at each point are in the directions of the principal stresses at this point. Such curves are called the *trajectories of the stresses*. Fig. 113 shows the stress trajectories for a rectangular cantilever beam, loaded at the end. All these curves intersect the neutral surface at 45° and have horizontal or vertical tangents at points where the shearing stress τ_{xy} is zero, i.e., at the top and at the bottom surfaces of the beam. The trajectories giving the directions of σ_{max} (tension) are represented by full lines and the other system of trajectories by dotted lines. Fig. 114 gives the trajectories and the stress distribution diagrams for σ_x and τ_{xy} over several cross sections of

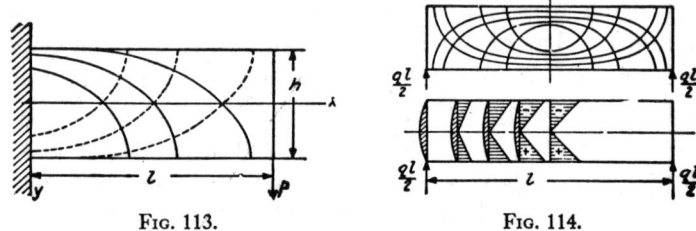

FIG. 113.   FIG. 114.

a simply supported rectangular beam under uniform load. It is clearly seen that σ_x has a maximum value at the middle, where the bending moment M is a maximum, and τ_{xy} is maximum at the supports, where the maximum shearing force acts.[9]

In the design of beams the concern is for the numerically maximum values of σ. From eq. (72) it can be seen that for the most remote fibers in tension, where the shear is zero, the longitudinal normal stress σ_x becomes the principal stress, i.e., $\sigma_{max} = (\sigma_x)_{max}$. For fibers nearer to the neutral axis the longitudinal fiber stress σ_x is smaller than at the extreme fiber. However, we now have a shear stress τ_{xy} also and the stresses σ_x and τ_{xy} acting together at this point may produce a principal stress, given by eq. (72), which will be numerically larger than that at the extreme fiber. In the case of beams of rectangular or circular cross section, in which the shearing stress τ_{xy} varies continuously down the depth of the beam, this is

---

[9] Several examples of construction of the trajectories of stresses are discussed by I. Wagner, Z. österr. *Ing. u. Architekt. Ver.*, p. 615, 1911.

not usually the case, i.e., the stress $(\sigma_x)_{max}$ calculated for the most remote fiber at the section of maximum bending moment is the maximum stress acting in the beam. However, in such a case as an I beam, where a sudden change occurs in the magnitude of shearing stress at the junction of flange and web (see p. 124), the maximum stress calculated at this joint from eq. (72) may be larger than the tensile stress $(\sigma_x)_{max}$ in the most remote fiber and should be taken into account in design. To illustrate this fact, consider the case represented in Fig. 112a with a beam of I section and the same dimensions as in Prob. 1, p. 124, the length $l = 2$ ft and $P = 60,000$ lb. Then $M_{max} = 30,000$ ft lb and $V_{max} = 30,000$ lb. From eq. (57) the tensile stress in the most remote fiber is

$$(\sigma_x)_{max} = \frac{30,000 \times 12 \times 6}{286} = 7,550 \text{ lb per sq in.}$$

Now for a point at the junction of flange and web we obtain the following values of normal and shearing stresses:

$$\sigma_x = \frac{7,550 \times 10\frac{1}{2}}{12} = 6,610 \text{ lb per sq in.,}$$

$$\tau_{xy} = 4,430 \text{ lb per sq in.}$$

Then, from eq. (72), the principal stress is

$$\sigma_{max} = 8,830 \text{ lb per sq in.}$$

It can be seen that σ_{max} at the joint between the flange and the web is larger than the tensile stress at the most remote fiber and therefore it must be considered in design.

Problems

1. Determine σ_{max} and σ_{min} at a point 2 in. below the neutral axis in the section 3 ft from the loaded end of the cantilever (Fig. 113) if the depth $h = 8$ in., the width $b = 4$ in. and $P = 2,000$ lb. Determine the angle between σ_{max} at this point and th

Answer. $(\sigma_x) = -844$ lb per sq in.; $(\tau_{xy}) = 70.3$ per sq. $\sigma_{max} = 5.7$ lb per sq in.; $\sigma_{min} = -849.7$ lb per sq in. The an between σ_{max} and the x axis is 85°16′ measured clockwise.

2. Determine σ_{max} and σ_{min} at the neutral axis and in the cross section 1 ft from the left support for the uniformly loaded rectangular beam supported at the ends (Fig. 114). The cross-sectional dimensions are the same as in the previous problem, and $q = 1,000$ lb per ft; $l = 10$ ft.

Answer. $\sigma_{max} = -\sigma_{min} = 187.5$ lb per sq in.

3. Determine the length of the I beam considered on p. 129 if $(\sigma_x)_{max}$ is equal to σ_{max} at the junction of flange and web.

Answer. $l = 39.8$ in.

30. Stresses in Built-up Beams.—In engineering practice built-up beams are frequently used and the stresses in them are usually calculated on the assumption that their parts are rigidly connected. The computation will then involve (*a*) the designing of the beam as a solid beam and (*b*) the designing and spacing of the elements which unite the parts of the beam. In the first case the formulas for solid beams are used, making an allowance for the effect of rivet holes, bolts, slots, etc., by the use of reduced sections. The design of keys and rivets will now be illustrated.

Let us discuss first a wooden beam built up as shown in Fig. 103. It is assumed that the keys used between the two parts of the beam are strong enough to resist the shearing forces S (Fig. 103*b*). Then eq. (57) can be used for calculating σ_x. In order to take into account the weakening of the section by the keyways and the bolt holes, only the shaded portion of the section, indicated in Fig. 103*c*, should be taken into consideration. Then

$$I_z = \frac{(b - d)}{12} [(2h)^3 - (2c)^3].$$

In calculating the shearing force S acting on each key we assume that this force is equal to the shearing force distributed in a solid beam over the area eb of the neutral surface, where b is the width of the beam and e is the distance between the middle points of the keys (see Fig. 103*a*). Then by using eq. (66) and considering that the depth of the beam is equal to $2h$ in this case, we obtain

$$S = eb \cdot \frac{3}{2} \frac{V}{2hb} = \frac{3}{2} \frac{Ve}{2h}. \tag{75}$$

The dimensions of the keys and the distance e between them should be chosen so as to insure sufficient strength against shearing of the key and against crushing of the wood on the lateral sides of the key and the keyway. In such calculations the rough assumption is usually made that the shearing stresses are uniformly distributed over the middle section $a \times b$ of the key and that the pressure on the lateral sides of the keys is uniformly distributed over the area $c \times b$. Then denoting by τ_W the working shearing stress for the keys, and by σ_W' the working stress in lateral compression of the wood of the keys or the keyways, the following equations for designing the keys are obtained:

$$\frac{S}{ab} \leqq \tau_W, \qquad \frac{S}{bc} \leqq \sigma_W'.$$

It is also necessary to insure sufficient strength against shearing of the wood of the beam along the fibers between two keys. The shearing force will be again equal to S and the resisting area is $b \times (e - a)$. Denoting with τ_W' the working stress in shear of the material of the beam along the fibers, the condition of strength becomes

$$\frac{S}{b(e - a)} \leqq \tau_W'.$$

In addition to keys there are bolts (Fig. 103) uniting the parts of the beam. By tightening them friction between the parts of the beam is produced. This friction is usually neglected in calculations and it is assumed that the total shearing force is taken by keys. Experiments show that such built-up wooden beams are weaker than solid beams of the same dimensions.[10]

In calculating the stresses σ_x in built-up steel beams or plate girders, the weakening effect of rivet holes is usually taken into account by assuming that all the holes are in the

---

[10] The experiments made by Prof. E. Kidwell at the Michigan College of Mines show that built-up wooden beams have about 75 per cent of the strength of the solid beam of the same dimensions.

same cross section (Fig. 115a) of the beam [11] and subtracting their diametral sections in calculating I_z in eq. (57).

In calculating the maximum shearing stress τ_{xy} it is also the practice to take into account the weakening effect of the rivet holes. It can be seen that the cross-sectional area of the web is diminished, by holes, in the ratio $(e - d)/e$, where e is the distance between the centers of the holes and d the diameter of the holes. Hence the factor $e/(e - d)$ is usually included in the right side of eq. (64) for calculating τ_{xy} in the

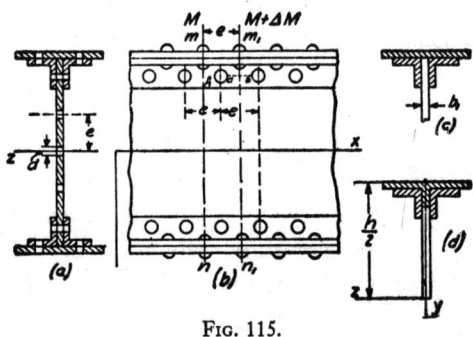

Fig. 115.

web of built-up beams. It should be noted that this manner of calculating the weakening effect of rivet holes is only a rough approximation. The actual distribution of stresses near the holes is very complicated. Some discussion of stress concentration near the edge of a hole will be given later (see Part II).

In calculating the shearing force acting on one rivet, such as rivet A (Fig. 115b), let us consider the two cross sections mn and m_1n_1. Due to the difference of bending moments in these two cross sections, the normal stresses σ_x on sections mn and m_1n_1 will be different and there is a tendency for the flange of the beam shaded in Fig. 115c to slide along the web. This sliding is prevented by friction forces and by the rivet A. Neglecting friction, the force acting on the rivet becomes equal to the difference of forces acting in sections mn and m_1n_1 of the flange. The force in the flange in the cross section mn is

----

[11] The holes in the vertical web are present in sections where vertical stiffeners are riveted to the girder.

(see eq. *a*, p. 117)

$$\frac{M}{I_z} \int y \, dA,$$

where the integration should be extended over the shaded cross-sectional area of the flange. In the same manner for the cross section $m_1 n_1$ we obtain

$$\frac{(M + \Delta M)}{I_z} \int y \, dA.$$

Then the force transmitted by the rivet A from the flange to the web will be

$$S = \frac{\Delta M}{I_z} \int y \, dA. \tag{a}$$

By using eq. (50) and substituting the distance e between the rivets instead of dx, we obtain

$$\Delta M = Ve,$$

where V is the shearing force in the cross section of the beam through the rivet A. Substituting in eq. (a), we obtain

$$S = \frac{Ve}{I_z} \int y \, dA. \tag{76}$$

The integral entering in this equation represents the moment of the shaded cross section (Fig. 115c) of the flange with respect to the neutral axis z.

It is easy to see that in order to have sliding of the flange along the web the rivet must be sheared through two cross sections. Assuming that the force S is uniformly distributed over these two cross sections, the shearing stress in the rivet will be

$$\tau = \frac{S}{2 \cdot \dfrac{\pi d^2}{4}} = \frac{2Ve}{\pi d^2 I_z} \int y \, dA. \tag{77}$$

The force S sometimes produces considerable shearing stress in the web of the beam along the plane ab (see Fig. 115b) which must be taken into consideration. Assuming that these stresses are uniformly distributed and dividing S by the area $b_1(e - d)$, we obtain

$$\tau' = \frac{V}{b_1 I_z} \cdot \frac{e}{e - d} \int y dA. \tag{b}$$

In addition to this stress produced by forces S transmitted from the flanges, there will act along the same plane ab shearing stresses τ'' due to bending of the web. The magnitude of these stresses will be obtained by using eq. (b) above and substituting for the integral $\int y dA$ the statical moment with respect to the neutral axis z of the portion of the rectangular cross section of the web above the plane ab. In this manner we arrive at the following equation for the shearing stress τ_{yx} in the web along the plane ab (Fig. 115b):

$$\tau_{yx} = \tau' + \tau'' = \frac{V}{b_1 I_z} \cdot \frac{e}{e - d} \int y dA, \tag{78}$$

in which the integral is extended over the shaded area of the cross section shown in Fig. 115d. Knowing σ_x and τ_{yx}, then σ_{max} and σ_{min} for the points in the plane ab can be calculated from eqs. (72) and (74), as was explained in the previous article, and the directions of principal stresses can be determined.

From the above discussion it is seen that in calculating stresses in built-up I beams several assumptions are made for simplifying the calculations. This to a certain extent reduces the accuracy of the calculated stresses, which fact should be considered in choosing the working stresses for built-up I beams.[12]

Problems

1. A built-up wooden beam (Fig. 103) consists of two parts of rectangular cross section connected by keys. Determine the shear-

---

[12] Experiments show that the failure of I beams usually occurs due to buckling of the compressed flanges or of the web (see H. F. Moore, Univ. of Illinois, *Bull. No. 68*, 1913). This question of buckling will be considered later. The effect of bending of rivets on the distribution of stresses in I beams has been discussed by I. Arnovlevic, *Z. Architekt. u. Ingenieurw.*, p. 57, 1910. He found that, due to this, bending stresses for usual proportions of beams increase about 6 per cent.

ing force acting on the keys, the shearing stress in the key and pressure per unit area on its lateral sides if the load $P = 5,000$ lb, the width of the beam $b = 5$ in., the depth $2h = 16$ in., the width of the key $a = 3$ in., the depth of the key $2c = 2\frac{1}{2}$ in. and the distance between centers of the keys $e = 11$ in

Answer. $S = \dfrac{3}{2} \cdot \dfrac{2,500 \times 11}{16} = 2,580$ lb.

Shearing stress in the key is

$$\tau = \frac{2,580}{5 \times 3} = 172 \text{ lb per sq in.}$$

The pressure per unit area on the lateral side is

$$) = \frac{S}{bc} = \frac{2,580 \times 2}{2\frac{1}{2} \times 5} = 413 \text{ lb per sq in.}$$

2. Determine the shearing stress at the neutral axis of a plate girder, the web of which is $\frac{3}{4}$ in. thick and 50 in. high, the flanges consisting of two pairs of angles $6 \times 6 \times \frac{1}{2}$ in., when the total shearing force on the section is 150,000 lb. Determine also the shearing stresses in the rivets attaching the flanges to the web if the diameter of these rivets is 1 in. and the pitch $e = 4$ in. (Fig. 115).

Solution. For the dimensions given we have

$$I_z = \frac{3}{4} \times \frac{50^3}{12} + 4(19.9 + 5.75 \times 23.3^2) = 20,400 \text{ in.}^4$$

The static moment of half of the cross-sectional area with respect to the neutral axis is

$$\int_0^{h/2} ydA = \frac{3}{4} \frac{25 \times 25}{2} + 2 \times 5.75 \times 23.3 = 502 \text{ in.}^3$$

In this calculation 5.75 in.^{2} is the cross-sectional area of one angle, 19.9 in.^{4} is the moment of inertia of the cross section of one angle with respect to the axis through its centroid parallel to the neutral axis of the beam, 23.3 in. is the distance of the centroid of each angle from the neutral axis z of the beam. All such numerical data can be taken directly from a handbook or the Appendix. Now we obtain, from eq. (64),

$$(\tau_{yz})_{\max} = \frac{150,000 \times 502}{\frac{3}{4} \times 20,400} = 4,920 \text{ lb per sq in.}$$

If we consider weakening of the web by the rivet holes, then

$$(\tau_{yx})_{max} = \frac{e}{e-d} \cdot 4{,}920 = \frac{4}{3} 4{,}920 = 6{,}560 \text{ lb per sq in}$$

The force S transmitted by one rivet, from eq. (76),

$$S = \frac{150{,}000 \times 4 \times 268}{20{,}400} = 7{,}880 \text{ lb.}$$

The shearing stress in the rivet, from eq. (77),

$$\tau = \frac{7{,}880 \times 2}{3.14} = 5{,}020 \text{ lb per sq in.}$$

3. Determine σ_{max} in points of the plane ab (Fig. 115) a distance of 21.5 in. from the neutral axis if the dimensions of the beam are the same as in the previous problem, $V = 150{,}000$ lb and the bending moment $M = 3 \times 10^6$ in. lb.

Solution. From eq. (78),

$$\tau_{yx} = \frac{150{,}000}{\frac{3}{4} \times 20{,}400} \cdot \frac{4}{3} (268 + 61) = 4{,}300 \text{ lb per sq in.,}$$

$$\sigma_x = \frac{3 \times 10^6 \times 21.5}{20{,}400} = 3{,}160 \text{ lb per sq in.,}$$

$$\sigma_{max} = \frac{\sigma_x}{2} + \sqrt{\frac{\sigma_x^2}{4} + \tau_{yx}^2} = 6{,}160 \text{ lb per sq in.}$$

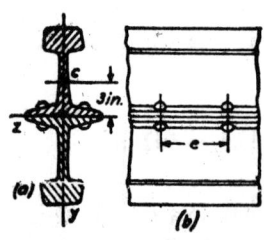

FIG. 116.

4. Determine the shearing force in the rivets connecting the two rails of the beam shown in Fig. 116 if the cross-sectional area of a rail is $A = 10$ sq in., the distance from the bottom of the rail to the centroid of its cross section $c = 3$ in., the moment of inertia of the cross section of the rail with respect to the axis through its centroid c and parallel to the z axis is 40 in.[4], the distance between the rivets $e = 6$ in., the shearing force $V = 5{,}000$ lb.

Solution. $S = \frac{1}{2} \cdot \frac{5{,}000 \times 6 \times 30}{2(40 + 10 \times 9)} = 1{,}730$ lb.

CHAPTER V

DEFLECTION OF LATERALLY LOADED SYMMETRICAL BEAMS

31. Differential Equation of the Deflection Curve.—In the design of a beam the engineer is usually interested not only in the stresses produced by the acting loads but also in the deflections produced by these loads. In many cases, furthermore, it is specified that the maximum deflection shall not exceed a certain small portion of the span.

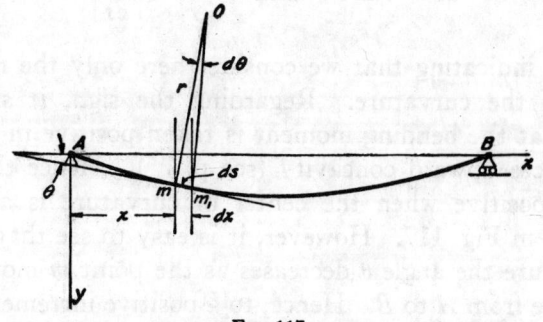

FIG. 117.

Let the curve AmB in Fig. 117 represent the shape of the axis of the beam after bending. Bending takes place in the plane of symmetry due to transverse forces acting in that plane. This curve is called the *deflection curve*. To derive the differential equation of this curve we take the coordinate axes as shown in the figure and assume that the curvature of the deflection curve at any point depends only on the magnitude of the bending moment M at that point.[1] In such a case the relation between the curvature and the bending moment is

---

[1] The effect of shearing force on the curvature will be discussed later (see Art. 39). It will be shown that this effect is usually small and can be neglected.

the same as in the case of pure bending (see eq. 56), and we obtain

$$\frac{1}{r} = \frac{M}{EI_z}.$$
(a)

To derive an expression for the relation between the curvature and the shape of the curve, we shall consider two adjacent points m and m_1, distance ds apart on the deflection curve. If the angle which the tangent at m makes with the x axis is denoted by θ, the angle between the normals to the curve at m and m_1 is $d\theta$. The intersection point O of these normals gives the center of curvature and defines the length r of the radius of the curvature. Then

$$ds = rd\theta \quad \text{and} \quad \frac{1}{r} = \left| \frac{d\theta}{ds} \right|,$$
(b)

the bars indicating that we consider here only the numerical value of the curvature. Regarding the sign, it should be noted that the bending moment is taken positive in eq. (a) if it produces upward concavity (see p. 75). Hence the curvature is positive when the center of curvature is above the curve as in Fig. 117. However, it is easy to see that for such a curvature the angle θ decreases as the point m moves along the curve from A to B. Hence, to a positive increment ds corresponds a negative $d\theta$. Thus to have the proper sign eq. (b) must be written in the form

$$\frac{1}{r} = -\frac{d\theta}{ds}.$$
(c)

In practical applications only very small deflections of beams are allowable and the deflection curves are very flat. In such cases we can assume with sufficient accuracy that

$$ds \approx dx \quad \text{and} \quad \theta \approx \tan\theta = dy/dx.$$
(d)

Substituting these approximate values for ds and θ in eq. (c), we obtain

$$\frac{1}{r} = -\frac{d^2y}{dx^2}.$$
(e)

Eq. (a) thus becomes

$$EI_z \frac{d^2y}{dx^2} = -M. \tag{79}$$

This is the differential equation of the deflection curve and must be integrated in each particular case to find deflections of beams.

It should be noted that the sign in eq. (79) depends upon the direction of the coordinate axes. For example, if we take y positive upwards, it is necessary to put

$$\theta \approx -dy/dx$$

in place of eq. (d), and we obtain plus instead of minus on the right side of eq. (79).

In the case of very slender bars, in which the deflection may be large, it is not permissible to use the simplifications (d), and we must have recourse to the exact expression

$$\theta = \arctan\left(\frac{dy}{dx}\right).$$

Then

$$\frac{1}{r} = -\frac{d\theta}{ds} = -\frac{d \arctan\left(\frac{dy}{dx}\right)}{dx} \frac{dx}{ds}$$

$$= -\frac{\frac{d^2y}{dx^2}}{\left[1 + \left(\frac{dy}{dx}\right)^2\right]^{3/2}}.$$

Comparing this result with eq. (e), it can be concluded that the simplifications shown in eq. (d) are equivalent to assuming that the quantity $(dy/dx)^2$ in the denominator of the exact formula (f) is small in comparison with unity and can therefore be neglected.[2]

---

[2] The exact expression (f) for the curvature was used by the first investigators of deflection curves. It was used, e.g., by L. Euler in his famous work, *Elastic Curves*, an English translation of which was published in *Isis*, Vol. 20, p. 1, Nov. 1933. See also S. Timoshenko, *History of Strength of Materials*. New York, p. 32, 1953.

By differentiating eq. (79) with respect to x and using eqs. (50) and (51), we obtain

$$EI_z \frac{d^3y}{dx^3} = -V$$

and

$$EI_z \frac{d^4y}{dx^4} = q. \qquad (80)$$

The last equation is sometimes used in considering the deflection of beams under a distributed load.

32. Bending of a Uniformly Loaded Beam.—In the case of a *simply supported and uniformly loaded beam*, Fig. 68, the bending moment at any cross section mn, a distance x from the left support, is

$$M = \frac{qlx}{2} - \frac{qx^2}{2},$$

and the differential eq. (79) becomes

$$EI_z \frac{d^2y}{dx^2} = -\frac{qlx}{2} + \frac{qx^2}{2}.$$

Multiplying both sides by dx and integrating, we obtain

$$EI_z \frac{dy}{dx} = -\frac{qlx^2}{4} + \frac{qx^3}{6} + C \qquad (a)$$

where C is the constant of integration which is to be adjusted so as to satisfy the conditions of this particular problem. To this end, we note that as a result of symmetry the slope at the middle of the span is zero. Setting $dy/dx = 0$ when $x = l/2$, we thus obtain

$$C = \frac{ql^3}{24},$$

and eq. (a) becomes

$$EI_z \frac{dy}{dx} = -\frac{qlx^2}{4} + \frac{qx^3}{6} + \frac{ql^3}{24}. \qquad (b)$$

A second integration gives

$$EI_z y = -\frac{qlx^3}{12} + \frac{qx^4}{24} + \frac{ql^3 x}{24} + C_1. \qquad (c)$$

The new constant of integration C_1 is determined from the condition that the deflection at the supports is zero. Substituting $y = 0$ and $x = 0$ into eq. (c), we find $C_1 = 0$. Eq. (c) then becomes

$$y = \frac{q}{24EI_z} (l^3x - 2lx^3 + x^4). \tag{81}$$

This is the deflection curve of a simply supported and uniformly loaded beam. The maximum deflection of this beam is evidently at the middle of the span. Substituting $x = l/2$ in eq. (81), we thus find

$$y_{max} = \frac{5}{384} \frac{ql^4}{EI_z}. \tag{82}$$

The maximum slope occurs at the left end of the beam where, by substituting $x = 0$ in eq. (b), we obtain

$$\left(\frac{dy}{dx}\right)_{max} = \frac{ql^3}{24EI_z}. \tag{83}$$

In the case of a *uniformly loaded cantilever beam*, Fig. 118a, the bending moment at a cross section mn a distance x from the left end is

$$M = -\frac{qx^2}{2},$$

and eq. (79) becomes

Fig. 118.

$$EI_z \frac{d^2y}{dx^2} = \frac{qx^2}{2}.$$

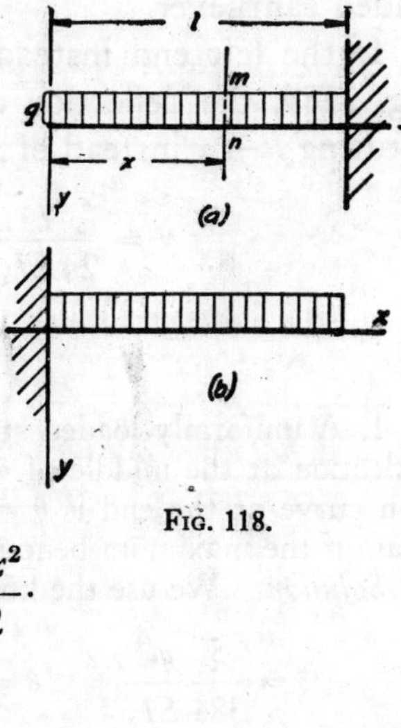

The first integration gives

$$EI_z \frac{dy}{dx} = \frac{qx^3}{6} + C. \tag{d}$$

The constant of integration is found from the condition that the slope at the built-in end is zero, that is $dy/dx = 0$ for $x = l$. Substituting these values in eq. (d), we find

$$C = -\frac{ql^3}{6}$$

The second integration gives

$$EI_z y = \frac{qx^4}{24} - \frac{ql^3 x}{6} + C_1. \tag{e}$$

The constant C_1 is found from the condition that the deflection vanishes at the built-in end. Thus, by substituting $x = l$, $y = 0$ in eq. (e), we obtain

$$C_1 = \frac{ql^4}{8}.$$

Substituting this value in eq. (e), we find

$$y = \frac{q}{24EI_z} (x^4 - 4l^3 x + 3l^4). \tag{84}$$

This equation defines the deflection curve of the uniformly loaded cantilever.

If the left end instead of the right end is built in, as in Fig. 118b, the deflection curve is evidently obtained by substituting $l - x$ instead of x into eq. (84). In this way we find

$$y = \frac{q}{24EI_z} (x^4 - 4lx^3 + 6l^2 x^2). \tag{85}$$

Problems

1. A uniformly loaded steel I beam supported at the ends has a deflection at the middle of $\delta = \frac{5}{16}$ in. while the slope of the deflection curve at the end is $\theta = 0.01$ radian. Find the depth h of the beam if the maximum bending stress is $\sigma = 18,000$ lb per sq in.

Solution. We use the known formulas

$$\delta = \frac{5}{384} \frac{ql^4}{EI_z}, \qquad \theta = \frac{ql^3}{24EI_z}, \qquad \sigma_{max} = \frac{ql^2}{8} \times \frac{h}{2I_z}.$$

From the first two formulas we find

$$\frac{5}{16} l = \frac{\delta}{\theta} = \frac{5}{16} \times 100 \text{ in.} \qquad \text{and} \qquad l = 100 \text{ in.}$$

The second formula then gives

$$\frac{ql^2}{8I_z} = \frac{3E\theta}{l} = \frac{3 \times 30 \times 10^6 \times 0.01}{100}.$$

Substituting this in the third formula, we obtain

$$h = \frac{2 \times 18{,}000 \times 100}{3 \times 30 \times 10^6 \times 0.01} = 4 \text{ in.}$$

2. A simply supported and uniformly loaded wooden beam of square cross section has a span $l = 10$ ft. Find the maximum deflection if $(\sigma_x)_{\max} = 1{,}000$ lb per sq in., $E = 1.5 \times 10^6$ lb per sq in. and $q = 400$ lb per ft.
Answer. $\delta = 0.281$ in.

3. Find the depth of a uniformly loaded and simply supported steel I beam having a span of 10 ft, if the maximum bending stress is 16,000 lb per sq in. and the maximum deflection $\delta = 0.1$ in.
Answer. $h = 16$ in.

4. A uniformly loaded cantilever beam of span l has a deflection at the end equal to $0.01l$. What is the slope of the deflection curve at the end?
Answer. $\theta = 0.0133$ radian.

5. What is the length of a uniformly loaded cantilever beam if the deflection at the free end is 1 in. and the slope of the deflection curve at the same point is 0.01?
Answer. $l = 11.1$ ft.

33. Deflection of a Simply Supported Beam Loaded by a Concentrated Load.—In this case there are two different expressions for the bending moment (see Art. 22) corresponding to the two portions of the beam, Fig. 119. Eq. (79) for the

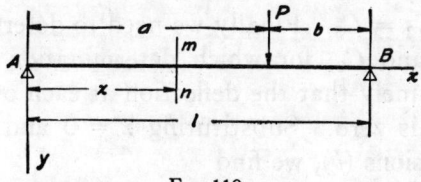

Fig. 119.

deflection curve must therefore be written for each portion. In this way we obtain

$$EI_z \frac{d^2y}{dx^2} = -\frac{Pb}{l}x \quad \text{for} \quad x \leqq a,$$

and

$$EI_z \frac{d^2y}{dx^2} = -\frac{Pb}{l}x + P(x - a) \quad \text{for} \quad x \geqq a.$$

By integrating these equations we obtain

$$EI_z \frac{dy}{dx} = -\frac{Pbx^2}{2l} + C \qquad \text{for} \qquad x \leqq a,$$

and

$$EI_z \frac{dy}{dx} = -\frac{Pbx^2}{2l} + \frac{P(x-a)^2}{2} + C_1 \qquad \text{for} \qquad x \geqq a. \qquad (a)$$

Since the two branches of the deflection curve must have a common tangent at the point of application of the load P, the above expressions (a) for the slope must be equal for $x = a$. From this we conclude that the constants of integration are equal, i.e., $C = C_1$. Performing the second integration and substituting C for C_1, we obtain

$$EI_z y = -\frac{Pbx^3}{6l} + Cx + C_2 \qquad \text{for} \quad x \leqq a,$$

and

$$EI_z y = -\frac{Pbx^3}{6l} + \frac{P(x-a)^3}{6} + Cx + C_3 \qquad \text{for} \quad x \geqq a. \qquad (b)$$

Since the two branches of the deflection curve have a common deflection at the point of application of the load, the two expressions (b) must be identical for $x = a$. From this it follows that $C_2 = C_3$. Finally we need to determine only two constants C and C_2, for which determination we have two conditions, namely that the deflection at each of the two ends of the beam is zero. Substituting $x = 0$ and $y = 0$ in the first of expressions (b), we find

$$C_2 = C_3 = 0. \qquad (c)$$

Substituting $y = 0$ and $x = l$ in the second of expressions (b), we obtain

$$C = \frac{Pbl}{6} - \frac{Pb^3}{6l} = \frac{Pb(l^2 - b^2)}{6l} \qquad (d)$$

Substituting the values (c) and (d) of the constants into eqs. (b) for the deflection curve, we obtain

$$EI_z y = \frac{Pbx}{6l}(l^2 - b^2 - x^2) \quad \text{for} \quad x \leqq a, \quad (86)$$

and

$$EI_z y = \frac{Pbx}{6l}(l^2 - b^2 - x^2) + \frac{P(x - a)^3}{6} \quad \text{for} \quad x \geqq a. \quad (87)$$

The first of these equations gives the deflections for the left portion of the beam and the second gives the deflections for the right portion.

Substituting the value (d) into eqs. (a), we obtain

$$\left.\begin{array}{l} EI_z \dfrac{dy}{dx} = \dfrac{Pb}{6l}(l^2 - b^2 - 3x^2) \quad \text{for} \quad x \leqq a, \\[3mm] \text{and} \\[2mm] EI_z \dfrac{dy}{dx} = \dfrac{Pb}{6l}(l^2 - b^2 - 3x^2) + \dfrac{P(x - a)^2}{2} \quad \text{for} \quad x \geqq a. \end{array}\right\} \quad (e)$$

From these equations the slope at any point of the deflection curve can readily be calculated. Often we need the values of the slopes at the ends of the beam. Substituting $x = 0$ in the first of eqs. (e), $x = l$ in the second, and denoting the slopes at the corresponding ends by θ_1 and θ_2, we obtain [*]

$$\theta_1 = \left(\frac{dy}{dx}\right)_{x=0} = \frac{Pb(l^2 - b^2)}{6lEI_z}, \quad (88)$$

$$\theta_2 = \left(\frac{dy}{dx}\right)_{x=l} = -\frac{Pab(l + a)}{6lEI_z}. \quad (89)$$

The maximum deflection occurs at the point where the tangent to the deflection curve is horizontal. If $a > b$ as in Fig. 119, the maximum deflection is evidently in the left portion of the beam. We can find the position of this point by equating the first of the expressions (e) to zero to obtain

$$l^2 - b^2 - 3x^2 = 0,$$

from which

$$x = \frac{\sqrt{l^2 - b^2}}{\sqrt{3}}.$$

---

[*] For flat curves, which we have in most cases, the slopes θ_1 and θ_2 can be taken numerically equal to the angles of rotation of the ends of the beam during bending, the angles being taken positive when the rotation is clockwise

This is the distance from the left support to the point of maximum deflection. To find the maximum deflection itself, we substitute expression (f) in eq. (86), which gives

$$y_{max} = \frac{Pb(l^2 - b^2)^{3/2}}{9\sqrt{3}lEI_z}. \tag{g}$$

If the load P is applied at the middle of the span the maximum deflection is evidently at the middle also. Its magnitude is obtained by substituting $b = l/2$ in eq. (g), which gives

$$(y)_{\substack{x=l/2 \\ a=b}} = \frac{Pl^3}{48EI_z}. \tag{90}$$

From eq. (f) it can be concluded that in the case of one concentrated force the maximum deflection is always near the middle of the beam. When $b = l/2$ it is at the middle; in the limiting case, when b is very small and P is near the support, the distance x as given by eq. (f) is $l/\sqrt{3}$, and the point of maximum deflection is only a distance

$$\frac{l}{\sqrt{3}} - \frac{l}{2} = 0.077l$$

from the middle. Due to this fact the deflection at the middle is a close approximation to the maximum deflection. To obtain the deflection at the middle we substitute $x = l/2$ in eq. (86), which gives

$$(y)_{\substack{x=l/2 \\ a>b}} = \frac{Pb}{48EI_z}(3l^2 - 4b^2) \tag{91}$$

The difference of the deflections (g) and (91) in the most unfavorable case, that is when b approaches zero, is only about 2.5 per cent of the maximum deflection.

Problems

1. Find the position of the load P, Fig. 119, if the ratio of the numerical values of the slopes at the ends of the beam is $|\theta_1/\theta_2| = \frac{3}{4}$.

Answer. $a = \frac{5}{7}l$.

2. Find the difference between the maximum deflection and the deflection at the middle of the beam in Fig. 119 if $b = 2a$.

Answer. $0.0046 \dfrac{Pl^3}{27EI_z}$.

3. Find the maximum deflection of the beam shown in Fig. 119 if the beam is an 8 I 18.4 section (see Appendix) with 8 in. depth and 5.34 sq in. cross-sectional area, and $a = 12$ ft, $b = 8$ ft and $P = 2,000$ lb.

4. What will be the maximum deflection if the I beam of the previous problem is replaced by a wooden beam having a cross section 10×10 in.? The modulus of elasticity for wood can be taken as $E = 1.5 \times 10^6$ lb per sq in.

34. Determination of Deflections by the Use of the Bending Moment Diagram.—*Area-Moment Method.*—In the preced-

ing articles it was shown how the deflection curve of a beam can be obtained by integration of the differential eq. (79). In many cases, however, especially if we need the deflection at a prescribed point rather than the general equation of the deflection curve, the calculation can be considerably simplified by the use of the bending moment diagram, as will be described in the following discussion.[4]

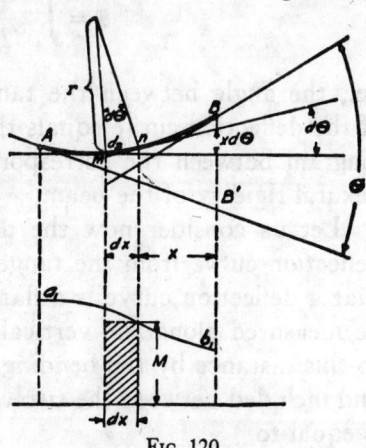

Fig. 120.

In Fig. 120 AB represents a portion of a deflection curve of a beam and $a_1 b_1$ is the corresponding portion of the bending moment diagram. Two adjacent cross sections of the beam at distance ds apart will intersect, after bending, at an angle $d\theta$, and, from eq. (56),

$$d\theta = \frac{1}{r} ds = \frac{M}{EI_z} ds.$$

---

[4] The use of the bending moment diagram in calculating deflections of beams was developed by O. Mohr, see *Z. Architekt. u. Ing.-Ver. (Hannover)*, p. 10, 1868; see also his *Abhandlungen*, p. 294; *loc. cit.*, p. 40. A similar method was developed independently by Prof. C. E. Green, Univ. of Michigan, 1874. It seems that the originator of the method was St.-Venant. See his notes in Navier's book, pp. 72 and 79; *loc. cit.*, p. 100.

For beams used in structures the curvature is very small, and we may use dx for ds. Then

$$d\theta = \frac{M}{EI_z} dx. \qquad (a)$$

Graphically interpreted, this means that the elemental angle $d\theta$ between two consecutive radii or two consecutive tangents to the deflection curve equals the shaded elemental area Mdx of the bending moment diagram, divided by the flexural rigidity.[5] This being valid for each element, the angle θ between the tangents at A and B will be obtained by summing up such elements as given by eq. (a). Then

$$\theta = \int_A^B \frac{1}{EI_z} Mdx \qquad (92)$$

i.e., the angle between the tangents at two points A and B of the deflection curve equals the area of the bending moment diagram between the corresponding verticals, divided by the flexural rigidity of the beam.

Let us consider now the distance of the point B of the deflection curve from the tangent AB' at point A. Recalling that a deflection curve is a flat curve, the above distance can be measured along the vertical BB'. The contribution made to this distance by the bending of an element mn of the beam and included between the two consecutive tangents at m and n is equal to

$$xd\theta = x \frac{Mdx}{EI_z}.$$

Interpreted graphically this is the moment of the shaded area Mdx with respect to the vertical through B, divided by EI_z. Integration gives the total deflection BB':

$$\overline{BB'} = \delta = \int_A^B \frac{1}{EI_z} xMdx, \qquad (93)$$

i.e., the distance of B from the tangent at A is equal to the

---

[5] By way of dimensional check: $d\theta$ is in radians, i.e., a pure number, Mdx is in in. lb $\times$ in., and EI_z is in lb per sq in. $\times$ in.[4]

moment with respect to the vertical through B of the area of the bending moment diagram between A and B, divided by the flexural rigidity EI_z. By using eqs. (92) and (93) the slope of the deflection curve and the magnitude of deflection at any cross section of the beam can easily be calculated in each particular case. We calculate first the absolute values of θ and δ. Then taking the positive directions of the coordinate axes as shown in Fig. 122, we consider the rotation of a tangent to the deflection curve as positive if it is in the clockwise direction, and deflection of the beam positive if it is in the direction of the positive y axis. This method of calculating deflections is called the *area-moment method*.

The calculation of the integrals in eqs. (92) and (93) can often be simplified by the use of known formulas concerning areas and centroids. Several formulas which are often encountered in applications are given in Fig. 121.

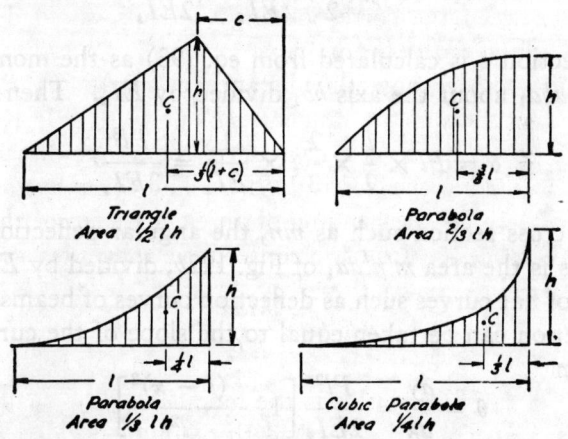

Triangle
Area $\frac{1}{2} l h$

Parabola
Area $\frac{2}{3} l h$

Parabola
Area $\frac{1}{3} l h$

Cubic Parabola
Area $\frac{1}{4} l h$

Fig. 121.

35. Deflection of a Cantilever Beam by the Area-Moment Method.—For the case of a cantilever beam with a concentrated load at the end (Fig. 122a) the bending moment diagram is shown in Fig. 122b. Since a tangent at the built-in end A remains fixed, the distances of points of the deflection curve from this tangent are actual deflections. The angle θ_b

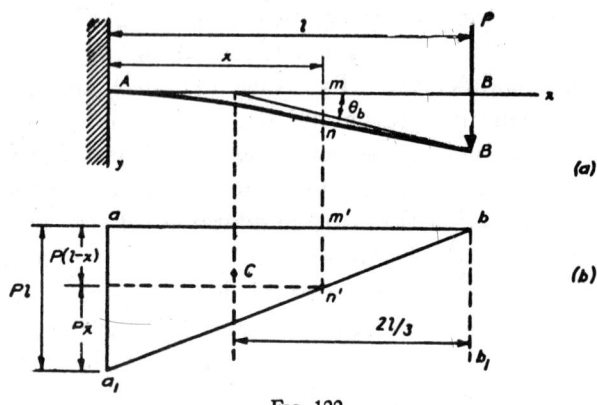

Fig. 122.

which the tangent to the deflection curve at B makes with the tangent at A is, from eq. (92),[6]

$$\theta_b = Pl \times \frac{l}{2} \times \frac{1}{EI_z} = \frac{Pl^2}{2EI_z}.$$ (94)

The deflection δ is calculated from eq. (93) as the moment of the area aba_1 about the axis bb_1 divided by EI_z. Then

$$\delta = Pl \times \frac{l}{2} \times \frac{2}{3}l \times \frac{1}{EI_z} = \frac{Pl^3}{3EI_z}.$$ (95)

For any cross section such as mn, the angular deflection from the x axis is the area $m'n'aa_1$ of Fig. 122b, divided by EI_z. In the case of flat curves such as deflection curves of beams, angular deflection can be taken equal to the slope of the curve and we obtain

$$\theta = \frac{dy}{dx} = \frac{Pl^2}{2EI_z}\left[1 - \frac{(l-x)^2}{l^2}\right].$$ (96)

The deflection y at the same cross section is the moment of the area $m'n'aa_1$ about $m'n'$ divided by EI_z (see eq. 93). Separating this area into the rectangle and the triangle indicated in the figure, this is

$$y = \frac{1}{EI_z}\left[P(l-x)\frac{x^2}{2} + \frac{Px^2}{2}\frac{2x}{3}\right] = \frac{P}{EI_z}\left(\frac{lx^2}{2} - \frac{x^3}{6}\right).$$ (97)

---

[6] It is taken positive since the rotation is clockwise.

For a cantilever with a concentrated load P at a cross section distance c from the support (Fig. 123a) the bending moment diagram is shown in Fig. 123b. The slope and the deflection for any section to the left of the point of application of the load are determined from eqs. (96) and (97) with c in place of l. For any cross section to the right of the load the bending moment and the curvature

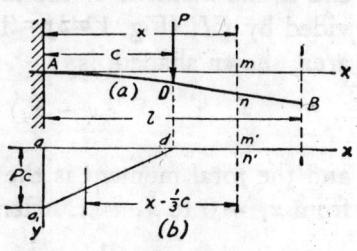

FIG. 123.

are zero, hence this portion of the beam remains straight. The slope is constant and equal to the slope at D, i.e., from eq. (94), $Pc^2/2EI_z$. The deflection at any cross section mn is the moment of the area of the triangle aa_1d about the vertical $m'n'$ divided by EI_z, which gives

$$v = \frac{1}{EI_z}\frac{Pc^2}{2}\left(x - \frac{1}{3}c\right). \qquad (98)$$

In the case of a cantilever with a uniform load of intensity q (Fig. 124a), the bending moment at any cross section mn distant x_1 from the built-in end is

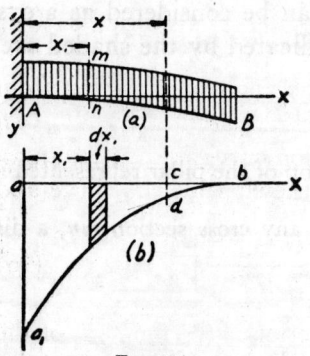

$$M = -\frac{q(l - x_1)^2}{2},$$

and is represented by the parabola a_1db in Fig. 124b. The slope at any cross section a distance x from the support is, from eq. (92),

$$\theta = \frac{dy}{dx} = \frac{1}{EI_z}\int_0^x \frac{q(l - x_1)^2}{2}\,dx_1$$

FIG. 124.

$$= \frac{q}{2EI_z}\left(l^2 x - lx^2 + \frac{x^3}{3}\right). \qquad (99)$$

The slope at the end is obtained by substituting l for x in the above equation, giving

$$\left(\frac{dy}{dx}\right)_{x=l} = \frac{ql^3}{6EI_z}. \qquad (100)$$

The deflection at any section a distance x from the built-in end is the moment of the area aa_1cd about the vertical cd divided by EI_z (Fig. 124b). The moment of the element of this area, shown shaded, is

$$(x - x_1) \frac{q(l - x_1)^2}{2} dx_1,$$

and the total moment is the integral of this with respect to x_1 from $x_1 = 0$ to $x_1 = x$. Hence

$$y = \frac{1}{EI_z} \frac{q}{2} \int_0^x (x - x_1)(l - x_1)^2 dx_1.$$

The deflection at any point a distance x from the support is then, after integration,

$$y = \frac{q}{2EI_z} \left(\frac{l^2 x^2}{2} - \frac{l x^3}{3} + \frac{x^4}{12} \right). \qquad (101)$$

For the deflection at the end $x = l$,

$$\delta = (y)_{x=l} = \frac{q l^4}{8EI_z}. \qquad (102)$$

The same problem can be solved by using the method of superposition. The uniform load can be considered as a system of infinitesimal loads qdc, as indicated by the shaded area in Fig. 140, p. 164.

Problems

1. Determine the deflection of the top of the pillar represented in Fig. 98.

Solution. The bending moment at any cross section mn, a distance x from the top, is

$$M = - \frac{W x^3}{3 l^2},$$

where $W = \frac{1}{2} d l^2 \times 62.4$ lb is the total hydrostatic pressure transmitted to one pillar. Using eq. (93), the deflection of the top of the pillar is

$$\delta = \frac{W}{EI_z} \int_0^l \frac{x^4 dx}{3 l^2}$$

$$= \frac{W l^3}{15 EI_z} = \frac{3 \times 6^2 \times 62.4 \times 6^3 \times 12^3 \times 12}{2 \times 15 \times 1.5 \times 10^6 \times 9.9^4} = 0.070 \text{ in.}$$

2. Determine the deflection and the slope at the end of a cantilever bent by a couple M (Fig. 125).

Answer. $(y)_{x=l} = -\dfrac{Ml^2}{2EI_z}$, $\left(\dfrac{dy}{dx}\right)_{x=l} = -\dfrac{Ml}{EI_z}$.

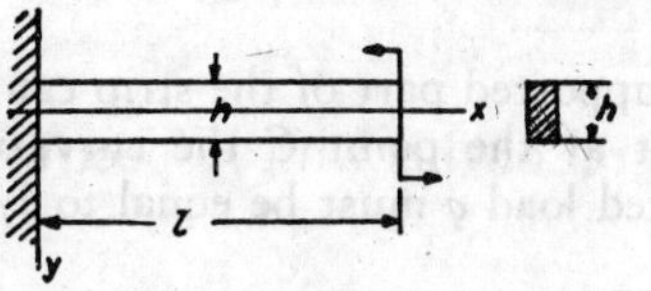

FIG. 125. 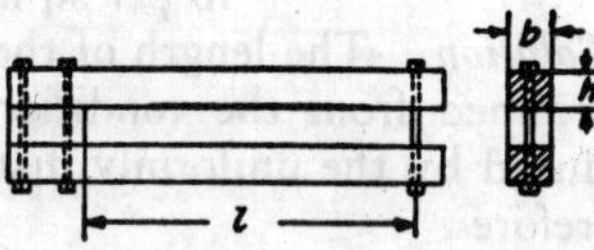   FIG. 126.

3. Two wooden rectangular beams clamped at the left end (Fig. 126) are bent by tightening the bolt at the right end. Determine the diameter d of the bolt to make the factors of safety for the wooden beams and for the steel bolt the same. The length of the beams $l = 3$ ft, the depth $h = 8$ in., the width $b = 6$ in., working stress for steel $\sigma_W = 12,000$ lb per sq in., for wood $\sigma_W = 1,200$ lb per sq in. Determine the deflection of the beams when the tensile stress in the bolt is 12,000 lb per sq in.

Solution. If P is the force in the bolt, the equation for determining the diameter d will be

$$\frac{4P}{\pi d^2} \div \frac{6Pl}{bh^2} = \frac{12,000}{1,200} = 10,$$

from which

$$d = 0.476 \text{ in.} \quad \text{and} \quad P = 12,000 \times \frac{\pi d^2}{4} = 2,130 \text{ lb.}$$

Then from eq. (95), by taking $E = 1.5 \times 10^6$ lb per sq in. the deflection $\delta = 0.0864$ in.

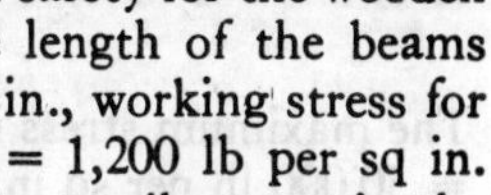

4. What must be the equation of the axis of the curved bar AB before it is bent if the load P, moving along the bar, remains always on the same level (Fig. 127)?

FIG. 127.

Answer. $y = -\dfrac{Px^3}{3EI_z}$.

5. Determine the safe deflection of the beam shown in Fig. 125 when the working stress σ_W is given. Determine this also for a cantilever loaded at the end (Fig. 122).

Answer. (1) $\delta = \dfrac{\sigma_W l^2}{Eh}$, (2) $\delta = \dfrac{2}{3}\dfrac{\sigma_W l^2}{Eh}$.

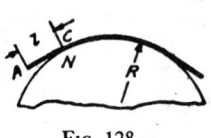

FIG. 128.

6. A circular disc N of radius R (Fig. 128) produces on a thin steel strip of thickness h an attraction of q lb per sq in. uniformly distributed. Determine the length l of the unsupported part AC of the strip and the maximum stress in it if $h = 0.01$ in., $R = 3$ in. and $q = 15$ lb per sq in.

Solution. The length of the unsupported part of the strip can be determined from the condition that at the point C the curvature produced by the uniformly distributed load q must be equal to $1/R$. Therefore

$$\frac{ql^2}{2} = \frac{EI_z}{R},$$

from which

$$l = \sqrt{\frac{2EI_z}{qR}} = \frac{1}{3} \text{ in.}$$

The maximum stress is determined from the equation $\sigma_{max} = Eh/2R$ = 50,000 lb per sq in.

36. Deflection of a Simply Supported Beam by the Area-Moment Method.—Let us consider the case of a simply supported beam with a load P applied at point F, Fig. 129. The

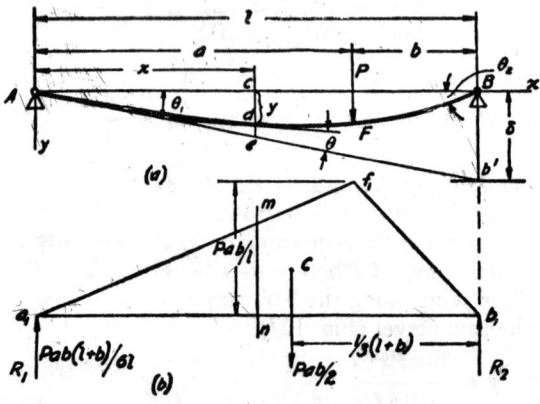

FIG. 129.

bending moment diagram is the triangle $a_1b_1f_1$, Fig. 129b. Its area is $Pab/2$, and its centroid C is at distance $(l + b)/3$ from the vertical Bb_1. The vertical distance δ from the end

E to the line Ab' which is tangent to the deflection curve at A is obtained from eq. (93) and is

$$\delta = \frac{1}{EI_z} \frac{Pab}{2} \times \frac{l+b}{3} = \frac{Pab(l+b)}{6EI_z}.$$

By using this value the slope θ_1 at the left end of the beam is found to be

$$\theta_1 = \frac{\delta}{l} = \frac{Pab(l+b)}{6lEI_z}, \qquad (a)$$

which coincides with previously obtained formula (88).[7] In calculating the angle of rotation θ_2 of the end B of the beam (Fig. 129a) we observe that rotation of the end B with respect to the end A, from eq. (92), is

$$\theta = \frac{Pab}{2EI_z}.$$

Hence

$$\theta_2 = \theta_1 - \theta = \frac{Pab(l+b)}{6lEI_z} - \frac{Pab}{2EI_z} = -\frac{Pab(l+a)}{6lEI_z}. \qquad (b)$$

A simple interpretation of formulas (a) and (b) is obtained if we consider a_1b_1 in Fig. 129b as a simply supported beam, carrying the triangular load represented by the triangle $a_1f_1b_1$. The reaction at the left support a_1 of this imaginary beam is

$$R_1 = \frac{Pab}{2} \times \frac{l+b}{3} \times \frac{1}{l} = \frac{Pab(l+b)}{6l}.$$

Similarly,

$$R_2 = \frac{Pab(l+a)}{6l}.$$

By comparing these results with formulas (a) and (b), it can be concluded that the angles of rotation θ_1 and θ_2 of the ends of the actual beam AB are obtained by dividing by the flexural rigidity EI_z the shearing forces at the ends of the imaginary beam a_1b_1. The imaginary beam a_1b_1 is called the *conjugate beam*.

---

[7] Note that $a = l - b$.

To calculate the slope at any point d of the deflection curve, Fig. 129a, it is necessary to subtract the angle θ between the tangents at A and at d from the angle θ_1 at the support. Using eq. (92) for the calculation of the angle θ, we obtain, from Fig. 129b,

$$\frac{dy}{dx} = \theta_1 - \theta = \frac{1}{EI_z} (R_1 - \Delta a_1 mn).$$

The first term in the parentheses is the reaction at the left support of the conjugate beam $a_1 b_1$ and the second is the load on the conjugate beam to the left of the cross section mn. The expression in the parentheses therefore represents the shearing force at the cross section mn of the conjugate beam. Consequently the slope of the actual beam at a point d can be obtained by dividing the shearing force at the corresponding cross section of the conjugate beam by the flexural rigidity EI_z.

Considering next the deflection y at a point d, it may be seen from Fig. 129a that

$$y = \overline{ce} - \overline{de}. \tag{c}$$

From the triangle Ace we obtain the relation

$$\overline{ce} = \theta_1 x = \frac{R_1 x}{EI_z}, \tag{d}$$

where R_1 is the reaction at the left support of the conjugate beam. The second term on the right side of eq. (c) represents the distance of the point d of the deflection curve from the tangent Ae and is obtained from eq. (93) as

$$\overline{de} = \frac{1}{EI_z} (\text{area } \Delta a_1 mn) \times \frac{x}{3}. \tag{e}$$

Substituting expressions (d) and (e) in eq. (c), we obtain

$$v = \frac{1}{EI_z} \left(R_1 x - \Delta a_1 mn \times \frac{x}{3} \right). \tag{f}$$

The expression in parentheses is seen to be the bending moment at the cross section mn of the conjugate beam. Thus the deflection at any point of a simply supported beam is obtained by dividing the bending moment at the corresponding cross section of the conjugate beam by the flexural rigidity EI_z. Substituting the value of R_1 in eq. (f) and noting that

$$\text{area } \Delta a_1 mn = \frac{Pbx^2}{2l},$$

we obtain

$$y = \frac{1}{EI_z}\left[\frac{Pabx(l+b)}{6l} - \frac{Pbx^3}{6l}\right] = \frac{Pbx}{6lEI_z}(l^2 - b^2 - x^2).$$

This checks with eq. (86), which was previously obtained by integration of the differential equation of the deflection curve. The deflection for a point to the right of the load P can be calculated in a similar manner. The result will, of course, be the same as eq. (87). It is seen that by using the area-moment method we eliminate the process of integration which was applied in Art. 33.

In the case of the uniformly loaded beam, Fig. 130a, we consider the conjugate beam ab, Fig. 130b, loaded by the parabolic segment acb, which is the bending moment diagram in this case. The total fictitious load on the conjugate beam is

FIG. 130.

$$\frac{2}{3} \times \frac{ql^2}{8} \times l,$$

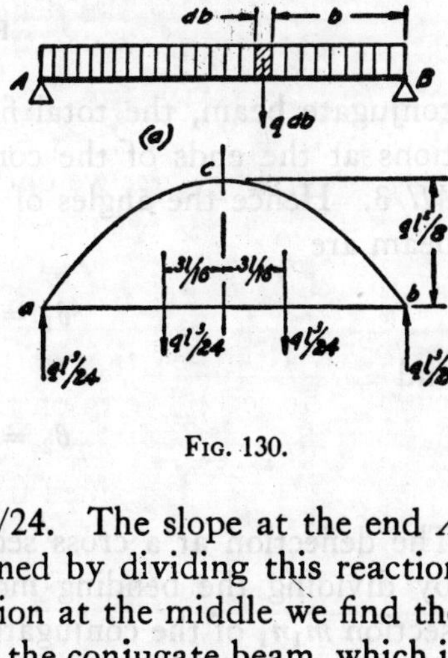

and each reaction is equal to $ql^3/24$. The slope at the end A of the actual beam is then obtained by dividing this reaction by EI_z. To calculate the deflection at the middle we find the bending moment at the middle of the conjugate beam, which is

$$\frac{ql^3}{24}\left(\frac{l}{2} - \frac{3l}{16}\right) = \frac{5ql^4}{384}.$$

The deflection is then obtained by dividing this moment by EI_z.

In the case of a simply supported beam AB with a couple M acting at the end, Fig. 131, the bending moment diagram is a triangle abd, as shown in Fig. 131b. Considering ab as the

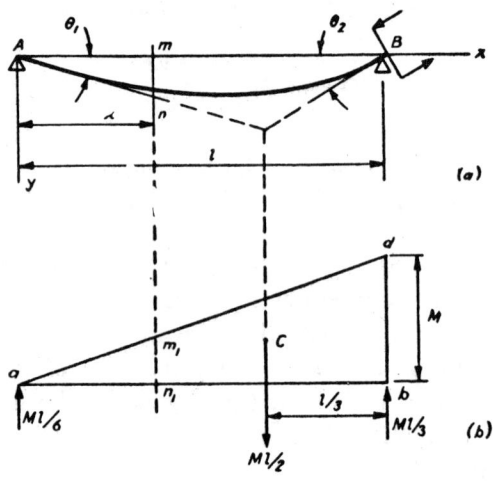

Fig. 131.

conjugate beam, the total fictitious load is $Ml/2$. The reactions at the ends of the conjugate beam are thus $Ml/6$ and $Ml/3$. Hence the angles of rotation of the ends of the actual beam are

$$\theta_1 = \frac{Ml}{6EI_z} \tag{103}$$

and

$$\theta_2 = -\frac{Ml}{3EI_z}. \tag{104}$$

The deflection at a cross section mn of the beam is obtained by dividing the bending moment at the corresponding cross section m_1n_1 of the conjugate beam by EI_z, which gives

$$y = \frac{1}{EI_z}\left(\frac{Ml}{6}x - \frac{Ml}{2}\cdot\frac{x^2}{l^2}\cdot\frac{x}{3}\right) = \frac{Mlx}{6EI_z}\left(1 - \frac{x^2}{l^2}\right). \tag{105}$$

Problems

1. Determine the angles at the ends and the deflection under the loads and at the middle of the beam shown in Fig. 132.

Solution. The conjugate beam will be loaded by the trapezoid $adeb$, the area of which is $Pc(l - c)$. The angles at the ends are

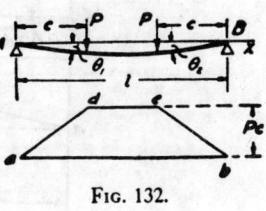

Fig. 132.

$$\theta_1 = -\theta_2 = \frac{1}{EI_z} \frac{Pc(l - c)}{2}.$$

The deflection under the loads is

$$(y)_{x=c} = \frac{1}{EI_z} \left[\frac{Pc^2(l - c)}{2} - \frac{Pc^2}{2} \cdot \frac{c}{3} \right] = \frac{Pc^2}{EI_z} \left(\frac{l}{2} - \frac{2}{3}c \right).$$

The deflection at the middle, from eq. (91), is

$$(y)_{x=l/2} = \frac{Pc}{24EI_z} (3l^2 - 4c^2).$$

2. Determine the slope at the ends of the beam shown in Fig. 92. *Answer.*

$$\left(\frac{dy}{dx} \right)_{x=0} = \frac{7}{180} \frac{Wl^2}{EI_z}, \qquad \left(\frac{dy}{dx} \right)_{x=l} = -\frac{2}{45} \frac{Wl^2}{EI_z},$$

where W is the total pressure on the beam.

3. A simply supported beam AB is loaded as shown in Fig. 133. Find the deflection at the center of the beam and the maximum deflection; find the slopes at the ends of the beam. *Answer.*

$$(\delta)_{x=l/2} = 0, \quad \delta_{max} = \frac{2\sqrt{2}Pl^3}{(27)^2\sqrt{3}EI_z}, \quad \left(\frac{dy}{dx} \right)_{x=0} = \left(\frac{dy}{dx} \right)_{x=l} = \frac{Pl^2}{81EI_z}.$$

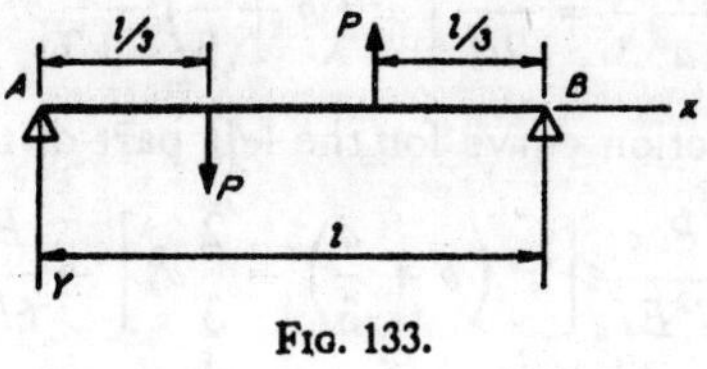

Fig. 133.

4. Determine the angles θ_1 and θ_2 and the deflection at any cross section mn of a beam simply supported at the ends and bent by a couple Pc (Fig. 134).

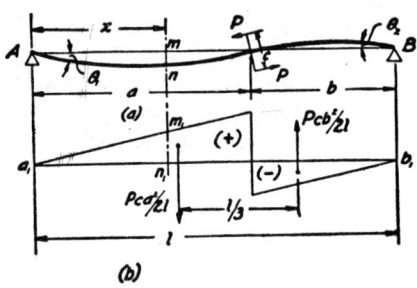

Fig. 134.

Solution. The loading of the conjugate beam is indicated in Fig. 134b. The reactions at a_1 and b_1 are

$$R_a = \frac{1}{l}\left[\frac{Pca^2}{2l}\left(b + \frac{a}{3}\right) - \frac{Pcb^2}{2l} \cdot \frac{2}{3}b\right],$$

$$R_b = \frac{1}{l}\left[\frac{Pca^2}{2l}\frac{2a}{3} - \frac{Pcb^2}{2l}\left(a + \frac{b}{3}\right)\right].$$

Therefore

$$\theta_1 = \frac{Pc}{2l^2 EI_z}\left[a^2\left(b + \frac{a}{3}\right) - \frac{2}{3}b^3\right] = \frac{Pc}{2lEI_z}\left(\frac{l^2}{3} - b^2\right),$$

$$\theta_2 = -\frac{Pc}{2l^2 EI_z}\left[\frac{2}{3}a^3 - b^2\left(a + \frac{b}{3}\right)\right] = \frac{Pc}{2lEI_z}\left(\frac{l^2}{3} - a^2\right).$$

If $a = b = l/2$, we obtain

$$\theta_1 = \theta_2 = \frac{Pcl}{24EI_z}.$$

If $a > l/\sqrt{3}$ the angle θ_2 changes its sign and the deflection is everywhere downward. The bending moment at the cross section m_1n_1 of the conjugate beam is

$$R_a x - \frac{Pca^2}{2l}\frac{x^2}{a^2}\frac{x}{3} = \frac{Pcx}{2l^2}\left[a^2\left(b + \frac{a}{3}\right) - \frac{2}{3}b^3\right] - \frac{Pcx^3}{6l}.$$

Therefore the deflection curve for the left part of the actual beam is

$$y = \frac{Pcx}{2l^2 EI_z}\left[a^2\left(b + \frac{a}{3}\right) - \frac{2}{3}b^3\right] - \frac{Pcx^3}{6lEI_z}.$$

5. A beam is bent by two couples as shown in Fig. 135. Determine the ratio $M_1:M_2$ if the point of inflection is at a distance $l/3$ from the left support.

Answer. $M_2 = 2M_1$.

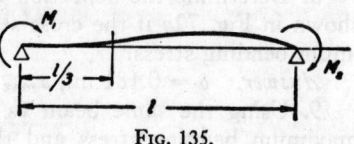

6. Two planks of different thicknesses h_1 and h_2, resting one

Fig. 135.

upon the other, support a uniformly distributed load as shown in Fig. 136. Determine the ratio of the maximum stresses occurring in each.

Solution. Both planks have the same deflection curves and curvature, hence their bending moments are in the same ratio as the moments of inertia of their cross sections, i.e., in the ratio $h_1^3:h_2^3$. The section moduli are in the ratio $h_1^2:h_2^2$, hence the maximum stresses are in the ratio $h_1:h_2$.

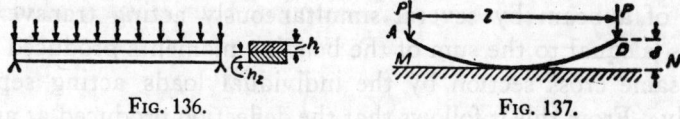

Fig. 136.   Fig. 137.

7. A steel bar AB has such an initial curvature that after being straightened by the forces P (Fig. 137) it produces a uniformly distributed pressure along the length of the rigid plane surface MN. Determine the forces P necessary to straighten the bar and the maximum stress produced in it if $l = 20$ in., $\delta = 0.1$ in. and the cross section of the bar is a square having 1 in. sides.

Solution. To obtain a uniformly distributed pressure, the initial curvature of the bar must be the same as the deflection curve of a simply supported beam carrying a uniformly distributed load of intensity $2P/l$. Then we obtain

$$M_{\max} = \frac{2P}{l}\frac{l^2}{8} = \frac{Pl}{4},\qquad (g)$$

$$\delta = \frac{5}{384} \times \frac{2P}{l} \times \frac{l^4}{EI_z}.\qquad (h)$$

The maximum stress will be

$$\sigma_{\max} = \frac{M_{\max}}{Z} = \frac{Plh}{8I_z}.\qquad (i)$$

Now from (h) and (i)

$$\sigma_{\max} = \frac{24 E\delta h}{5l^2} = \frac{24 \times 30 \times 10^6 \times 0.1 \times 1}{5 \times 20^2} = 36{,}000 \text{ lb per sq in.,}$$

and from (i)

$$P = 1{,}200 \text{ lb.}$$

8. Determine the deflection δ at the middle of the wooden beam shown in Fig. 72a if the cross section is 10×10 in. Find the maximum bending stress.

Answer. $\delta = 0.181$ in., $\sigma_{max} = 494$ lb per sq in.

9. Using the same beam as in the preceding problem, find the maximum bending stress and the deflection at the middle for the loading condition shown in Fig. 72e.

Answer. $\delta = 0$, $\sigma_{max} = 123$ lb per sq in.

37. Method of Superposition.—From the discussion of the area-moment method (Art. 34) it is seen that the deflections of a beam are entirely defined by the bending moment diagram. From the definition of the bending moment (Art. 20) it follows that the bending moment produced at any cross section of a beam by several simultaneously acting transverse loads is equal to the sum of the bending moments produced at the same cross section by the individual loads acting separately. From this it follows that the deflection produced at any point of a beam by a system of simultaneously acting transverse loads can be obtained by summing up the deflections produced at that point by the individual loads. Having, for example, deflection curves for the cases illustrated in Figs. 123 and 119, we can obtain by simple summation the deflections for a cantilever or a simply supported beam carrying any transverse load.

Taking as an example the case shown in Fig. 138 and using eqs. (97) and (98), we conclude that the deflection at B is

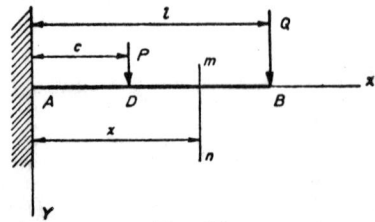

FIG. 138.

$$\delta_b = \frac{Ql^3}{3EI_z} + \frac{Pc^2}{2EI_z}\left(l - \frac{c}{3}\right).$$

Similarly the deflection at D is

$$\delta_d = \frac{Q}{EI_z}\left(\frac{lc^2}{2} - \frac{c^3}{6}\right) + \frac{Pc^3}{3EI_z}.$$

The deflection at any cross section mn for the portion DB of the cantilever beam will be

$$y = \frac{Pc^2}{2EI_z}\left(x - \frac{c}{3}\right) + \frac{Q}{EI_z}\left(\frac{lx^2}{2} - \frac{x^3}{6}\right).$$

In the case of distributed load the summation will naturally be replaced by integration. As an example we shall take the case of a simply supported beam under a uniformly distributed load, Fig. 130, and calculate the slopes at the ends and the deflection at the middle. From eq. (a), Art. 36, the increment of slope $d\theta_1$ produced at the left end of the beam by the element of load qdb shown in Fig. 130 is

$$d\theta_1 = \frac{qab(l + b)db}{6lEI_z} = \frac{qb(l^2 - b^2)db}{6lEI_z}.$$

The slope θ_1 produced by the total load is then the summation of the increments of slope produced by all the elements qdb from $b = 0$ to $b = l$. Thus

$$\theta_1 = \int_0^l \frac{qb(l^2 - b^2)db}{6lEI_z} = \frac{ql^3}{24EI_z}. \qquad (a)$$

The deflection at the middle is obtained from eq. (91), which was derived on the assumption that the load is to the right of the middle. Any element of load qdb to the right of the middle produces at the middle a deflection

$$(dy)_{x=l/2} = \frac{qbdb}{48EI_z} (3l^2 - 4b^2).$$

Summing up the deflections produced by all such elements of load to the right of the middle, and noting that the load on the left half of the beam produces the same deflection at the middle as the load on the right half, we obtain for the total deflection

$$\delta = (y)_{x=l/2} = 2 \int_0^{l/2} \frac{qbdb}{48EI_z} (3l^2 - 4b^2) = \frac{5}{384} \frac{ql^4}{EI_z}. \qquad (b)$$

The results (a) and (b) coincide with formulas (83) and (82) previously obtained by integration of the differential equation of the deflection curve.

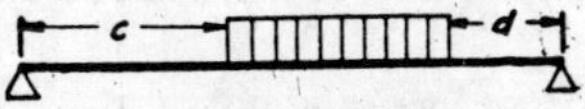

The method of superposition is especially useful if the distributed load covers only a part of the span as in Fig. 139. Using the expression developed above for

Fig. 139.

$(dy)_{x=l/2}$, the deflection produced at the middle by the load to the right of the middle is

$$\delta_1 = \int_d^{l/2} \frac{qbdb}{48EI_z} (3l^2 - 4b^2).$$

The load to the left of the middle produces the deflection

$$\delta_2 = \int_c^{l/2} \frac{qbdb}{48EI_z} (3l^2 - 4b^2).$$

The total deflection at the middle is therefore

$$\delta = \delta_1 + \delta_2 = \int_d^{l/2} \frac{qbdb}{48EI_z} (3l^2 - 4b^2) + \int_c^{l/2} \frac{qbdb}{48EI_z} (3l^2 - 4b^2).$$

Let us consider now a uniformly loaded cantilever, Fig. 140. The deflection produced at the cross section mn by each elemental load qdc to its left can be found from eq. (98) by substituting qdc for P. The deflection y_1 produced by the total load to the left of mn is the summation of the deflections produced by all such elemental loads with c varying from $c = 0$ to $c = x$:

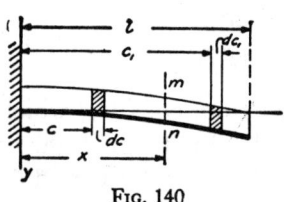

FIG. 140

$$v_1 = \frac{1}{EI_z} \int_0^x \frac{qc^2}{2} \left(x - \frac{1}{3}c \right) dc = \frac{q}{2EI_z} \frac{x^4}{4}.$$

The deflection produced at the cross section mn by an elemental load qdc_1 to its right is found from eq. (97) by substituting qdc_1 for P and c_1 for l. The deflection y_2 produced at mn by the total load to the right is the summation of the deflections due to all such elemental loads, with c_1 varying from $c_1 = x$ to $c_1 = l$:

$$y_2 = \frac{1}{EI_z} \int_x^l q \left(\frac{c_1 x^2}{2} - \frac{x^3}{6} \right) dc_1 = \frac{q}{2EI_z} \left(-\frac{x^4}{6} + \frac{x^2 l^2}{2} - \frac{lx^3}{3} \right).$$

Then the total deflection at the section mn is

$$y = y_1 + y_2 = \frac{q}{2EI_z} \left(\frac{l^2 x^2}{2} - \frac{lx^3}{3} + \frac{x^4}{12} \right),$$

which agrees with eq. (101) found previously.

Problems

1. Determine the deflection at the middle of the beam AB, shown in Fig. 141, when $I_z = 91.4$ in.[4], $q = 500$ lb per ft, $l = 24$ ft, $a = 12$ ft, $b = 8$ ft, $E = 30 \times 10^6$ lb per sq in.

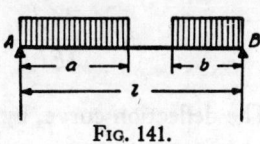

Solution. Due to the fact that $a = l/2$, the deflection produced at the middle by the load acting on the left half of the beam, from eq. (82), is

FIG. 141.

$$(y_1)_{x=l/2} = \frac{1}{2}\frac{5}{384}\frac{ql^4}{EI_z}.$$

The deflection produced at the middle by the load on the right half of the beam is

$$(y_2)_{x=l/2} = \int_0^b \frac{qcdc}{48EI_z}(3l^2 - 4c^2) = \frac{25}{48 \times 162} \times \frac{ql^4}{EI_z}.$$

The total deflection is

$$(y)_{x=l/2} = (y_1)_{x=l/2} + (y_2)_{x=l/2}$$

$$= \left(\frac{1}{2}\frac{5}{384} + \frac{25}{48 \times 162}\right)\frac{ql^4}{EI_z} = 1.02 \text{ in.}$$

2. Determine the deflection at the middle of the beam shown in Fig. 95 when the load is in a position to produce the maximum bending moment.

Suggestion. The deflection can be obtained by using eq. (91) together with the method of superposition and substituting $b = l/2 - d/4$ in this equation for one load and $b = l/2 - \frac{3}{4}d$ for the other.

3. Determine the deflections at the middle and the angles of rotation of the ends of the beams shown in Figs. 72b and 72d. Assume in these calculations a standard steel I beam, 8 I 23.0, with $I_z = 64.2$ in.[4]

Answer. For the beam in Fig 72b, $\delta = 0.11$ in., $\theta_1 = 0.00223$ radian, $\theta_2 = -0.00255$ radian.

4. A beam with supported ends is bent by two couples M_1 and M_2, applied at the ends (Fig. 142). Determine the angles of rota-

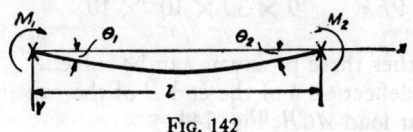

FIG. 142

tion of the ends and the position of the cross section in which the deflection is a maximum.

Solution. The absolute values of the angles from eqs. (103) and (104) are

$$\theta_1 = \frac{M_1 l}{3EI_z} + \frac{M_2 l}{6EI_z}, \qquad |\theta_2| = \frac{M_2 l}{3EI_z} + \frac{M_1 l}{6EI_z}.$$

The deflection curve, by using eq. (105), is

$$y = \frac{M_1 l(l - x)}{6EI_z}\left[1 - \left(\frac{l - x}{l}\right)^2\right] + \frac{M_2 lx}{6EI_z}\left(1 - \frac{x^2}{l^2}\right).$$

The position of maximum deflection can be found from this equation by equating the first derivative to zero.

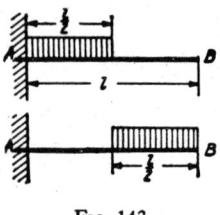

5. What is the ratio of the deflections at the ends of the cantilevers shown in Fig. 143 if the intensity of uniform load is the same in both cases?

Answer. 7:41.

6. Determine the deflections at the ends of the cantilever beams shown in Fig. 73, assuming that the material is steel, the depth h of each beam is 10 in., and the maximum bending stress is 16,000 lb per sq in. Use the method of superposition.

Fig. 143.

Solution. Taking, for example, the case represented in Fig. 73b and observing that the total uniform load is $ql = 4,000$ lb and the load at the end is $P = 1,000$ lb $= \frac{1}{4}ql$, we have

$$\delta = \frac{ql^4}{8EI_z} + \frac{Pl^3}{3EI_z} = \frac{5ql^4}{24EI_z},$$

$$\sigma_{max} = \frac{M_{max}h}{2I_z} = \frac{h}{2I_z}\left(Pl + \frac{ql^2}{2}\right) = \frac{3ql^2 h}{8I_z}.$$

Eliminating I_z, we obtain

$$\delta = \frac{5l^2 \sigma_{max}}{9Eh} = \frac{5 \times (120)^2 \times 16,000}{9 \times 30 \times 10^6 \times 10} = 0.427 \text{ in.}$$

Similarly, the other three problems can be solved.

7. Find the deflection δ of the end B of the cantilever AB loaded by the triangular load ACB, Fig. 144.

Solution. Applying the method of superposition and using eq. (102) and the result of Prob. 1, Art. 35, we obtain

$$\delta = \frac{q l^4}{8 E I_z} - \frac{q l^4}{30 E I_z} = \frac{11 q l^4}{120 E I_z}.$$

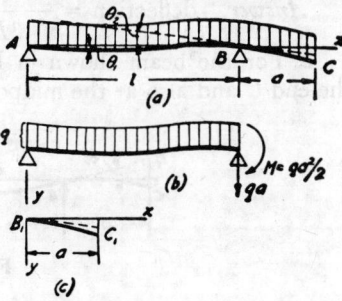

Fig. 144.

38. Deflection of Beams with Overhangs.—A beam with an overhang can be divided into two parts: the part between the supports which is to be treated as a beam with supported ends, and the overhang which is to be treated as a cantilever. As an illustration, we consider the bending of a beam with an overhang under the action of a uniformly distributed load q (Fig. 145). The beam is divided into the parts AB and BC and the action of the overhang on the portion of the beam between the supports is replaced by a shearing force qa and a couple $M = qa^2/2$. We find that the shearing force is directly transmitted to the support and that only the couple $qa^2/2$ need be considered. Then the deflection at any cross section between the supports is obtained by subtracting the deflection produced by the couple $qa^2/2$ from the deflections produced by the uniform load q (Fig. 145b). Using eqs. (81) and (105), we obtain

$$y = \frac{q}{24 E I_z} (l^3 x - 2 l x^3 + x^4) - \frac{qa^2 l x}{12 E I_z} \left(1 - \frac{x^2}{l^2}\right).$$

The angle of rotation of the cross section at B is obtained by using eqs. (83) and (104), from which, by considering rotation positive when in the clockwise direction, we have

$$\theta_2 = \frac{qa^2l}{6EI_z} - \frac{ql^3}{24EI_z}.$$

The deflection at any cross section of the overhang (Fig. 145c) is now obtained by superposing the deflection of a cantilever (eq. 101) on the deflection,

$$\theta_2 x = \left(\frac{qa^2l}{6EI_z} - \frac{ql^3}{24EI_z}\right) x,$$

due to the rotation of the cross section B.

Problems

1. Determine the deflection and the slope at the end C of the beam shown in Fig. 147a.

   Answer. deflection $= \dfrac{Pa^2(l + a)}{3EI_z}$, slope $= \dfrac{Pa(2l + 3a)}{6EI_z}$.

2. For the beam shown in Fig. 146 determine the deflection at the end C and also at the midpoint between the supports.

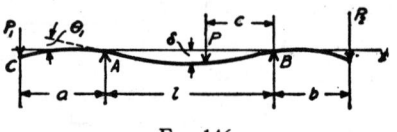

FIG. 146.

Solution. That part of the beam between the supports will be in the condition of a beam loaded by the force P and by the couples P_1a and P_2b at the supports. By using eqs. (91) and (105) and the method of superposition, the deflection at the middle is

$$\delta = \frac{Pc}{48EI_z}(3l^2 - 4c^2) - \frac{P_1al^2}{16EI_z} - \frac{P_2bl^2}{16EI_z}.$$

The angle θ_1 at the support A is obtained from eqs. (88), (103) and (104),

$$\theta_1 = \frac{Pc(l^2 - c^2)}{6lEI_z} - \frac{P_1al}{3EI_z} - \frac{P_2bl}{6EI_z}.$$

From eq. (95) the deflection at the end C is

$$\frac{P_1 a^3}{3EI_z} - a\theta_1.$$

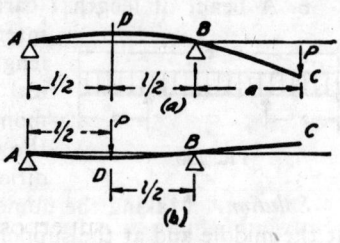

FIG. 147.

3. A beam with an overhang is bent in one case by the force P at the end (Fig. 147a), and in another case by the same force applied at the middle of the span (Fig. 147b).

Prove that the deflection at the point D in the first case is equal to the deflection at the end C in the second case.

Answer. In each case the deflection is

$$\frac{Pl^2 a}{16EI_z}.$$

4. A beam of length l with two equal overhangs is loaded by two equal forces P at the ends (Fig. 148). Determine the ratio x/l at which (1) the deflection at the middle is equal to the deflection at either end, (2) the deflection at the middle has its maximum value.

Answer. (1) $x = 0.152l$; (2) $x = l/6$.

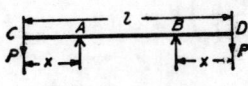

FIG. 148.

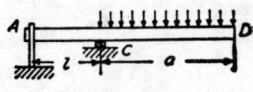

FIG. 149.

5. A wooden beam of circular cross section supported at C, with the end attached at A, carries a uniformly distributed load q on the overhang CD (Fig. 149). Determine the diameter of the cross section and the deflection at D if $l = 3$ ft, $a = 6$ ft, $q = 300$ lb per ft, $\sigma_W = 1,200$ lb per sq in.

Solution. The diameter d is found from the equation

$$\frac{qa^2}{2} \div \frac{\pi d^3}{32} = \sigma_W.$$

Then the deflection at the end D is found from the equation

$$\delta = \frac{qa^4}{8EI_z} + \frac{qa^3 l}{6EI_z} = \frac{qa^3}{24EI_z}(3a + 4l).$$

6. A beam of length l carries a uniformly distributed load of intensity q (Fig. 150). Determine the length of overhangs to make the numerical maximum value of the bending moment as small as possible. Determine the deflection at the middle for this condition.

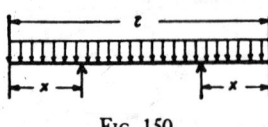

Fig. 150.

Solution. Making the numerical values of the bending moments at the middle and at the supports equal, we obtain

$$x = 0.207l.$$

The deflection at the middle is determined from the equation

$$\delta = \frac{5}{384} \cdot \frac{q(l - 2x)^4}{EI_z} - \frac{qx^2(l - 2x)^2}{16EI_z},$$

in which the first term on the right side represents the deflection produced by the load between the support (eq. 82) and the second, the deflection produced by the load on the overhangs (eq. 105).

7. Determine the deflections at the ends of the overhangs for the beams represented in Fig. 79a, b, c. Assume a standard 8 I 23.0 beam, $I_z = 64.2$ in.[4], $E = 30 \times 10^6$ lb per sq in.

Answer. If l is the length of the beam between supports, and a is the length of the overhang, then

$$\delta_a = \frac{qa^4}{8EI_z} + \left(\frac{qa^2l}{6EI_z} - \frac{ql^3}{24EI_z} \right) a$$

$$\delta_b = \frac{Pa^2l}{4EI_z} - \frac{117Pa^3}{384EI_z},$$

$$\delta_c = \frac{qa^3}{8EI_z} (a + 2l).$$

39. Effect of Shearing Force on the Deflection of Beams.—In the previous discussion (see p. 137) only the action of the bending moment in causing deflection was considered. An additional deflection will be produced by the shearing force, in the form of a mutual sliding of adjacent cross sections along each other. As a result of the nonuniform distribution of the shearing stresses, the cross sections, previously plane, become curved as in Fig. 151, which shows the bending due to shear alone.[8] The elements

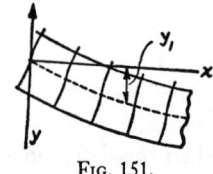

Fig. 151.

[8] The deformation produced by the bending moment and consisting of a mutual rotation of adjacent cross sections has been subtracted.

of the cross sections at the centroids remain vertical and slide along one another. Therefore the slope of the deflection curve, due to the shear alone, is equal at each cross section to the shearing strain at the centroid of this cross section. Denoting by y_1 the deflections due to shear, we obtain for any cross section the following expression for the slope:

$$\frac{dy_1}{dx} = \frac{(\tau_{xy})_{y=0}}{G} = \frac{\alpha V}{AG}, \qquad (a)$$

in which V/A is the average shearing stress τ_{xy}, G is the modulus in shear and α is a numerical factor with which the average shearing stress must be multiplied in order to obtain the shearing stress at the centroid of the cross sections. For a rectangular cross section $\alpha = \frac{3}{2}$ (see eq. 66, p. 118); for a circular cross section $\alpha = \frac{4}{3}$ (see eq. 68, p. 122). With a continuous load on the beam, the shearing force V is a continuous function which may be differentiated with respect to x. The curvature caused by the shear alone is then

$$\frac{d^2 y_1}{dx^2} = \frac{\alpha}{AG} \frac{dV}{dx} = -\frac{\alpha}{AG} q,$$

where q is the intensity of the load. The sum of this and the curvature produced by the bending moment (see eq. 79) gives the complete expression for the curvature:

$$\frac{d^2 y}{dx^2} = -\frac{1}{EI_z} \left(M + \frac{\alpha EI_z}{AG} q \right). \qquad (106)$$

This equation must be used instead of eq. (79) to determine deflections in all cases in which the effect of the shearing force should be taken into consideration.[9] Knowing M and q as functions of x, eq. (106) can be integrated in the same manner as has been shown in Art. 32.

The conjugate beam method (see p. 155) may also be applied to good advantage in this case by taking as ordinates of the imaginary load diagram

$$M + \alpha \frac{EI_z}{AG} q, \qquad (b)$$

instead of only M.

Let us consider, for example, the case of a simply supported beam carrying a uniform load (Fig. 152). The bending moment at any cross section x is

$$M = \frac{ql}{2} x - \frac{qx^2}{2}. \qquad (c)$$

---

[9] Another way of determining additional deflection due to shear is discussed on p. 318.

The load on the conjugate beam consists of two parts: (1) that represented by the first term of (b) and given by the parabolic bending moment diagram (Fig. 152b), and (2) that represented by the second term of (b), which is $\alpha(EI_z/AG)q$. Since q is constant, the second term is a uniformly distributed load shown in Fig. 152c.

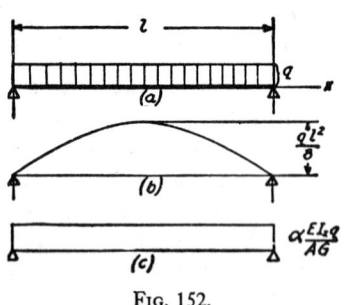

FIG. 152.

The additional deflection at any cross section, due to the shearing force, is the bending moment produced at this cross section of the conjugate beam by the load shown in Fig. 152c, divided by EI_z. At the middle of the beam the additional deflection is consequently

$$\delta_1 = \frac{1}{EI_z}\left(\alpha \frac{EI_z}{AG} q\right)\frac{l^2}{8} = \frac{\alpha l^2 q}{8AG}.$$

Adding this to the deflection due to the bending moment (see eq. 82, p. 141), we obtain the total deflection

$$\delta = \frac{5}{384}\frac{ql^4}{EI_z} + \frac{\alpha l^2 q}{8AG} = \frac{5}{384}\frac{ql^4}{EI_z}\left(1 + \frac{48\alpha}{5}\frac{k_z^2}{l^2}\frac{E}{G}\right), \qquad (d)$$

in which $k_z = \sqrt{I_z/A}$ is the radius of gyration of the cross section with respect to the z axis.

For a rectangular cross section of depth h, $k_z^2 = \frac{1}{12}h^2$, $\alpha = \frac{3}{2}$. Putting $E/G = 2(1 + \mu) = 2.6$, we obtain from (d)

$$\delta = \frac{5}{384}\frac{ql^4}{EI_z}\left(1 + 3.12\frac{h^2}{l^2}\right).$$

It may be seen that for $l/h = 10$ the effect of the shearing force on the deflection is about 3 per cent. As the ratio l/h decreases this effect increases.

The factor α is usually larger than 2 for I beams and when these beams are short the effect of the shearing force may be comparatively great. Using eq. (70) and Fig. 110, we have

$$\frac{\alpha V}{A} = \frac{V}{b_1 I_z}\left[\frac{bh^2}{8} - \frac{h_1^2}{8}(b - b_1)\right],$$

from which

$$\alpha = \frac{A}{b_1 I_z}\left[\frac{bh^2}{8} - \frac{h_1^2}{8}(b - b_1)\right]. \qquad (e)$$

For example, suppose $h = 24$ in., $A = 31.0$ sq in., $I_z = 2{,}810$ in.[4], the thickness of the web $b_1 = \frac{5}{8}$ in., $l = 6h$. Then eq. (e) gives $\alpha = 2.42$. Substituting in eq. (d), we find

$$\delta = \frac{5}{384}\frac{ql^4}{EI_z}\left(1 + \frac{48}{5} \times 2.42 \times \frac{2,810}{31 \times 144^2} \times 2.6\right) = 1.265\frac{5ql^4}{384EI_z}.$$

The additional deflection due to shear in this case is equal to 26.5 per cent of the deflection produced by the bending moment and must therefore be considered.

In the case of a concentrated load P (Fig. 153) such a load can be considered as the limiting case of a load distributed over a very short portion e of the beam. The amount of the imaginary loading P_1 on the conjugate beam A_1B_1, corresponding to the second term in expression (b), will be

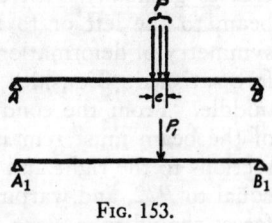

Fig. 153.

$$P_1 = \alpha\frac{EI_z}{AG}P. \qquad (f)$$

The additional deflection due to shearing forces is obtained by dividing by EI_z the bending moment produced in the conjugate beam by the imaginary concentrated load given by eq. (f). For instance, for central loading of a beam the bending moment at the middle of the conjugate beam produced by the load (f) will be $\alpha(EI_z/AG)Pl/4$ and the additional deflection at the middle due to shearing forces is

$$\delta_1 = \frac{\alpha}{AG}\frac{Pl}{4}. \qquad (g)$$

Adding this to the deflection produced by the bending moment alone (see eq. 90, p. 146), the following expression for the complete deflection is obtained:

$$\delta = \frac{Pl^3}{48EI_z} + \frac{\alpha}{AG}\cdot\frac{Pl}{4} = \frac{Pl^3}{48EI_z}\left(1 + \frac{12\alpha k_z^2}{l^2}\frac{E}{G}\right).$$

For a beam of rectangular cross section of depth h we have

$$\frac{k_z^2}{l^2} = \frac{h^2}{12l^2}, \qquad \alpha = \frac{3}{2},$$

and we obtain

$$\delta = \frac{Pl^3}{48EI_z}\left(1 + 3.90\frac{h^2}{l^2}\right). \qquad (h)$$

For $h/l = \frac{1}{10}$ the additional effect of the shearing force is about 4 per cent.

It has been assumed throughout the above discussion that the cross sections of the beam can warp freely as shown in Fig. 151. The uniformly loaded beam is one case in which this condition is approximately satisfied. The shearing force at the middle of such a beam is zero and there will be no warping here. The warping increases gradually with the shearing force as we proceed along the beam to the left or to the right of the middle. The condition of symmetry of deformation with respect to the middle section is therefore satisfied. Consider now bending by a concentrated load at the middle. From the condition of symmetry the middle cross section of the beam must remain plane. At the same time, adjacent cross sections to the right and to the left of the load carry a shearing force equal to $P/2$, and warping of cross sections caused by these shearing forces should take place. From the condition of continuity of deformation, however, there can be no abrupt change from a plane middle section to warped adjacent sections. There must be a continuous increase in warping as we proceed along the beam in either direction from the middle, and only at some distance from the load can the warping be such as a shearing force $P/2$ produces under conditions of freedom in warping. From this discussion it must be concluded that in the neighborhood of the middle cross section the stress distribution will not be that predicted by the elementary theory of bending (see p. 94). Warping will be partially prevented and the additional deflection due to shearing forces will be somewhat less than that found above (see eq. g). A more detailed investigation [10] shows that in the case of a concentrated load at the middle the deflection at the middle is

$$\delta = \frac{Pl^3}{48EI}\left[1 + 2.85\,\frac{h^2}{l^2} - 0.84\left(\frac{h}{l}\right)^3\right]. \qquad (i)$$

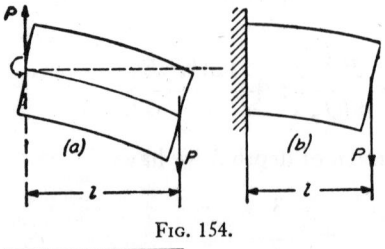

FIG. 154.

We have an analogous condition also in the case of a cantilever beam. If the built-in cross section can warp freely as shown in Fig. 154a, the conditions will be as assumed in the derivation of eq. (h). The deflection of a cantilever of rectangular cross

[10] See L. N. G. Filon, *loc. cit.*, p. 119; and S. Timoshenko, *Phil. Mag.*, Vol. 47, p. 1095, 1924. See also Th. Kármán, *Scripta Universitatis atque Bibliothecae Hierosolmitanarum*, 1923; and writer's *Theory of Elasticity*, p. 95, 1934.

section will be obtained by substituting l for $l/2$ and P for $P/2$ in this equation, giving

$$\delta = \frac{Pl^3}{3EI}\left(1 + 0.98\frac{h^2}{l^2}\right). \qquad (j)$$

When the built-in cross section is completely prevented from warping (Fig. 154b), the conditions will be the same as assumed in the derivation of eq. (i) and the deflection will be

$$\delta = \frac{Pl^3}{3EI}\left[1 + 0.71\frac{h^2}{l^2} - 0.10\left(\frac{h}{l}\right)^3\right], \qquad (k)$$

which is less than the deflection given by (j).

CHAPTER VI

STATICALLY INDETERMINATE PROBLEMS IN BENDING

40. Redundant Constraints.—In our previous discussion three types of beams have been considered: (1) a cantilever beam, (2) a beam simply supported at the ends and (3) a beam with overhangs. In all three cases the reactions at the supports can be determined from the fundamental equations of statics so that the problems are *statically determinate*. We will now consider problems of bending of beams in which the equations of statics are not sufficient to determine all the reactive forces at the supports, so that additional equations, based on a consideration of the deflection of the beams, must be derived. Such problems are called *statically indeterminate*.

Let us consider the various types of supports which a beam may have. The support represented in Fig. 155a is a

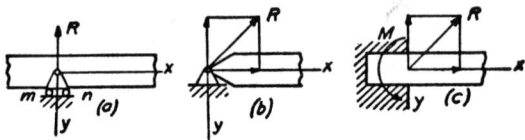

Fig. 155.

hinged movable support. Neglecting the friction in the hinge and in the rollers, it is evident that in this type of support the reaction must act through the center of the hinge and must be perpendicular to the plane *mn* on which the rollers are moving. Hence we know the point of application of the reaction and its direction. There remains only one unknown element, the magnitude of the reaction.

In Fig. 155b a *hinged immovable* support is shown. In this case the reaction must go through the center of the hinge, but it may have any direction in the plane of the figure. We have two unknowns to determine from the equations of statics, the

176

direction of the reaction and its magnitude, or, if we like, the vertical and horizontal components of the reaction.

In Fig. 155c a *built-in support* is represented. In this case, not only are the direction and the magnitude of the reaction unknown, but also the point of application. The reactive forces distributed over the built-in cross section can be replaced by a force R applied at the centroid of the cross section and a couple M. We then have three unknowns to determine from the equations of statics, the two components of the reactive force R and the magnitude of the couple M.

For beams loaded by transverse loads in the plane of symmetry we have, for determining the reactions at the supports, the three equations of statics, namely

$$\Sigma X = 0, \qquad \Sigma Y = 0, \qquad \Sigma M = 0. \qquad (a)$$

If the beam is supported so that there are only three unknown *reactive elements*, they can be determined from eqs. (a), hence the problem is statically determinate. These three elements are just sufficient to assure the immovability of the beam. When the number of reactive elements is larger than three, we say there are *redundant constraints* and the problem is statically indeterminate.

A cantilever is supported by a built-in support. In this case, as was explained above, the number of unknown reactive elements is three and they can be determined from the equations of statics, (a). For beams supported at both ends and beams with overhangs it is usually assumed that one of the supports is an immovable and the other a movable hinge. In such a case we again have three unknown reactive elements, which can be determined from the equations of statics.

If the beam has immovable hinges at both ends (Fig. 156),

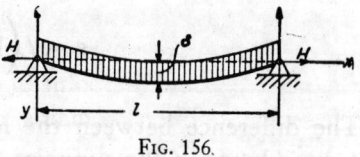

FIG. 156.

the problem becomes statically indeterminate. At each end we have two unknown elements, the two components of the corresponding reaction, and for determining these four unknowns we have only the three eqs. (a). Hence we have one redundant constraint and a consideration of the deformation of the beam becomes necessary to determine the reactions. The vertical components of the reactions can be calculated from the equations of statics. In the case of vertical loads it can be concluded also from statics that the horizontal components H are equal and opposite in direction. To find the magnitude of H let us consider the elongation of the axis of the beam during bending. A good approximation to this elongation can be obtained by assuming that the deflection curve of the beam is a parabola,[1] the equation of which is

$$y = \frac{4\delta x(l - x)}{l^2}, \qquad (b)$$

where δ is the deflection at the middle. The length of the curve is

$$s = 2\int_0^{l/2} \sqrt{dx^2 + dy^2} = 2\int_0^{l/2} dx \sqrt{1 + \left(\frac{dy}{dx}\right)^2}. \qquad (c)$$

In the case of a flat curve the quantity $(dy/dx)^2$ is small in comparison with unity and, neglecting small quantities of order higher than the second, we obtain approximately

$$\sqrt{1 + \left(\frac{dy}{dx}\right)^2} \approx 1 + \frac{1}{2}\left(\frac{dy}{dx}\right)^2.$$

Substituting this expression in eq. (c) and using eq. (b), we find the length of the curve to be

$$s = l\left(1 + \frac{8}{3}\frac{\delta^2}{l^2}\right).$$

The difference between the length of the curve and the distance l between the supports represents the total axial elongation of the beam and is $(\frac{8}{3})(\delta^2/l)$. The unit elongation is then $(\frac{8}{3})(\delta^2/l^2)$. Knowing this and denoting by E the modulus of

---

[1] The exact expression for the deflection curve will be given later (see Part II).

elasticity of the material of the beam and by A its cross-sectional area, we obtain the horizontal reactions from the equation

$$H = \frac{8}{3} \frac{\delta^2}{l^2} EA. \tag{d}$$

It is important to note that in practice the deflection δ for most beams is very small in comparison with the length l and the tensile stress $(\frac{8}{3})(\delta^2/l^2)E$ produced by the forces H is usually small in comparison with bending stresses and can be neglected. This justifies the usual practice of calculating beams with supported ends by assuming that one of the two supports is a movable hinge, although the special provisions for permitting free motion of the hinge are actually used only in cases of large spans such as bridges.

In the case of bending of flexible bars and thin metallic strips, where the deflection δ is no longer very small in comparison with l, the tensile stresses produced by the longitudinal forces H cannot be neglected. Such problems will be discussed later (see Part II).

In the following discussion of statically indeterminate problems of bending, the method of superposition will be used and the solutions will be obtained by combining the previously investigated statically determinate cases in such a manner as to satisfy the conditions at the supports.

41. Beam Built In at One End and Supported at the Other. —In this case we have three unknown reactive elements at one end and one unknown at the other end. Hence the problem is statically indeterminate with one redundant constraint. Starting with the case of a single transverse load P, Fig. 157a, let us consider as redundant the constraint which prevents the left end A of the beam from rotating during bending. Removing this constraint, we obtain the statically determinate problem shown in Fig. 157b. The bending produced by the statically indeterminate couple M_a is now studied separately as shown in Fig. 157c.[2] It is evident that the bending of the beam represented in Fig. 157a can be obtained by the com-

---

[2] Deflection curves and bending moment diagrams are shown together.

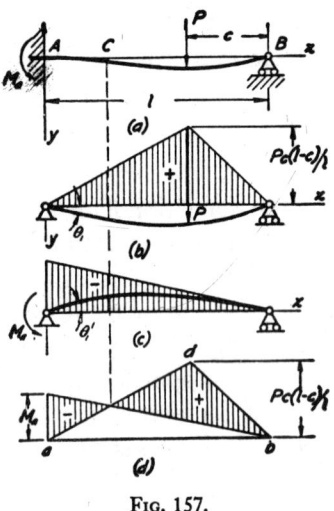

FIG. 157.

bination of cases (b) and (c). It is only necessary to adjust the magnitude of the couple M_a at the support A in such a manner as to satisfy the condition

$$\theta_1 = -\theta_1'. \tag{a}$$

Thus, the rotation of the left end of the beam, due to the force P, will be eliminated by the couple M_a and the condition of a built-in end with zero slope will be satisfied. To obtain the statically indeterminate couple M_a it is necessary only to substitute in eq. (a) the known values for the angles θ_1 and θ_1' from eqs. (88), p. 145, and (104), p. 158. Then

$$\frac{Pc(l^2 - c^2)}{6lEI_z} = -\frac{M_a l}{3EI_z},$$

from which

$$M_a = -\frac{Pc(l^2 - c^2)}{2l^2}. \tag{107}$$

The bending moment diagram can now be obtained by combining the diagrams for cases (b) and (c) as shown by the shaded area in Fig. 157d. The maximum bending moment will be either at a or d.

The deflection of the beam at any point can easily be obtained by subtracting from the deflection at this point produced by the load P (Fig. 157b) the deflection produced by the couple M_a, Fig. 157c. The equations of the deflection curves for both these cases have already been given in (86) and (87), p. 145, and in (105), p. 158. Let us take, for example, the case $c < \frac{1}{2}l$ and calculate the deflection δ at the middle of the span. From eqs. (91) and (105) we have

$$\delta = \frac{Pc}{48EI_z}(3l^2 - 4c^2) + \frac{M_a l^2}{16EI_z},$$

or, by using eq. (107),

$$\delta = \frac{Pc}{96EI_z}(3l^2 - 5c^2).$$

At the point C, where the bending moment becomes zero, the curvature of the deflection curve is also zero and we have a point of inflection, i.e., a point where the curvature changes sign.

It may be seen from eq. (107) that the bending moment at the built-in end depends on the position of the load P. If we equate to zero the derivative of eq. (107) with respect to c, we find that the moment M_a has its numerical maximum value when $c = l/\sqrt{3}$. Then

$$|M_a|_{\text{max}} = \frac{Pl}{3\sqrt{3}} = 0.192Pl. \qquad (108)$$

The bending moment under the load, from Fig. 157d, is

$$M_d = \frac{Pc(l - c)}{l} - \frac{c}{l}\frac{Pc(l^2 - c^2)}{2l^2} = \frac{Pc}{2l^3}(l - c)^2(2l + c). \qquad (b)$$

If we take the derivative of (b) with respect to c and equate it to zero, we find that M_d becomes a maximum when

$$c = \frac{l}{2}(\sqrt{3} - 1) = 0.366l.$$

Substituting this in eq. (b), we obtain

$$(M_d)_{\text{max}} = 0.174Pl.$$

Comparing this with eq. (108), we find that in the case of a moving load the maximum normal stresses σ_x are at the built-in section and occur when

$$c = \frac{l}{\sqrt{3}}.$$

Having the solution for the single concentrated load and using the method of superposition, the problem can be solved

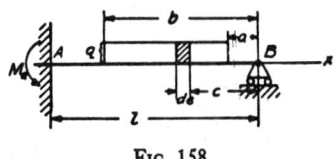

FIG. 158.

for other types of transverse loading by simple extension of the above theory. Take, for example, the case represented in Fig. 158. The moment at the support A, produced by an element qdc of the load, is obtained from eq. (107) by substituting qdc for P. The total moment M_a at the support will be

$$M_a = -\int_a^b \frac{qcdc(l^2 - c^2)}{2l^2}$$

$$= -\frac{q}{2l^2}\left[\frac{l^2(b^2 - a^2)}{2} - \frac{b^4 - a^4}{4}\right]. \quad (c)$$

If the load be distributed along the entire length of the beam, Fig. 159a, then substituting in eq. (c) $a = 0$, $b = l$, we obtain

$$M_a = -\frac{ql^2}{8} \quad (109)$$

The bending moment diagram is obtained by subtracting the triangular diagram due to the couple M_a (Fig. 159) from the parabolic diagram, due to uniform loading. It can be seen that the maximum bending stresses will be at the built-in section.

The deflection at any point of the beam is obtained by subtracting the deflection at this point produced by the couple M_a (see eq. 105, p. 158) from the deflection at the same point produced by the uniform load (see eq. 81, p. 141).

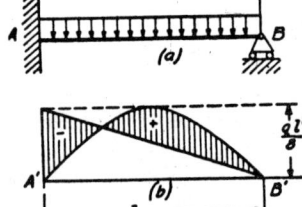

FIG. 159.

For the middle of the span we will then obtain

$$\delta = \frac{5}{384}\frac{ql^4}{EI_z} + \frac{M_a l^2}{16EI_z} = \frac{ql^4}{192EI_z}. \quad (110)$$

Problems

1. Draw shearing force diagrams for the cases shown in Figs. 157 and 159.

2. Determine the maximum deflection for the case of a uniformly distributed load shown in Fig. 159.

Solution. Combining eqs. (81) and (105), the following equation for the deflection curve is obtained:

$$y = \frac{q}{48EI_z} (3l^2x^2 - 5lx^3 + 2x^4). \tag{d}$$

Setting the derivative dy/dx equal to zero, we find the point of maximum deflection at $x = (l/16)(15 - \sqrt{33}) = 0.579l$. Substituting in (d), we obtain

$$\delta_{max} = \frac{ql^4}{185EI_z}.$$

3. Determine the reaction at the right-hand support of the beam shown in Fig. 159, considering this reaction as the redundant constraint.

Solution. Removing support B, the deflection of this end of the beam, considered as a cantilever, will be $ql^4/8EI_z$ from eq. (84). Reaction R_b at B (Fig. 159a) must be such as to eliminate the above deflection. Then by using eq. (95) we obtain the equation

$$\frac{ql^4}{8EI_z} = \frac{R_b l^3}{3EI_z},$$

from which

$$R_b = \tfrac{3}{8}ql.$$

4. A beam is loaded as shown in Fig. 160. Determine the bending moment M_a and the reactions R_a and R_b at the supports.

Answer.

$$-M_a = \frac{ql^2}{8} + \frac{7}{120}q_1 l^2, \qquad R_a = \tfrac{5}{8}ql + \tfrac{9}{40}q_1 l, \qquad R_b = \tfrac{3}{8}ql + \tfrac{11}{40}q_1 l.$$

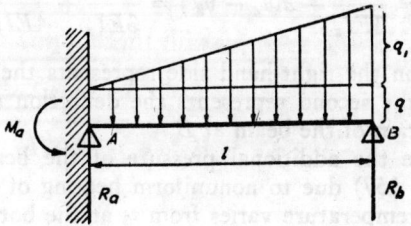

Fɪɢ. 160.

5. Determine the reaction R_b at the support B of the uniformly loaded beam shown in Fig. 159 if the support B is elastic, so that a downward force of magnitude k lowers the support a unit distance.

Solution. Using the same method as in Prob. 3 above, the equation for determining R_b will be

$$\frac{ql^4}{8EI_z} - \frac{R_b l^3}{3EI_z} = \frac{R_b}{k},$$

from which

$$R_b = \frac{3}{8} ql \frac{1}{1 + \dfrac{3EI_z}{kl^3}}.$$

6. Construct the bending moment and shearing force diagrams or a uniformly loaded beam supported at the middle and at the ends.

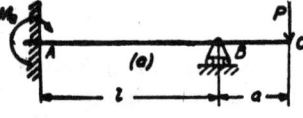

(a)

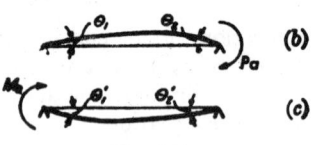

(b)

(c)

Fɪɢ. 161.

Suggestion. From the condition of symmetry the middle cross section does not rotate during bending and each half of the beam will be in the condition of a beam built in at one end and supported at the other.

7. Determine the deflection of the end C of the beam shown in Fig. 161.

Solution. Replacing the action of the overhang by a couple Pa, the bending of the beam between the supports will be obtained by superposing cases (b) and (c) in Fig. 161. Using formulas (103) and (104) (p. 158), the statically indeterminate couple M_a will be found from the equation $\theta_1 = -\theta_1'$, or

$$\frac{Pal}{6EI_z} = \frac{M_a l}{3EI_z},$$

from which $M_a = Pa/2$. The deflection at C will be

$$\delta = \frac{Pa^3}{3EI_z} + a(\theta_2 - \theta_2') = \frac{Pa^3}{3EI_z} + \frac{Pa^2 l}{4EI_z}.$$

The first term on the right-hand side represents the deflection of a cantilever and the second represents the deflection due to rotation of the cross section of the beam at B.

8. Determine the additional pressure of the beam AB on the support B (Fig. 157) due to nonuniform heating of the beam, provided that the temperature varies from t_0 at the bottom to t at the top of the beam according to a linear law ($t > t_0$).

Solution. If the support at B is removed, the nonuniform heating will cause the beam to deflect in an arc of a circle. The radius of this circle can be determined from the equation $1/r = \alpha(t - t_0)/h$, in which $h =$ the depth of the beam and $\alpha =$ the coefficient of thermal expansion. The corresponding deflection at B can be found as in Prob. 2, p. 97, and is

$$\delta = \frac{l^2}{2r} = \frac{l^2 \alpha(t - t_0)}{2h}.$$

This deflection is eliminated by the reaction of the support B. Letting R_b denote this reaction, we obtain

$$\frac{R_b l^3}{3EI_z} = \frac{l^2 \alpha(t - t_0)}{2h},$$

from which

$$R_b = \frac{3EI_z}{2hl} \cdot \alpha(t - t_0).$$

9. A cantilever AB, Fig. 162, loaded at the end B, is supported by a shorter cantilever CD of the same cross section as cantilever AB. Determine the pressure X between the two beams at C.

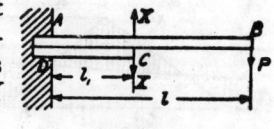

Fig. 162.

Solution. Pressure X will be found from the condition that at C both cantilevers have the same deflection. Using eq. (95) for the lower cantilever and eq. (97) together with eq. (95) for the upper, we obtain

$$\frac{X l_1^3}{3EI_z} = \frac{P}{EI_z}\left(\frac{l l_1^2}{2} - \frac{l_1^3}{6}\right) - \frac{X l_1^3}{3EI_z},$$

from which

$$X = \frac{3P}{4}\left(\frac{l}{l_1} - \frac{1}{3}\right).$$

From a consideration of the bending moment diagrams for the upper and lower cantilevers it can be concluded that at C the upper cantilever has a larger angular deflection than the lower. This indicates that there will be contact between the two cantilevers only at points D and C.

10. Solve Prob. 7 assuming, instead of a concentrated load P, a uniform load of intensity q to be distributed (1) along the length a of the overhang, and, (2) along the entire length of the beam. Draw bending moment and shearing force diagrams for these two cases.

11. Draw the bending moment and shearing force diagrams for the case shown in Fig. 158 if $a = 4$ ft, $b = 12$ ft, $l = 15$ ft and $q = 400$ lb per ft.

42. Beam with Both Ends Built In.—In this case (Fig. 163) we have six reactive elements (three at each end), so that the problem has three statically indeterminate elements. However, for ordinary beams, the horizontal components of the reactions can be neglected (see p. 179), which reduces the number of statically indeterminate quantities to two. Let us take the bending moments M_a and M_b at the supports for the statically indeterminate quantities. Then for the case of a single concentrated load P (Fig. 163a) the solution can be obtained by combining the two statically determinate problems shown in Figs. 163b and 163c. It is evident that the conditions at the built-in ends

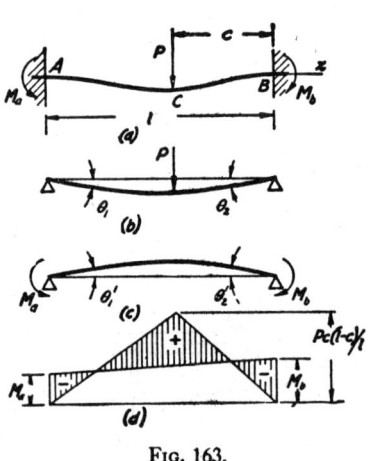

FIG. 163.

of the beam AB will be satisfied if the couples M_a and M_b are adjusted so as to make

$$\theta_1 = -\theta_1', \qquad \theta_2 = -\theta_2'. \qquad (a)$$

From these two equations the two statically indeterminate couples are obtained. Using eqs. (88) and (89) for a concentrated load and eqs. (103) and (104) for the couples, eqs. (a) become

$$-\frac{Pc(l^2 - c^2)}{6lEI_z} = \frac{M_a l}{3EI_z} + \frac{M_b l}{6EI_z},$$

$$-\frac{Pc(l - c)(2l - c)}{6lEI_z} = \frac{M_a l}{6EI_z} + \frac{M_b l}{3EI_z},$$

from which

$$M_a = -\frac{Pc^2(l - c)}{l^2}, \qquad M_b = -\frac{Pc(l - c)^2}{l^2} \qquad (111)$$

Both bending moments are negative and produce bending convex upward. Combining the bending moment diagrams for cases (b) and (c), the diagram shown in Fig. 163d is obtained. We see that the maximum positive bending moment for the case in Fig. 163a is under the load at the point C. Its magnitude can be found from Fig. 163d and is given by the following:

$$M_c = \frac{Pc(l-c)}{l} + \frac{M_a c}{l} + \frac{M_b(l-c)}{l} = \frac{2Pc^2(l-c)^2}{l^3}. \qquad (112)$$

From Fig. 163d it may be seen that the numerically greatest bending moment is either at C or at the nearest support. For a moving load, i.e., when c varies, assuming $c < l/2$, the maximum numerical value of M_b is obtained by putting $c = l/3$ in eq. (111). This maximum is equal to $4Pl/27$. The bending moment under the load is a maximum when $c = l/2$ and this maximum, from eq. (112), is equal to $Pl/8$. Hence for a moving load the greatest moment is at the end.

By using the method of superposition, the deflection at any point of the beam can be obtained by combining the deflection produced by load P with that produced by couples M_a and M_b.

Having the solution for a single concentrated load P, any other type of transverse loading can easily be studied by using the method of superposition.

Problems

1. Draw the shearing force diagram for the case in Fig. 163a if $P = 1,000$ lb, $l = 12$ ft and $c = 4$ ft.

2. Find the bending moments at the ends of the beam loaded at the third points, Fig. 164. Draw the bending moment and shearing force diagrams.

Answer. $M_a = M_b = -\frac{2}{9}Pl$.

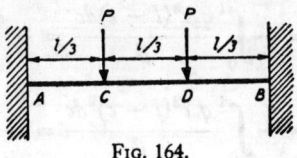

FIG. 164.

3. Solve the preceding problem when the force at D has the opposite direction

Answer. $M_a = -M_b = -\frac{2}{27}Pl.$

4. Construct the bending moment diagram for a uniformly loaded beam with built-in ends, Fig. 165.

Solution. The bending moment at A produced by one element qdc of the load (Fig. 165a) is, from eq. (111),

$$dM_a = -\frac{qc^2(l-c)dc}{l^2}.$$

The moment produced by the load over the entire span is then

$$M_a = -\int_0^l \frac{qc^2(l-c)dc}{l^2} = -\frac{ql^2}{12};$$

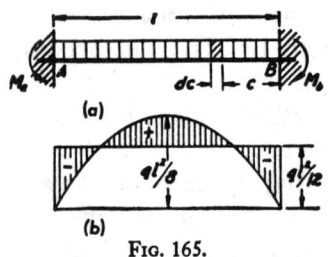

(a)

(b)

Fɪɢ. 165.

the moment at the support B will have the same magnitude. Combining the parabolic bending moment diagram produced by the uniform load with the rectangular diagram given by two equal couples applied at the ends, we obtain the diagram shown in Fig. 165b by the shaded area.

5. Determine the bending moments at the ends of a beam with built-in ends and loaded by the triangular load shown in Fig. 166.

Solution. The intensity of the load at distance c from the support B is q_ac/l and the load represented by the shaded element is q_acdc/l. The bending moments acting at the ends, produced by this elementary load, as given by eqs. (111), are

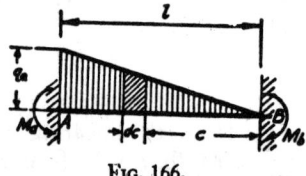

Fɪɢ. 166.

$$dM_a = -\frac{q_ac^3(l-c)dc}{l^3}, \qquad dM_b = -\frac{q_ac^2(l-c)^2dc}{l^3}.$$

Therefore

$$M_a = -\int_0^l \frac{q_ac^3(l-c)dc}{l^3} = -\frac{q_al^2}{20},$$

$$M_b = -\int_0^l \frac{q_ac^2(l-c)^2dc}{l^3} = -\frac{q_al^2}{30}.$$

6. Determine the bending moments M_a and M_b in a beam with built-in ends and bent by a couple Pc (Fig. 167).

Solution. By using the solution of Prob. 4, p. 160, and eqs. (103) and (104), the following equations are obtained:

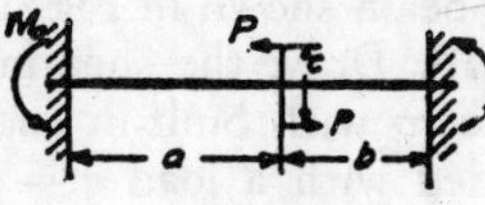

FIG. 167.

$$2M_a + M_b = -\frac{3Pc}{l^3}\left[a^2\left(b + \frac{a}{3}\right) - \frac{2}{3}b^3\right],$$

$$2M_b + M_a = -\frac{3Pc}{l^3}\left[\frac{2}{3}a^3 - b^2\left(a + \frac{b}{3}\right)\right],$$

from which M_a and M_b can easily be calculated.

7. Determine the bending moments at the ends of a built-in beam due to nonuniform heating of the beam if the temperature varies from t_0 at the bottom to t at the top of the beam according to a linear law.

Answer. $M_a = M_b = \dfrac{\alpha E I_z(t - t_0)}{h},$

where α is the coefficient of thermal expansion and h is the depth of the beam.

8. Determine the effect on the reactive force and reactive couple at A of a small vertical displacement δ of the built-in end A of the beam AB (Fig. 163).

Solution. Remove the support A; then the deflection δ_1 at A and the slope θ_1 at this point will be found as for a cantilever built in at B and loaded by P, i.e.,

$$\delta_1 = \frac{Pc^3}{3EI_z} + \frac{Pc^2}{2EI_z}(l - c), \qquad \theta_1 = \frac{Pc^2}{2EI_z}.$$

Applying at A an upward reactive force X and a reactive couple Y in the same direction as M_a, of such magnitude as to eliminate the slope θ_1 and to make the deflection equal to δ, the equations for determining the unknown quantities X and Y become

$$\frac{Xl^2}{2EI_z} - \frac{Yl}{EI_z} = \frac{Pc^2}{2EI_z},$$

$$\frac{Xl^3}{3EI_z} - \frac{Yl^2}{2EI_z} = \delta_1 - \delta.$$

9. Draw the shearing force and bending moment diagrams for the beam shown in Fig. 166 if $q_a = 400$ lb per ft and $l = 15$ ft.

10. Draw the shearing force and bending moment diagrams for a beam with built-in ends if the left half of the beam is uniformly loaded with a load $q = 400$ lb per ft. The length of the beam is $l = 16$ ft.

43. Frames.—The method used in the preceding article for statically indeterminate beams can be applied also to the study of frames. Take, as a simple example, the symmetrical frame, Fig. 168, hinged at C and D, and loaded symmetrically.

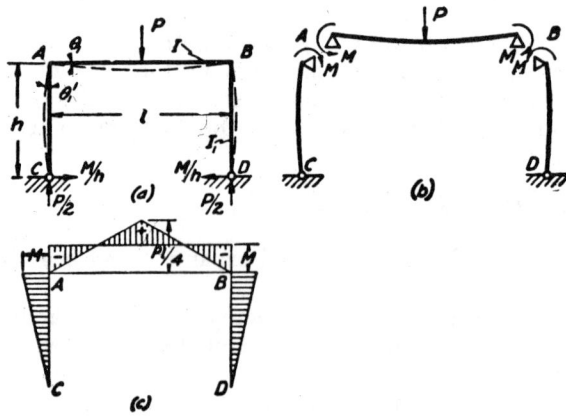

Fig. 168.

The shape of the frame after deformation is shown by the dotted lines. Neglecting the change in the length of the bars and the effect of axial forces on the bending of bars,[3] the frame can be considered to be made up of three beams as shown in Fig. 168b. It is evident that there will be couples M at the ends of the horizontal beam AB which oppose the free rotation of these ends and represent the action of the vertical bars on the horizontal beam. These couples M can be considered as the statically indeterminate quantity. Knowing M, the bending of all three bars can be investigated without any difficulty.

---

[3] Simultaneous action of bending and thrust will be discussed later (see Part II).

For determining M we have the condition that at A and at B there are rigid joints between the bars so that the rotation of the top end of the vertical bar AC must be equal to the rotation of the left end of the horizontal bar. Hence the equation for determining M is

$$\theta_1 = \theta_1'. \qquad (a)$$

θ_1 must be determined from the bending of the horizontal beam AB. Denoting by l the length of this beam and by EI its flexural rigidity, the rotation of the end A due to the load P, by eq. (88) with $b = l/2$, is $Pl^2/16EI$. The couples at the ends resist this bending and give a rotation in the opposite direction, which, from eqs. (103) and (104), equals $Ml/2EI$. The final value of the angle of rotation will be

$$\theta_1 = \frac{Pl^2}{16EI} - \frac{Ml}{2EI}.$$

Considering now the vertical bar AC as a beam with supported ends, bent by a couple M, and denoting by h its length and by EI_1 its flexural rigidity, the angle at the top, from eq. (104) will be

$$\theta_1' = \frac{Mh}{3EI_1}.$$

Substituting in eq. (a), we obtain

$$\frac{Pl^2}{16EI} - \frac{Ml}{2EI} = \frac{Mh}{3EI_1},$$

from which

$$M = \frac{Pl}{8} \frac{1}{1 + \frac{2}{3} \frac{h}{l} \frac{I}{I_1}}. \qquad (113)$$

This is the absolute value of M. Its direction is shown in Fig. 168b. Knowing M, the bending moment diagram can be constructed as shown in Fig. 168c. The reactive forces at the

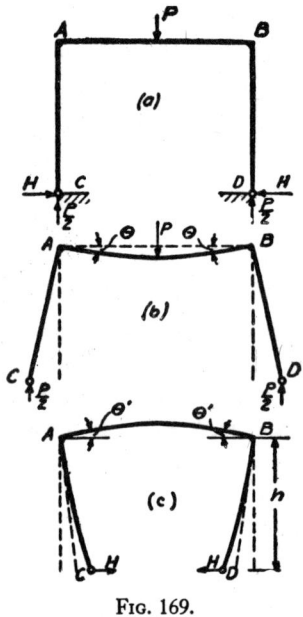

FIG. 169.

hinges C and D are also shown (Fig. 168a). The vertical components of these forces, from considerations of symmetry, are each equal to $P/2$. As regards the horizontal components, their magnitude M/h is obtained by considering the vertical bars as simply supported beams loaded at the top by the couples M.

The same problem can be solved in another way by taking the horizontal reactions H at the hinges C and D as the statically indeterminate quantity, Fig. 169, instead of M. The statically indeterminate problem is solved by superposing the two statically determinate problems shown in Figs. 169b and 169c. In case (b) the redundant constraint preventing the horizontal motion of the hinges C and D is removed. The vertical bars no longer have any bending. The horizontal bar AB is in the condition of a bar with simply supported ends whose angles of rotation are equal to $Pl^2/16EI$, and the horizontal motion of each hinge C and D is therefore $h(Pl^2/16EI)$. In case (c) the effect of the forces H is studied. These forces produce bending couples on the ends of the horizontal bar AB equal to $H \cdot h$, so that the angles of rotation of its ends θ' will be $Hh \cdot l/2EI$. The deflection of each hinge C and D consists of two parts, (1) the deflection $\theta'h = Hh^2l/2EI$ due to rotation of the upper end and, (2) the deflection $Hh^3/3EI_1$ of the vertical bars as cantilevers. In the actual case (Fig. 169a) the hinges C and D do not move. Hence the horizontal displacements produced by the force P (Fig. 169b) must be counteracted by the forces H (Fig. 169c), i.e.,

$$\frac{Pl^2}{16EI} h = \frac{Hh^2l}{2EI} + \frac{Hh^3}{3EI_1},$$

from which

$$H = \frac{1}{h}\frac{Pl}{8}\frac{1}{1 + \frac{2}{3}\frac{h}{l}\frac{I}{I_1}}.$$

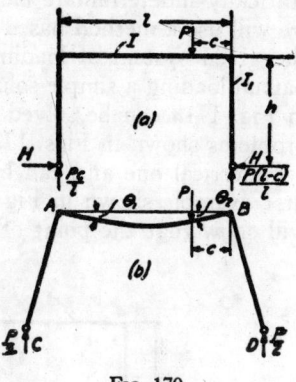

(a)

(b)

Fig. 170.

Observing that $Hh = M$, this result agrees with eq. (113) above.

This latter method of analysis is
especially useful for nonsymmetrical
loading such as shown in Fig. 170.
Removing the constraint preventing
the hinges C and D from horizontal
motion, we have the condition rep-
represented in Fig. 170b. It is evident that the increase in distance between C and D may be obtained by multiplying by h
the sum of the angles θ_1 and θ_2. Using eqs. (88) and (89), this
increase in distance becomes

$$h\left[\frac{Pc(l^2 - c^2)}{6lEI} + \frac{Pc(l - c)(2l - c)}{6lEI}\right] = \frac{Pc(l - c)h}{2EI}.$$

It must be eliminated by the horizontal reactions H (Fig.
169c). Then, using the results obtained in the previous problem, we obtain the following equation for determining H:

$$2\left(\frac{Hh^2l}{2EI} + \frac{Hh^3}{3EI_1}\right) = \frac{Pc(l - c)h}{2EI},$$

from which

$$H = \frac{Pc(l - c)}{2hl}\frac{1}{1 + \frac{2}{3}\frac{h}{l}\frac{I}{I_1}}. \tag{114}$$

Having the solution for one concentrated load, any other case
of loading of the beam AB of the frame can easily be studied
by the method of superposition.

Let us consider now a frame with built-in supports and an un-
symmetrical loading as shown in Fig. 171. In this case we have
three reactive elements at each support and the system has three

statically indeterminate elements. In the solution of this problem
we will use a method based on the method of superposition in which
the given system of loading is split into parts such that for each
partial loading a simple solution can be found.[4] The problem shown
in Fig. 171a can be solved by superposing the solutions of the two
problems shown in Figs. 171b and 171c. The case shown in (b) is a
symmetrical one and can be considered in the same manner as the
first example shown in Fig. 168. A study of the case shown in (c)
will show that the point of inflection O of the horizontal bar AB is

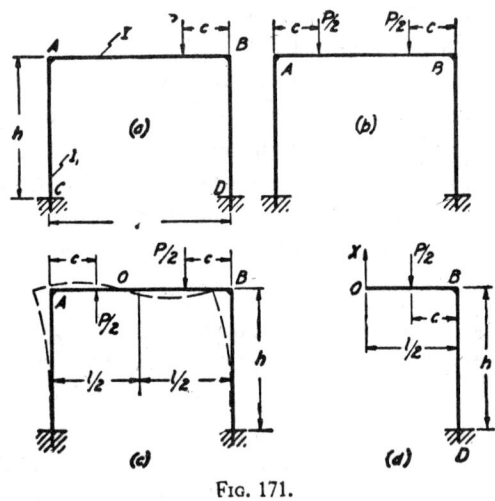

FIG. 171.

located at the middle of the bar. This follows from the condition
that the loads P/2 are equally distant from the vertical axis of symmetry of the frame and are opposite in sense. The moment, the
deflection and the axial force produced at the mid point O of the
horizontal beam AB by one of the loads P/2 will be removed by the
action of the other load P/2. Hence there will be no bending moment, no vertical deflection and no axial force at O. The magnitude
of the shearing force at the same point X can be found from the
condition that the vertical deflection of O is equal to zero (Fig. 171d).
This deflection consists of two parts, a deflection δ_1 due to the bending of the cantilever OB and a deflection δ_2 due to rotation of the
end B of the vertical bar BD. Using the known equations for a
cantilever (eq. 98), and using the notations given in the figure, the
following equations are obtained:

---

[4] Such a method was extensively used by W. L. Andrée; see his book,
Das B-U Verfahren, Berlin, 1919.

$$\delta_1 = \frac{P}{2}\frac{c^3}{3EI} + \frac{P}{2}\frac{c^2}{2EI}\left(\frac{l}{2} - c\right) - \frac{X\left(\frac{l}{2}\right)^3}{3EI}.$$

$$\delta_2 = \left(\frac{Pc}{2} - X\frac{l}{2}\right)\frac{h}{EI_1}\frac{l}{2}.$$

Substituting this in the equation $\delta_1 + \delta_2 = 0$, the magnitude X of the shearing force can be found. Having determined X, the bending moment at every cross section of the frame for case (c) can be calculated. Combining this with the bending moments for the symmetrical case (b), the solution of the problem (a) is obtained.[5]

Problems

1. Find the axial forces in all bars of the frame in Fig. 168a.
Answer. Compression in the vertical bars $= P/2$; compression in the horizontal bar $= M/h$.
2. Draw the bending moment diagram for the frame in Fig. 170a.
3. Determine the bending moments at the corners of the frame shown in Fig. 172.

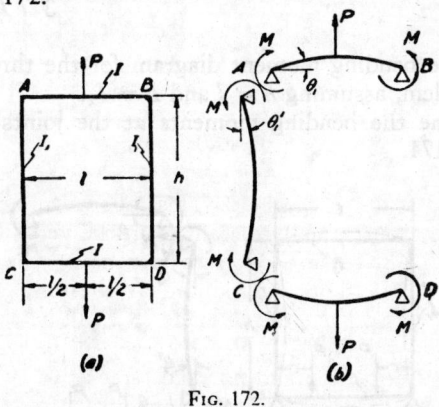

FIG. 172.

Solution. Considering the bar AB as a beam supported at the ends (Fig. 172b) and denoting by M the moments at the corners, the angle θ_1 will be

$$\frac{Pl^2}{16EI} - \frac{Ml}{2EI}.$$

[5] Solutions of many important problems on frames can be found in the book by Kleinlogel, *Mehrstielige Rahmen*, Berlin, 1927; English translation, New York, 1952.

Putting this equal to the angle θ_1' at the ends of the vertical bars which are bent by the couples M only, the following equation for M is obtained:

$$\frac{Pl^2}{16EI} - \frac{Ml}{2EI} = \frac{Mh}{2EI_1},$$

from which

$$M = \frac{Pl}{8} \frac{1}{1 + \frac{h}{l}\frac{I}{I_1}}.$$

4. Draw the bending moment diagram for the frame of the preceding problem.

5. Determine the horizontal reactions H for the case shown in Fig. 173.

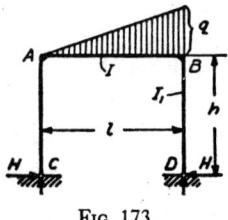

FIG. 173.

Suggestion. By using eq. (114) and applying the method of superposition, we obtain

$$H = \frac{ql^2}{24h} \frac{1}{1 + \frac{2}{3}\frac{h}{l}\frac{I}{I_1}}.$$

6. Draw the bending moment diagram for the three bars of the preceding problem, assuming $h = l$ and $I = I_1$.

7. Determine the bending moments at the joints of the frame shown in Fig. 174.

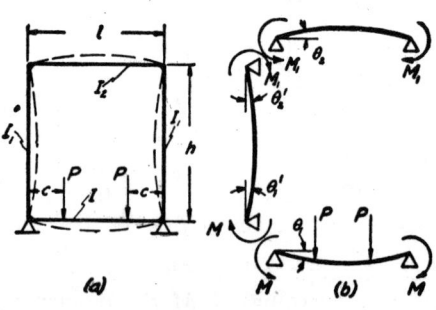

FIG. 174.

Solution. Disjointing the frame as shown in Fig. 174b, the equations for determining the couples M and M_1 are

$$\theta_1 = \theta_1' \quad \text{and} \quad \theta_2 = \theta_2'.$$

Substituting in these equations

$$\theta_1 = \frac{Pc(l-c)}{2EI} - \frac{Ml}{2EI}, \qquad \theta_1' = \frac{Mh}{3EI_1} - \frac{M_1h}{6EI_1},$$

$$\theta_2 = \frac{M_1l}{2EI_2}, \qquad \theta_2' = \frac{Mh}{6EI_1} - \frac{M_1h}{3EI_1},$$

we obtain two equations for determining M and M_1.

8. Draw the bending moment diagram and determine the axial forces in all bars of the frame in Fig. 174a, if $h = l$ and $I = I_1 = I_2$.

9. A symmetrical rectangular frame is submitted to the action of a horizontal force H as shown in Fig. 175. Determine the bending moments M and M_1 at the joints.

Solution. The deformed shape of the frame is shown in Fig. 175a. Disjointing the frame as shown in Fig. 175b and applying moments in directions which comply with the distorted shape of the frame, Fig. 175a, we have for the bar CD,

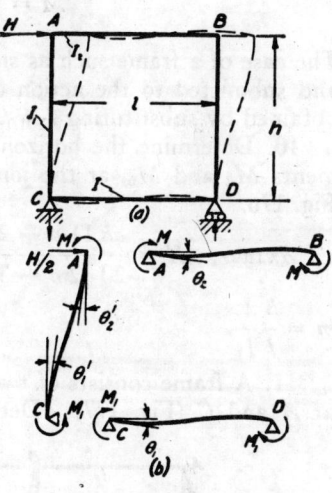

Fig. 175.

$$\theta_1 = \frac{M_1l}{6EI} = \left(\frac{Hh}{2} - M\right)\frac{l}{6EI}. \qquad (b)$$

Considering now the vertical bar AC as a cantilever built in at the end C at an angle θ_1, the slope at the end A will be

$$\theta_2' = \theta_1 + \frac{H}{2}\frac{h^2}{2EI_1} - \frac{Mh}{EI_1}. \qquad (c)$$

Finally, due to bending of the bar AB,

$$\theta_2 = \theta_2' = \frac{Ml}{6EI_2}. \qquad (d)$$

Then, from eqs. (b), (c), and (d),

$$M = \frac{Hh}{2}\left(1 + \frac{3h}{l}\frac{I}{I_1}\right)\frac{1}{1 + \frac{I}{I_2} + 6\frac{h}{l}\frac{I}{I_1}}. \qquad (e)$$

Substituting in eq. (b), the bending moment M_1 can be found. When the horizontal bar CD has very great rigidity, we approach the condition of the frame shown in Fig. 171a, submitted to a lateral load H. Substituting $I = \infty$ in (e), we obtain for this case

$$M = \frac{Hh}{4} \cdot \frac{1}{1 + \frac{1}{6} \frac{l}{h} \frac{I_1}{I_2}}. \qquad (f)$$

The case of a frame such as shown in Fig. 168 with hinged supports and submitted to the action of a lateral load applied at A can be obtained by substituting $I = 0$ in eq. (e).

10. Determine the horizontal reactions H and the bending moments M_a and M_b, at the joints A and B, for the frame shown in Fig. 176.

Answer. $H = \dfrac{qh}{20} \dfrac{11m + 20}{2m + 3}$, $M_a = M_b = -\dfrac{qh^2}{60} \dfrac{7m}{2m + 3}$, where $m = \dfrac{h}{l} \dfrac{I}{I_1}$.

11. A frame consists of two bars joined rigidly at B and built in at A and C (Fig. 177). Determine the bending moment M at B

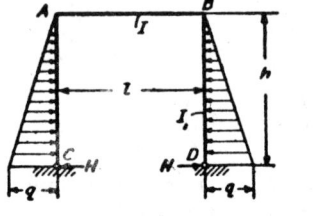

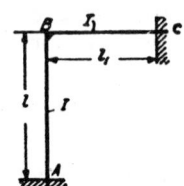

Fig. 176.   Fig. 177.

and the compressive force P in AB when, due to a rise in temperature, the bar AB increases in length by $\Delta = \alpha l(t - t_0)$.

Answer. P and M can be found from the equations:

$$\frac{Pl_1^3}{3EI} - \frac{Ml_1^2}{2EI} = \Delta,$$

$$\frac{Pl_1^2}{2EI} - \frac{Ml_1}{EI} = \frac{Ml}{4EI}.$$

44. Beams on Three Supports.—In the case of a beam on three supports (Fig. 178a) there is one statically indeterminate reactive element. Let the reaction of the intermediate support be this element. Then by using the method of superposition the solution of case (a) may be obtained by combining the cases represented in (b) and (c), Fig. 178. The intermediate

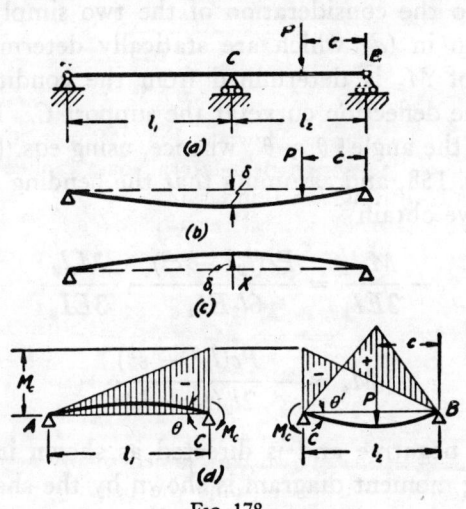

Fig. 178.

reaction X is found by using the condition that the deflection δ produced at C by the load P must be eliminated by the reaction X. Using eq. (86), we get the following equation for determining X:

$$\frac{Pcl_1[(l_1 + l_2)^2 - c^2 - l_1^2]}{6(l_1 + l_2)EI_z} = \frac{Xl_1^2 l_2^2}{3(l_1 + l_2)EI_z},$$

from which

$$X = \frac{Pc[(l_1 + l_2)^2 - c^2 - l_1^2]}{2l_1 l_2^2}. \tag{115}$$

If P is acting on the left span of the beam, the same equation can be used, but the distance c must be measured from the support A and l_1 and l_2 must be interchanged. For $l_1 = l_2 = l$, from eq. (115),

$$X = \frac{Pc(3l^2 - c^2)}{2l^3}. \tag{116}$$

Having the solution for a single load P, any other loading can easily be studied by using the method of superposition.

The same problem can be solved in another manner. Imagine the beam cut into two parts at C (Fig. 178d) and let M_c denote the magnitude of the bending moment of the original beam at this cross section. In this manner the problem is reduced to the consideration of the two simply supported beams shown in (d) which are statically determinate. The magnitude of M_c is determined from the condition of continuity of the deflection curve at the support C. From this it follows that the angle [6] $\theta = \theta'$, whence, using eqs. (88), p. 145, and (104), p. 158, and assuming that the bending moment M_c is positive, we obtain

$$-\frac{M_c l_1}{3EI_z} = \frac{Pc(l_2{}^2 - c^2)}{6l_2 EI_z} \neg \frac{M_c l_2}{3EI_z},$$

from which

$$M_c = -\frac{Pc(l_2{}^2 - c^2)}{2l_2(l_1 + l_2)}. \tag{117}$$

Thus M_c is negative and is directed as shown in Fig. 178d. The bending moment diagram is shown by the shaded area in Fig. 178d.

Problems

1. For the example in Fig. 178 prove that the magnitude of the bending moment M_c given by eq. (117) is the same as that obtained for the cross section C by using eq. (115).

2. Draw the shearing force diagram for the beam of the preceding problem if $l_1 = l_2$, $c = l_2/2$ and $P = 1,000$ lb.

3. A beam on three supports (Fig. 178a) carries a uniformly distributed load of intensity q. Determine the bending moment at the support C.

Solution. By the method of superposition, substituting $q\,dc$ for P in eq. (117) and integrating along both spans, we obtain

$$M_c = -\int_0^{l_2} \frac{qc(l_2{}^2 - c^2)dc}{2l_2(l_1 + l_2)} - \int_0^{l_1} \frac{qc(l_1{}^2 - c^2)dc}{2l_1(l_1 + l_2)} = -\frac{q}{8}\frac{l_2{}^3 + l_1{}^3}{l_1 + l_2},$$

and when

$$l_1 = l_2 = l, \qquad M_c = -\frac{ql^2}{8}.$$

The direction of this moment is the same as shown in Fig. 178d.

---

[6] The angle is taken positive if rotation is in the clockwise direction.

4. Draw the shearing force diagram for the preceding problem, assuming $l_1 = l_2$ and $q = 500$ lb per ft.

5. Determine the numerically maximum bending moment in the beam ACB (Fig. 178a) if $P = 10,000$ lb, $l_1 = 9$ ft, $l_2 = 12$ ft, $c = 6$ ft. *Answer.* $M_{max} = 23,600$ ft lb.

6. A beam on three equidistant supports carries a uniformly distributed load of intensity q. What effect will there be on the middle reaction if the middle support is lowered a distance δ?

Solution. Using the method shown in Figs. 178b and 178c, the middle reaction X is found from the equation

$$\frac{5}{384}\frac{q(2l)^4}{EI} = \frac{X(2l)^3}{48EI} + \delta,$$

from which

$$X = \frac{5}{8}2ql - \frac{6\delta EI}{l^3}.$$

7. Determine the additional pressure of the beam AB on the support C (Fig. 178a) due to nonuniform heating of the beam if the temperature varies from t at the bottom to t_1 at the top of the beam according to a linear law, assuming $t > t_1$ and $l_1 = l_2 = l$.

Solution. If the support at C were removed, then, due to the nonuniform heating, the deflection curve of the beam would become the arc of a circle. The radius of this circle is determined by the equation

$$\frac{1}{r} = \frac{\alpha(t - t_1)}{h},$$

in which $h =$ the depth of the beam and $\alpha =$ the coefficient of thermal expansion. The corresponding deflection at the middle is $\delta = l^2/2r$ and the reaction X at C can be found from the equation

$$\frac{X(2l)^3}{48EI} = \delta.$$

8. Determine the bending moment diagram for the beam ACB supported by three pontoons (Fig. 179) if the horizontal cross-sec-

FIG. 179.

tional area of each pontoon is A and the weight per unit volume of water is γ.

Solution. Removing the support at C, the deflection δ produced at this point by the load P consists of two parts: (1) the deflection due to bending of the beam, and (2) the deflection due to sinking of pontoons A and B. From eq. (91) we obtain

$$\delta = \frac{Pc}{48EI_z} [3(2l)^2 - 4c^2] + \frac{P}{2A\gamma}. \qquad (a)$$

The reaction X of the middle support diminishes the above deflection by

$$\frac{X(2l)^3}{48EI_z} + \frac{X}{2A\gamma}. \qquad (b)$$

The difference between (a) and (b) represents the distance $X/A\gamma$ which the middle pontoon sinks, from which we obtain the following equation for determining X:

$$\frac{Pc}{48EI_z} [3(2l)^2 - 4c^2] + \frac{P}{2A\gamma} - \frac{X(2l)^3}{48EI_z} - \frac{X}{2A\gamma} = \frac{X}{A\gamma}.$$

Knowing X, the bending moment diagram can readily be obtained.

45. Continuous Beams.—In the case of a continuous beam on many supports (Fig. 180) one support is usually considered as an immovable hinge while the other supports are hinges on rollers. In this arrangement every intermediate support has only one unknown reactive element, namely the magnitude of the vertical reaction. Hence the number of statically indeterminate elements is equal to the number of intermediate supports. For instance, in the case shown in Fig. 180a the number of statically indeterminate elements is five. Both methods shown in the previous article can be also used here. But if the number of supports is large, the second method, in which the bending moments at the supports are taken as the statically indeterminate elements, is by far the simpler method. Let Fig. 180b represent two adjacent spans n and $n + 1$ of a continuous beam cut at supports $n - 1$, n and $n + 1$. Let M_{n-1}, M_n and M_{n+1} denote the bending moments at these supports. The directions of these moments depend on the loads on the beam. We will assume them to be in the positive directions shown in the figure.[7] It is evident that if the bending moments at the supports are known the problem of the continuous beam will be reduced to that of calculating as many simply supported beams as there are spans in the continuous beam. For calculating the bending moments M_{n-1}, M_n, M_{n+1} the condition of continuity of the deflection curve at the supports will be used. For any support n this condition of

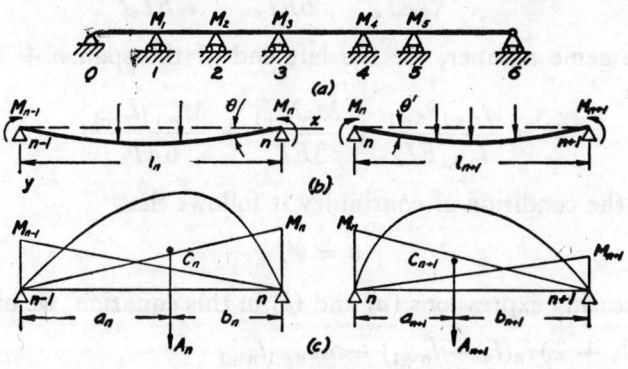

FIG. 180.

continuity is satisfied if the deflection curves of the two adjacent spans have a common tangent at the support n, i.e., if the slope at the right end of span n is equal to the slope at the left end of span $n + 1$. To calculate these slopes the area-moment method will be used. Let A_n denote the area of the bending moment diagram for the span n, Fig. 180c, considered as a simply supported beam, due to the actual load on this span. Let a_n and b_n represent the horizontal distances of the centroid C_n of the moment area from the supports $n - 1$ and n. Then the slope at the right end for this condition of loading is (see Art. 36)

$$-\frac{A_n a_n}{l_n E I_z}.$$

In addition to the deflection caused by the load on the span itself, the span n is also bent by the couples M_{n-1} and M_n.

---

[7] If finally we would obtain negative signs for some moments, this will indicate that the directions of these moments are opposite to that shown in the figure.

From eqs. (103) and (104) · slope produced at the support n by these couples is

$$-\left(\frac{M_n l_n}{3EI_z} + \frac{M_{n-1} l_n}{6EI_z}\right).$$

The total angle of rotation is then [8]

$$\theta = -\left(\frac{M_n l_n}{3EI_z} + \frac{M_{n-1} l_n}{6EI_z} + \frac{A_n a_n}{l_n EI_z}\right). \qquad (a)$$

In the same manner, for the left end of the span $n + 1$, we obtain

$$\theta' = \frac{A_{n+1} b_{n+1}}{l_{n+1} EI_z} + \frac{M_n l_{n+1}}{3EI_z} + \frac{M_{n+1} l_{n+1}}{6EI_z}. \qquad (b)$$

From the condition of continuity it follows that

$$\theta = \theta'. \qquad (c)$$

Substituting expressions (a) and (b) in this equation, we obtain

$$M_{n-1} l_n + 2M_n(l_n + l_{n+1}) + M_{n+1} l_{n+1}$$

$$= -\frac{6A_n a_n}{l_n} - \frac{6A_{n+1} b_{n+1}}{l_{n+1}}. \qquad (118)$$

This is the *equation of three moments.*[9] It is evident that the number of these equations is equal to the number of intermediate supports and thus the bending moments at the supports can be calculated without difficulty.

In the beginning it was assumed that the ends of the continuous beam were simply supported. If one or both ends are built in, then the number of statically indeterminate quantities will be larger than the number of intermediate supports and derivation of additional equations will be necessary to express the condition that no rotation occurs at the built-in ends (see Prob. 5 below).

Knowing the moments at the supports, there is no difficulty in calculating the reactions at the supports of a continuous beam. Taking, for example, the two adjacent spans n and

---

[8] The angle is taken positive if rotation is in the clockwise direction.
[9] This equation was established by Bertot, *Compt. rend. soc. ing. civils*, p. 278, 1855; see also Clapeyron, *Compt. rend.*, Vol. 45, 1857.

$n + 1$ (Fig. 180b), and considering them as two simply supported beams, the reaction R_n' at the support n, due to the loads on these two spans, can easily be calculated. In addition to this there will be a reaction due to the moments M_{n-1}, M_n and M_{n+1}. Taking the directions of these moments as indicated in Fig. 180b, the additional pressure on the support n will be

$$\frac{M_{n-1} - M_n}{l_n} + \frac{- M_n + M_{n+1}}{l_{n+1}}.$$

Adding this to the above reaction R_n', the total reaction will be

$$R_n = R_n' + \frac{M_{n-1} - M_n}{l_n} + \frac{- M_n + M_{n+1}}{l_{n+1}}. \tag{119}$$

If concentrated forces are applied at the supports, they will be transmitted directly to the corresponding supports and must be added to the right side of eq. (119).

The general equation of continuity (c) can also be used for those cases where, by misalignment or by settlement, the supports are not situated on the same level (Fig. 181). Let β_n and β_{n+1} denote the angles of inclination to the horizontal of the straight

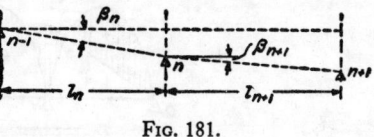

Fig. 181.

lines connecting the points of supports in the nth and $(n + 1)$th spans. The angle of rotation given by eqs. (a) and (b) was measured from the line connecting the centers of the hinges. Hence the angle θ between the tangent at n and the horizontal line will be, for the span n,

$$\theta = - \left(\frac{M_n l_n}{3EI_z} + \frac{M_{n-1} l_n}{6EI_z} + \frac{A_n a_n}{l_n EI_z} - \beta_n \right).$$

In the same manner, for the span $n + 1$,

$$\theta' = \frac{A_{n+1} b_{n+1}}{l_{n+1} EI_z} + \frac{M_n l_{n+1}}{3EI_z} + \frac{M_{n+1} l_{n+1}}{6EI_z} + \beta_{n+1}.$$

Equating these angles we obtain

$$M_{n-1} l_n + 2M_n (l_n + l_{n+1}) + M_{n+1} l_{n+1}$$

$$= - \frac{6A_n a_n}{l_n} - \frac{6A_{n+1} b_{n+1}}{l_{n+1}} - 6EI_z (\beta_{n+1} - \beta_n). \tag{120}$$

If h_{n-1}, h_n, h_{n+1} denote the vertical heights of the supports $n - 1$, n and $n + 1$ above a horizontal reference line, we have

$$\beta_n = \frac{h_{n-1} - h_n}{l_n}, \qquad \beta_{n+1} = \frac{h_n - h_{n+1}}{l_{n+1}}.$$

Substituting in eq. (120), the bending moments at the supports due to misalignment or settlement can be calculated.

Problems

1. Determine the bending moment and shearing force diagrams for a continuous beam with three equal spans carrying a uniformly distributed load of intensity q (Fig. 182).

Solution. For a simply supported beam and a uniformly distributed load the bending moment diagram is a parabola with maximum ordinate $ql_n^2/8$. The area of the parabolic segment is

$$A_n = \frac{2}{3} l_n \frac{ql_n^2}{8} = \frac{ql_n^3}{12}.$$

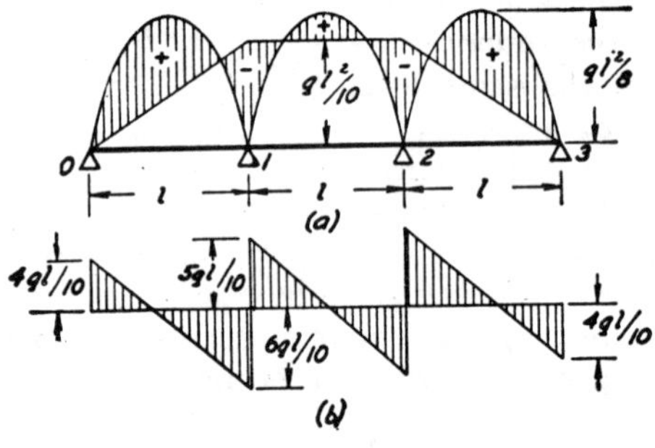

FIG. 182.

The centroid is above the middle of the span, so $a_n = b_n = l_n/2$. Substituting in eq. (118), we obtain

$$M_{n-1}l_n + 2M_n(l_n + l_{n+1}) + M_{n+1}l_{n+1} = -\frac{ql_n^3}{4} - \frac{q(l_{n+1})^3}{4}. \qquad (118')$$

Applying this equation to our case (Fig. 182) for the first and the second span and noting that at the support 0 the bending moment is zero, we obtain

$$4M_1l + M_2l = -\frac{ql^3}{2}. \qquad (d)$$

From the condition of symmetry it is evident that $M_1 = M_2$. Then, from (d), $M_1 = -(ql^2/10)$. The bending moment diagram is shown in Fig. 182a by the shaded area. The reaction at support 0 is

$$R_0 = \frac{ql}{2} - \frac{ql^2}{10}\frac{1}{l} = \frac{4}{10}ql$$

The reaction at support 1 is

$$R_1 = ql + \frac{ql^2}{10}\frac{1}{l} = \frac{11}{10}ql.$$

The shearing force diagram is shown in Fig. 182b. The maximum moment will evidently be at a distance $4l/10$ from the ends of the beam, where the shearing force is zero. The numerically maximum bending moment is at the inter-mediate supports.

2. Set up the expression for the right side of eq. (118) when there is a concentrated force in the span n and no load in the span $n+1$ (Fig. 183).

Solution. In this case A_n is the area of the triangle of height

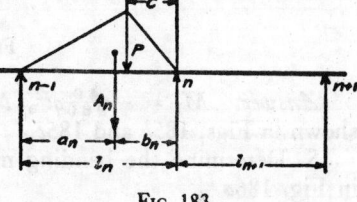

Fig. 183.

$Pc(l_n - c)/l_n$ and with the base l_n, hence $A_n = Pc(l_n - c)/2$ and $a_n = l_n - b_n = l_n - (l_n + c)/3$ Substituting in (118), we get

$$M_{n-1}l_n + 2M_n(l_n + l_{n+1}) + M_{n+1}l_{n+1} = -\frac{Pc(l_n - c)(2l_n - c)}{l_n}.$$

3. Determine the bending moments at the supports and the reactions for the continuous beam shown in Fig. 184.

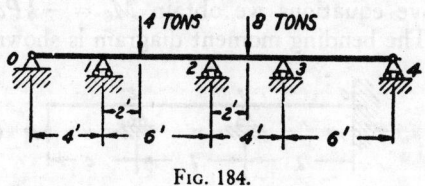

Fig. 184.

Answer. $M_1 = -1.54$ ft tons; $M_2 = -3.74$ ft tons; $M_3 = -1.65$ ft tons. The reactions are $R_0 = -0.386$ ton; $R_1 = 2.69$ tons; $R_2 = 6.22$ tons; $R_3 = 3.75$ tons; $R_4 = -0.275$ ton. The moments at the supports are negative and produce bending convex upward.

4. Construct the bending moment and shearing force diagrams for the continuous beam shown in Fig. 185a if $P = ql$, $c = l/4$.

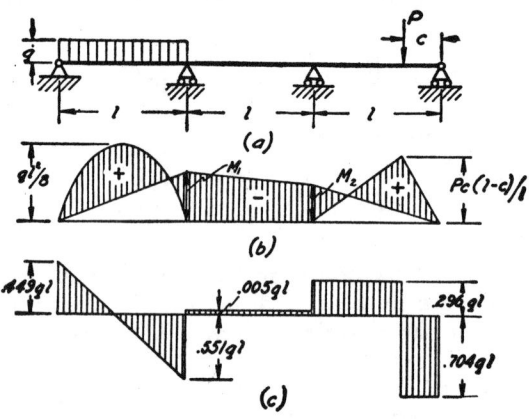

Fig. 185.

Answer. $M_1 = -\frac{49}{960}ql^2$, $M_2 = -\frac{11}{240}ql^2$. The diagrams are shown in Figs. 185b and 185c.

5. Determine the bending moment diagram for the case shown in Fig. 186a.

Solution. Eq. (118) for this case becomes

$$M_0l + 4M_1l + M_2l = 0.$$

It is evident that $M_2 = -Pc$, while the condition at the built-in end (support 0) gives (from eqs. 103 and 104)

$$\frac{M_0l}{3EI} + \frac{M_1l}{6EI} = 0.$$

From the above equations we obtain $M_0 = -\frac{1}{7}Pc$; $M_1 = +\frac{2}{7}Pc$; $M_2 = -Pc$. The bending moment diagram is shown in Fig. 186b.

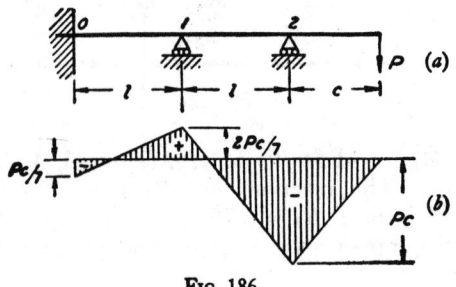

Fig. 186.

6. Determine the bending moments at the supports of a continuous beam with seven equal spans when the middle span alone is loaded by a uniformly distributed load q.

Answer. $M_3 = M_4 = -\frac{15}{284}ql^2$, $M_2 = M_5 = -\frac{4}{15}M_3$, $M_1 = M_6 = \frac{1}{15}M_3$.

7. A continuous beam having four equal spans, each of length 16 ft, is uniformly loaded over the last span. Draw the shearing force and bending moment diagrams if $q = 400$ lb per ft.

8. Solve Prob. 5, assuming that a uniform load of intensity q is distributed along the entire length of the beam and that $c = l/2$. Draw the shearing force diagram for this loading condition.

Answer. $M_0 = -\frac{5}{56}ql^2$, $M_1 = -\frac{1}{14}ql^2$, $M_2 = -\frac{1}{8}ql^2$.

CHAPTER VII

SYMMETRICAL BEAMS OF VARIABLE CROSS SECTION.
BEAMS OF TWO MATERIALS

46. Beams of Variable Cross Section.—In the preceding discussion all of the beams considered were of prismatic form. More elaborate investigation shows that eqs. (56) and (57), which were derived for prismatic bars, can also be used with sufficient accuracy for bars of variable cross section, provided the variation is not too extreme. Cases of abrupt changes in cross section, in which considerable stress concentration takes place, will be discussed in Part II.

As a first example of a beam of variable cross section, let us consider the deflection of a cantilever beam of *uniform strength*, i.e., a beam in which the section modulus varies along the beam in the same proportion as the bending moment. Then, as is seen from eqs. (60), $(\sigma_x)_{\max}$ remains constant along the beam and can be taken equal to σ_W. Such a condition is favorable as regards the amount of material used, because each cross section will have the minimum area necessary to satisfy the conditions of strength.

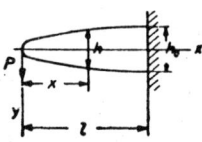

FIG. 187.

For a cantilever with an end load (Fig. 187), the bending moment at any cross section at distance x from the load is numerically equal to Px. In order to have a beam of uniform strength the section modulus must also be proportional to x. This condition can be fulfilled in various ways.

Let us take as a first example the case of a rectangular cross section of constant width b and variable depth h, Fig. 187. From the definition of the beam of equal strength it follows that

$$\frac{M}{Z} = \frac{6Px}{bh^2} = \frac{6Pl}{bh_0^2} = \text{const.},$$

in which h_0 is the depth of the beam at the built-in end. Then

$$h^2 = \frac{h_0{}^2 x}{l}.$$

It may be seen that in this case the depth of the beam varies following a parabolic law. At the loaded end the cross-sectional area is zero. This result is obtained because the shearing stress was neglected in the derivation of the shape of the beam of uniform strength. In practical applications this stress must be taken into account by making certain changes in the shape at the loaded end in order to have sufficient cross-sectional area to transmit the shearing force. The deflection of the beam at the end is found from eq. (93):

$$\delta = \int_0^l \frac{12Px^2 dx}{Ebh^3} = \frac{12Pl^{3/2}}{Ebh_0{}^3} \int_0^l \sqrt{x}\, dx = \frac{2}{3} \frac{Pl^3}{EI_0}, \quad (121)$$

where $I_0 = bh_0{}^3/12$ represents the moment of inertia of the cross section at the built-in end. Comparison with eq. (95) shows that this deflection is twice that of a prismatic bar having the flexural rigidity EI_0 and subjected to the same load, i.e., the bar has the same strength but not the same stiffness as the prismatic bar.

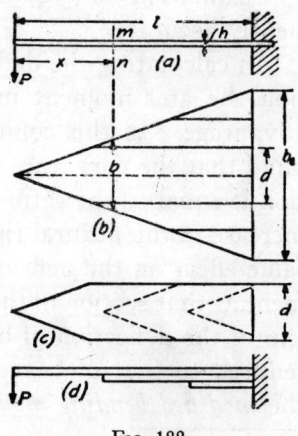

Fig. 188.

As a second example we consider a cantilever of rectangular cross section of constant depth h and variable width b (Figs. 188a and 188b). As the section modulus and moment of inertia I_z of a beam of triangular shape increases with x in the same proportion as the bending moment, the maximum stress $(\sigma_x)_{max}$ and the curvature (see eq. 56) remain constant along the beam and the magnitude of the radius of curvature can be determined from the equation (see eq. 55):

$$(\sigma_x)_{max} = \frac{hE}{2r}. \quad (a)$$

The deflection at the end of a circular arc can be taken, for small deflections, equal to

$$\delta = \frac{l^2}{2r} = \frac{Pl^3}{2EI_0},\tag{122}$$

or, by using (a),

$$\delta = (\sigma_x)_{\max}\frac{l^2}{hE}.\tag{123}$$

It is seen from this equation that for this type of cantilever of uniform strength the deflection at the end varies as the square of the length and inversely as the depth.

These results may be used to compute the approximate stresses and deflections in a *spring of leaf type*. The triangular plate considered above is thought of as divided into strips, arranged as shown in Fig. 188b, c, d. The initial curvature and the friction between the strips are neglected for a first approximation and eq. (123) can then be considered as sufficiently accurate.[1]

In calculating the deflection of beams of variable cross section the area-moment method (see Art. 34) can be used to advantage. In this connection it is only necessary to bear in mind that the curvature of the deflection line at any cross section is equal to the ratio M/EI_z (eq. 56, p. 95). Therefore an increase in the flexural rigidity at a given section will have the same effect on the deflection as a decrease of the bending moment at that section in the same ratio. Consequently the problem of the deflection of beams of variable cross section can be reduced to that of beams of constant cross section, by using the *modified bending moment diagram*. The modified moment

---

[1] This solution was obtained by E. Phillips, *Ann. mines*, Vol. 1, pp. 195–336, 1852. See also Todhunter and Pearson, *History of the Theory of Elasticity*, Vol. 2, p. 330, 1893; and A. Castigliano, *Theorie der Biegungs- und Torsions-Federn*, Vienna, 1888. The effect of friction between the leaves was discussed by G. Marié, *Ann. mines*, Vols. 7–9, 1905–6. D. Landau and P. H. Parr investigated the distribution of load between the individual leaves of the spring, *J. Franklin Inst.*, Vols. 185–7. A complete bibliography on mechanical springs was published by the Am. Soc. Mech. Engrs., New York, 1927. See also the book by S. Gross and E. Lehr, *Die Federn*, V. D. I. Verlag, 1938. A very complete study of various types of mechanical springs is given in the book by A. M. Wahl, *Mechanical Springs*, Cleveland, 1944.

diagram is obtained by multiplying the ordinates of the actual moment diagram by the ratio I_0/I, where I is the moment of inertia at any cross section and I_0 is a constant moment of inertia. This reduces the deflection of the bar of variable cross section to the deflection of a bar of constant cross section with moment of inertia I_0.

For example, the problem of the deflection of a circular shaft (Fig. 189) which has sections of two different diameters, with moments of inertia I_0 and I_1 and loaded by P, can be reduced to that for a circular shaft having a constant moment of inertia I_0 as follows. In considering the conjugate beam A_1B_1, we use the loading represented by the shaded area instead of the triangular load-

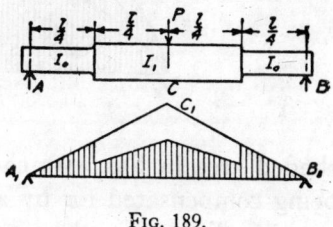

FIG. 189.

ing $A_1C_1B_1$ representing the actual bending moment diagram. This area is obtained by reducing the ordinates of the diagram along the middle portion of the shaft in the ratio I_0/I_1. Determination of deflections and slopes can now be made as in the case of prismatic bars, the magnitude of the deflection and slope at any cross section of the beam being equal to the bending moment and shearing force in the conjugate beam, divided by EI_0. It should be noted that in the case represented in Fig. 189 an abrupt change in the diameter of the shaft takes place at a distance $l/4$ from the supports, producing local stresses at these points. These have no substantial effect upon the deflection of the shaft, provided the difference in diameter of the two portions is small in comparison with the lengths of these portions.

The method used for a shaft of variable cross section can also be applied to built-up I beams or plate girders of variable cross section. An example of a plate girder supported at the ends and uniformly loaded is shown in Fig. 190. The bending moment decreases from the middle towards the ends of the girder and the weight of the girder can be reduced by diminishing the number of plates in the flanges as shown schematically in the figure. The deflection of such a beam may be calculated on the basis of the moment of inertia of the middle cross section. The load on the conjugate beam, instead of being a single parabola indicated by the dotted line, is represented by the

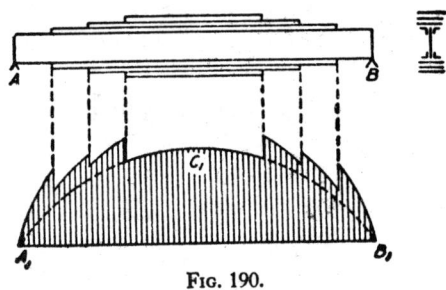

FIG. 190.

shaded area in Fig. 190, each diminution in the cross section being compensated for by an increase of the ordinates of the moment diagram in the ratio I_{middle}/I.

Problems

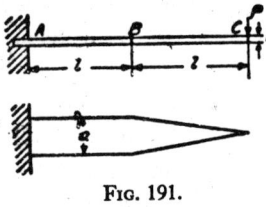

FIG. 191.

1. A steel plate of the form shown in Fig. 191 is built in at one end and loaded by a force P at the other. Determine the deflection at the end if the length is $2l$, a is the width and h the thickness of the plate and P is the load at the end.

Solution. The deflection will consist of three parts:

$$\delta_1 = \frac{Pl^3}{3EI_z} + \frac{Pl^3}{2EI_z} = \text{deflection at } P$$

$$\delta_2 = \frac{3Pl^3}{2EI_z} = \text{deflection at } C \text{ due to the slope at } B,$$

$$\delta_3 = \frac{Pl^3}{2EI_z} = \text{deflection due to bending of part } BC \text{ of the plate.}$$

The complete deflection is given by $\delta = \delta_1 + \delta_2 + \delta_3$.

2. Solve the previous problem, assuming $l = 10$ in., $a = 3$ in., $P = 1,000$ lb and $\sigma_{\max} = 70,000$ lb per sq in.

3. Determine the width d of a leaf spring (Fig. 188) and its deflection if $P = 6,000$ lb, $h = \frac{1}{2}$ in., $l = 24$ in., $\sigma_W = 70,000$ lb per sq in. and the number of leaves $n = 10$

Solution. Considering the leaves of the spring as cut out of a triangular plate (Fig. 188*b*), the maximum stress will be

$$\sigma_{max} = \frac{6Pl}{ndh^2},$$

from which

$$d = \frac{6Pl}{n\sigma_w h^2} = \frac{6 \times 6,000 \times 24 \times 4}{10 \times 70,000} = 4.94 \text{ in.}$$

The deflection is determined from eq. (123),

$$\delta = \frac{70,000 \times 24^2}{\frac{1}{2} \times 30 \times 10^6} = 2.69 \text{ in.}$$

4. Compare the deflection at the middle and the slope at the ends of the shaft shown in Fig. 189 with those of a shaft of the same length but of constant cross section whose moment of inertia is equal to I_0. Take $I_1/I_0 = 2$.

Solution. Due to the greater flexural rigidity at the middle, the slopes at the ends of the shaft shown in Fig. 189 will be less than those at the ends of the cylindrical shaft in the ratio of the shaded area to the total area of the triangle $A_1C_1B_1$. The total area is the loading for the case of the cylindrical shaft. For the values given, this ratio is $\frac{5}{8}:1$.

The deflections at the middle for the two types of shaft are in the ratio given by the bending moment produced by the shaded area, divided by that produced by the area of the triangle $A_1C_1B_1$. This will be $\frac{9}{16}:1$.

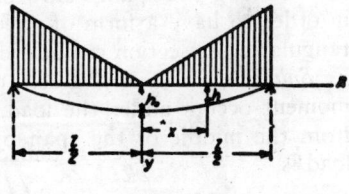

FIG. 192.

5. A beam supported at the ends is loaded as shown in Fig. 192. How should the depth h of the beam vary in order to have a form of equal strength if the width b of the rectangular cross section remains constant along the beam?

Answer. $h^2 = h_0{}^2 \left(1 - 8\frac{x^3}{l^3}\right).$

6. Determine the deflection of a steel plate $\frac{1}{2}$ in. thick shown in Figs. 193*a* and 193*b* under the action of the load $P = 20$ lb at the middle.

Solution. Reducing the problem to the deflection of a plate of constant width = 4 in., the transformed moment diagram for this case will be represented by the trapezoid $adeb$, Fig. 193c, and we obtain

$$\delta = \frac{11}{8} \cdot \frac{Pl^3}{48EI_z},$$

where I_z is the moment of inertia at the middle of the span. The numerical value of the deflection can now be easily calculated.

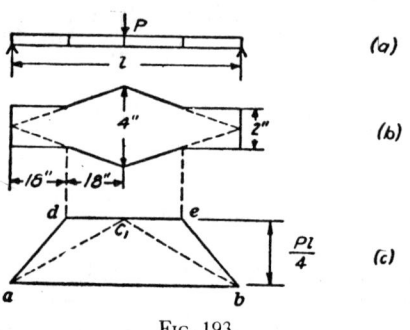

(a)

(b)

(c)

Fig. 193.

7. Determine the maximum deflection of a leaf spring (Fig. 188a) if $l = 36$ in., $h = \frac{1}{2}$ in., $E = 30 \times 10^6$ lb per sq in., $\sigma_W = 60,000$ lb per sq in.

Answer. $\delta = 5.18$ in.

8. A simply supported rectangular beam carries a load P which moves along the span. How should the depth h of the beam vary in order to have a form of equal strength if the width b of the rectangular cross section remains constant along the beam?

Solution. For any given position of the load the maximum moment occurs under the load. Denoting the distance of the load from the middle of the span by x, the bending moment under the load is

$$M = \frac{P\left(\dfrac{l}{2} - x\right)\left(\dfrac{l}{2} + x\right)}{l}.$$

The required depth h of the beam under the load is obtained from the equation

$$\sigma_W = \frac{6M}{bh^2},$$

from which

$$h^2 = \frac{6M}{b\sigma_W} = \frac{6P}{lb\sigma_W}\left(\frac{l^2}{4} - x^2\right)$$

and

$$\frac{h^2}{6Pl/4b\sigma_W} + \frac{x^2}{l^2/4} = 1.$$

It may be seen that in this case the depth of the beam varies following an elliptical law, the semi-axes of the ellipse being

$$l/2 \quad \text{and} \quad \sqrt{6Pl/4b\sigma_W}.$$

9. Determine the bending moments at the ends of the beam AB with built-in ends and centrally loaded, Fig. 194. Take $I_1/I_0 = 2$.

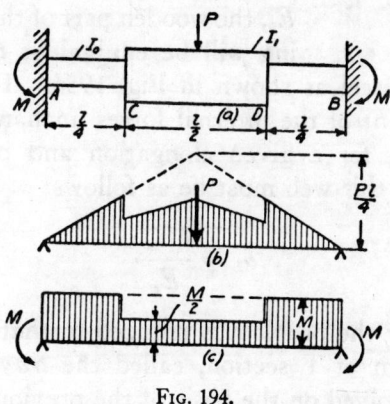

FIG. 194.

Solution. A solution is obtained by combining the two simple cases shown in (b) and (c). It is clear that the condition at the built-in ends will be satisfied if the slopes at the ends are equal to zero, i.e., if the reactions due to the imaginary loading (see p. 155) represented by the shaded areas in (b) and (c) are equal. Therefore the equation for calculating the numerical value of M is

$$\frac{Pl}{4}\frac{l}{2} - \frac{3}{8}\frac{Pl}{4}\frac{l}{2} = Ml - \frac{Ml}{4},$$

from which

$$M = 5Pl/48.$$

10. Solve the above problem on the assumption that two equal loads P are applied at C and D.
Answer. $M = Pl/6$.

47. Symmetrical Beams of Two Different Materials.— There are cases when beams of two or more different materials are used. Fig. 195a represents a simple case, a wooden beam reinforced by a steel plate bolted to the beam at the bottom. Assuming that there is no sliding between the steel and wood during bending, the theory of solid beams can also be used

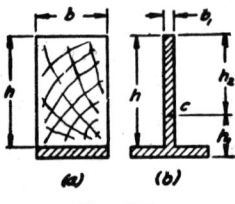

Fig. 195.

here. According to this theory elongations and contractions of longitudinal fibers are proportional to the distance from the neutral axis. Due to the fact that the modulus of elasticity of wood E_w is much smaller than that of steel E_s, the wooden part of the beam in bending will be equivalent to a much narrower web of steel as shown in Fig. 195b. To maintain the resisting moment of the internal forces unchanged for a given curvature, i.e., for a given elongation and contraction, the thickness b_1 of this web must be as follows:

$$b_1 = \frac{bE_w}{E_s}. \qquad (a)$$

In this manner the problem is reduced to that of the bending of a steel beam of T section, called the *transformed section*, which can be solved on the basis of the previous theory.

Consider, for instance, a simply supported beam 10 ft long loaded at the middle by 1,000 lb. The cross-sectional dimensions of the wooden part of the beam are $b = 4$ in. and $h = 6$ in. and on the convex side it is reinforced by a steel plate 1 in. wide and $\frac{1}{2}$ in. thick. Assuming $E_w/E_s = \frac{1}{20}$ and using eq. (a), the transformed section will have a web 6×0.20 and the flange 1×0.50. The distances of the outermost fibers from the neutral axis (Fig. 195b) are $h_1 = 2.54$ in. and $h_2 = 3.96$ in. The moment of inertia of the transformed section with respect to the neutral axis is $I_z = 7.37$ in.[4], whence the stresses in the outermost fibers of the transformed section are (from eqs. 61, p. 96):

$$\sigma_{max} = \frac{M_{max}h_1}{I_z} = \frac{30,000 \times 2.54}{7.37} = 10,300 \text{ lb per sq in.,}$$

$$\sigma_{min} = -\frac{M_{max}h_2}{I_z} = -\frac{30,000 \times 3.96}{7.37} = -16,000 \text{ lb per sq in.}$$

To obtain the maximum compressive stress in the wood of the

actual beam the stress σ_{min} obtained above for the transformed section (steel) must be multiplied by $E_w/E_s = \frac{1}{20}$. The maximum tensile stress for this case is the same in both the actual beam and the transformed beam.

As another example of the bending of a beam of two different materials let us consider the case of a bimetallic strip built up of nickel steel and monel metal (Fig. 196). The bending of such a strip

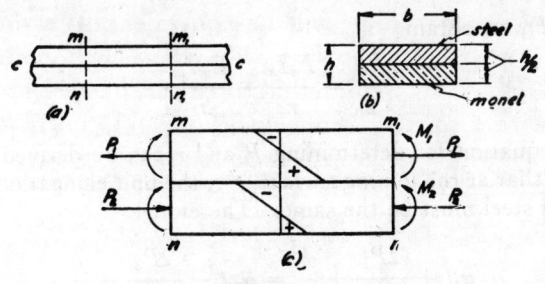

FIG. 196.

by external forces can be discussed in the same manner as in the above problem of wood and steel, provided we know the ratio E_m/E_s, in which E_m and E_s are the moduli of elasticity of monel metal and steel, respectively. Let us consider now the bending of such a strip due to a change in temperature. The coefficient of thermal expansion of monel metal is larger than that of nickel steel and, when the temperature rises, bending will occur with the concave side on the same side as the steel strip. This phenomenon of bending of bimetallic strips under varying temperatures is used in various automatic instruments such as thermostats for regulating temperature.[2] Let $h/2$ be the thickness and b the width of each metal strip, t the increase in temperature, r the radius of curvature, α_s and α_m the coefficients of thermal expansion of steel and monel, respectively, $E_sI_s =$ the flexural rigidity of the steel, $E_mI_m =$ the flexural rigidity of monel metal. When the temperature rises, the strip of monel metal, having a greater coefficient of expansion, will be subjected to both bending and compression and the steel will be subjected to bending and tension. Considering an element of the strip cut out by two adjacent cross sections mn and m_1n_1 (Fig. 196c), the internal forces over the cross section of the steel can be reduced to a tensile force P_1 and a couple M_1. In the same manner the internal forces in the monel metal can be reduced to a compressive force P_2 and a couple M_2. The internal forces over any

---

[2] See author's paper in *J. Opt. Soc. Amer.*, Vol. 11, p. 23.

cross section of the beam must be in equilibrium. Therefore

$$P_1 = P_2 = P$$

and

$$\frac{Ph}{2} = M_1 + M_2. \tag{b}$$

Substituting

$$M_1 = \frac{E_s I_s}{r}, \qquad M_2 = \frac{E_m I_m}{r},$$

into eq. (b), we obtain

$$\frac{Ph}{2} = \frac{E_s I_s}{r} + \frac{E_m I_m}{r}. \tag{c}$$

Another equation for determining P and r can be derived from the condition that at the joining surface, c–c, the unit elongation of monel metal and steel must be the same. Therefore

$$\alpha_s t + \frac{2P_1}{E_s h b} + \frac{h}{4r} = \alpha_m t - \frac{2P_2}{E_m h b} - \frac{h}{4r}$$

or

$$\frac{2P}{hb}\left(\frac{1}{E_s} + \frac{1}{E_m}\right) = (\alpha_m - \alpha_s)t - \frac{h}{2r}. \tag{d}$$

From eqs. (c) and (d) we obtain

$$\frac{4}{bh^2 r}(E_s I_s + E_m I_m)\left(\frac{1}{E_s} + \frac{1}{E_m}\right) = (\alpha_m - \alpha_s)t - \frac{h}{2r}. \tag{e}$$

Substituting in this equation

$$I_s = I_m = \frac{bh^3}{96} \qquad \text{and} \qquad E_s = 1.15 E_m,$$

the following approximate equation is obtained:

$$\frac{1}{r} = \frac{3}{2}\frac{(\alpha_m - \alpha_s)t}{h}. \tag{f}$$

Now, from eq. (c),

$$P = \frac{3}{h^2}(\alpha_m - \alpha_s)t(E_s I_s + E_m I_m) = \frac{bh}{32}(\alpha_m - \alpha_s)t(E_s + E_m) \tag{g}$$

and

$$M_1 = \frac{3}{2}\frac{(\alpha_m - \alpha_s)t}{h}E_s I_s, \qquad M_2 = \frac{3}{2}\frac{(\alpha_m - \alpha_s)t}{h}E_m I_m. \tag{h}$$

From eqs. (g) and (h) P, M_1, and M_2 can be determined. The maximum stress in the steel is obtained by adding to the tensile stress produced by the force P the tensile stress due to the curvature $1/r$:

$$\sigma_{max} = \frac{2P}{bh} + \frac{h}{4}\frac{E_s}{r} = \frac{4}{bh^2 r}\left(E_s I_s + E_m I_m + \frac{bh^3}{16}E_s\right).$$

Assuming, e.g., that both metals have the same modulus E, we obtain

$$\sigma_{max} = \frac{hE}{3r},$$

or. by using eq. (f) [3]

$$\sigma_{max} = \tfrac{1}{2}Et(\alpha_m - \alpha_s).$$

For $E = 27 \times 10^6$ lb per sq in., $t = 200°$ C and $\alpha_m - \alpha_s = 4 \times 10^{-6}$, we find

$$\sigma_{max} = 10,800 \text{ lb per sq in.}$$

The distribution of stresses due to heating is shown in Fig. 196c.

Problems

1. Find the safe bending moment for the wooden beam reinforced by a steel plate, Fig. 195, if $b = 6$ in., $h = 8$ in. and the thickness of the steel plate is $\frac{1}{2}$ in. Assume $E_w = 1.5 \times 10^6$ lb per sq in., $E_s = 30 \times 10^6$ lb per sq in., $\sigma_W = 1,200$ lb per sq in. for wood and $\sigma_W = 16,000$ lb per sq in. for steel.

2. Assume that the wooden beam of the preceding problem is reinforced at the top with a steel plate 2 in. wide and 1 in. thick and at the bottom with a steel plate 6 in. wide and $\frac{1}{2}$ in. thick. Calculate the safe bending moment if E and σ_W are the same as in the preceding problem.

Answer. $M = 308,000$ in. lb.

3. A bimetallic strip has a length $l = 1$ in. Find the deflection at the middle produced by a temperature increase equal to $200°$ C if $E_s = 1.15E_m$ and $\alpha_m - \alpha_s = 4 \times 10^{-6}$.

48. Reinforced-Concrete Beams.

It is well known that the strength of concrete is much greater in compression than in tension. Hence a rectangular beam of concrete will fail from the tensile stresses on the convex side. The beam can be greatly strengthened by the addition of steel reinforcing bars

---

[3] This equation also holds for $E_s = E_m$.

on the convex side as shown in Fig. 197. As concrete bonds to the steel strongly there will be no sliding of the steel bars with respect to the concrete during bending and the methods developed in the previous article can also be used here for calculating bending stresses. In practice the cross-sectional area of the steel bars is usually such that the tensile strength of the concrete on the convex side is overcome before yielding of the steel begins and at larger loads the steel alone takes practically all the tension. Hence it is the established practice in calculating bending stresses in reinforced-concrete beams to

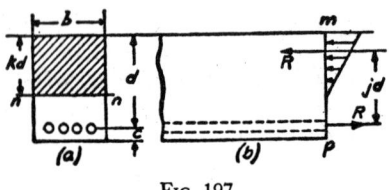

FIG. 197.

assume that all the tension is taken by the steel and all the compression by the concrete. Replacing the tensile forces in the steel bars by their resultant R, the distribution of internal forces over any cross section mp will be as shown in Fig. 197b. Assuming, as before, that cross sections remain plane during bending and denoting by kd the distance of the neutral axis nn from the top,[4] then the maximum longitudinal unit contraction ϵ_c in the concrete and the unit elongation ϵ_s of the axes of the steel bars are given by the following:

$$\epsilon_c = -\frac{kd}{r}, \qquad \epsilon_s = \frac{(1-k)d}{r} \qquad (a)$$

Concrete does not follow Hooke's law and a compression test diagram for this material has a shape similar to that for cast iron in Fig. 4b. As the compression stress increases, the slope of the tangent to the diagram decreases, i.e., the modulus of concrete decreases with an increase in stress. In calculating stresses in reinforced-concrete beams it is usual practice to assume that Hooke's law holds for concrete, and to compensate for the variable modulus by taking a lower value for this modulus than that obtained from compression tests when the stresses are small. In specifications for reinforced-concrete it is frequently assumed that $E_s/E_c = 15$. Then, from eqs. (a), the

[4] k is a numerical factor less than unity.

maximum compressive stress in the concrete and the maximum tensile stress in the steel [5] are, respectively,

$$\sigma_c = -\frac{kd}{r}E_c, \quad \sigma_s = \frac{(1-k)d}{r}E_s. \qquad (b)$$

We will now calculate the position of the neutral axis from the condition that the normal forces over the cross section mp must reduce to a couple equal to the bending moment at that cross section. The sum of the compressive forces in the concrete must equal the tensile force R in the steel bars, or

$$-\frac{bkd\sigma_c}{2} = \sigma_s A_s, \qquad (c)$$

where A_s is the total cross-sectional area of steel. Using the notation $A_s/bd = p$ and $E_s/E_c = n$, we obtain from (c) and (b)

$$k^2 = 2(1-k)pn, \qquad (d)$$

from which

$$k = \sqrt{(pn)^2 + 2pn} - pn. \qquad (124)$$

After determining the position of the neutral axis from eq. (124), the ratio between the maximum stress in the concrete and the stress in the steel becomes, from eqs. (b),

$$-\frac{\sigma_c}{\sigma_s} = \frac{k}{(1-k)n}. \qquad (125)$$

The distance jd between the resultants R of the compressive and tensile forces acting over the cross section of the beam (Fig. 197b) is [6]

$$jd = \frac{2}{3}kd + (1-k)d = \left(1 - \frac{k}{3}\right)d, \qquad (126)$$

and the moment of the internal forces equal to the bending moment M is

$$jdR = jdA_s\sigma_s = -\frac{jkbd^2}{2}\sigma_c = M,$$

---

[5] The cross-sectional dimensions of the steel bars are usually small and the average tensile stress is used instead of the maximum stress.

[6] j is a numerical factor less than unity.

from which

$$\sigma_s = \frac{M}{A_s j d},$$ (127)

$$\sigma_c = -\frac{2M}{j k b d^2}.$$ (128)

'By using eqs. (124)–(128), the bending stresses in reinforced-concrete beams are readily calculated.

Problems

1. If $E_s/E_c = 15$ and $A_s = 0.008bd$, determine the distance of the neutral axis from the top of the beam, Fig. 197.
Answer. $kd = 0.384d$.

2. Determine the ratio $p = A_s/bd$ if the maximum tensile stress in the steel is 12,000 lb per sq in., the maximum compressive stress in the concrete is 645 lb per sq in. and $E_s/E_c = n = 15$.
Solution. From eq. (125) $k = 0.446$. Then, from eq. (*d*),

$$p = \frac{k^2}{2(1-k)n} = 0.012.$$

3. Determine the ratio p if the maximum compressive stress in the concrete is $\frac{1}{20}$ of the tensile stress in the steel.
Answer. $p = 0.0107$.

4. If $n = 15$ and the working compressive stress for concrete is 650 lb per sq in., determine the safe load at the middle of a reinforced-concrete beam 10 ft long supported at the ends and having $b = 10$ in., $d = 12$ in., $A_s = 1.17$ sq in.
Answer. $P = 5,570$ lb.

5. Calculate the maximum bending moment which a concrete beam will safely carry if $b = 8$ in., $d = 12$ in., $A_s = 2$ sq in., $E_s/E_c = 12$ and the working stress for steel is 15,000 lb per sq in. and for concrete 800 lb per sq in.
Answer. $M = 16,000$ ft lb.

6. Determine the value of k for which the maximum permissible stresses in the concrete and the steel are realized simultaneously.
Solution. Let σ_c and σ_s be the allowable stresses for the concrete and the steel. Then taking the ratio of these stresses, as given by formulas (*b*), and considering only the absolute value of this ratio, we obtain

$$\frac{\sigma_c}{\sigma_s} = \frac{kE_c}{(1-k)E_s},$$

from which

$$k = \frac{\sigma_c}{\sigma_c + \sigma_s \dfrac{E_c}{E_s}}.$$

If this condition is satisfied the beam is said to have *balanced reinforcement*. Having k and using eq. (126) the depth is obtained from eq. (128) and the area A_s from eq. (127).

7. Determine the steel ratio $p = A_s/bd$ if $\sigma_s = 12,000$ lb per sq in., $\sigma_c = 645$ lb per sq in. and $n = E_s/E_c = 15$.

Solution. From the formula of the preceding problem we find

$$k = 0.446.$$

Then, substituting in eq. (d), we obtain

$$p = 0.012.$$

8. Design a beam 10 in. wide to withstand safely a bending moment of 22,500 ft lb if $\sigma_c = 750$ lb per sq in., $\sigma_s = 12,000$ lb per sq in. and $E_s/E_c = 12$. Find the depth d and the steel area A_s. Assume balanced reinforcement as in Prob. 6.

49. Shearing Stresses in Reinforced-Concrete Beams.—

Using the same method as in Art. 26 and considering an element mnm_1n_1 between the two adjacent cross sections mp and m_1p_1 (Fig. 198), it can be concluded that the maximum shearing stress τ_{yx} will act over the neutral surface nn_1. Denoting by dR the difference between the compressive forces on the concrete on cross sections mp

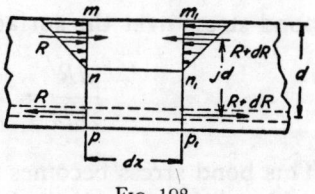

Fig. 198.

and m_1p_1, the shearing stress τ_{yx} over the neutral surface is found from the following:

$$(\tau_{yx})_{\max} b\,dx = dR,$$

from which

$$(\tau_{yx})_{\max} = \frac{1}{b}\frac{dR}{dx}. \qquad (a)$$

Since the bending moment is

$$M = Rjd,$$

eq. (*a*) becomes

$$(\tau_{yz})_{\text{max}} = \frac{1}{bjd}\frac{dM}{dx} = \frac{V}{bjd},\qquad(b)$$

in which V is the shearing force at the cross section considered. Using eq. (126), the above equation for shearing stresses becomes

$$(\tau_{yz})_{\text{max}} = \frac{3V}{bd(3 - k)}.\qquad(129)$$

In practical calculations, not only the shearing stresses over the neutral surface but also the shearing stresses over the surface of contact between the steel and concrete (bond stresses) are of importance. Considering again the two adjacent cross sections (Fig. 198), the difference between the tensile forces in the steel bars at these two sections is

$$dR = \frac{Vdx}{jd}.$$

This difference is balanced by the bond stresses distributed over the surface of the bars. Denoting by A the total lateral surface of all the steel bars per unit length of the beam, the bond stress over the surface of the bars is

$$\frac{dR}{Adx} = \frac{V}{Ajd} = \frac{3V}{A(3 - k)d}.\qquad(130)$$

This bond stress becomes larger than the stress on the neutral surface (eq. 129) if A is less than b. To increase A and at the same time keep the cross-sectional area of steel constant, it is only necessary to increase the number of bars and decrease their diameter.

CHAPTER VIII

BENDING OF BEAMS IN A PLANE WHICH IS NOT A PLANE OF SYMMETRY

50. Pure Bending in a Plane Which Is Not a Plane of Symmetry.—If a beam has a plane of symmetry, say plane xy, Fig. 199a, and couples acting in that plane are applied at the ends,

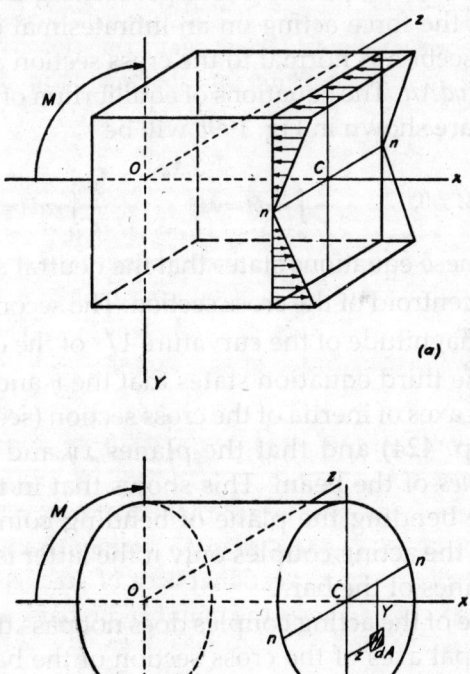

FIG. 199.

then the beam will be bent in the same plane and the neutral axis nn at each cross section of the beam (Fig. 199a) will pass through the centroid C and will be perpendicular to the plane

227

of the acting couples. From symmetry it can be concluded that the stresses corresponding to that direction of the neutral axis give a resultant couple in the xy plane, and with a proper selection of the magnitude of stresses (see eq. 56, p. 95) this couple will balance the external couple M.

Let us consider now the case in which the plane xy of the bending couples, Fig. 199b, is not a plane of symmetry, and investigate under what condition the neutral axis will be perpendicular to that plane. Assuming that the axis nn is perpendicular to the plane xy and proceeding as in Art. 23, we find that the force acting on an infinitesimal element dA of the cross section is normal to the cross section and has the magnitude $EydA/r$. The equations of equilibrium of the portion of the beam are shown in Fig. 199b will be [1]

$$\frac{E}{r}\int_A y dA = 0, \qquad \frac{E}{r}\int_A y^2 dA = M, \qquad \frac{E}{r}\int_A yz dA = 0 \qquad (a)$$

The first of these equations states that the neutral axis passes through the centroid of the cross section. The second equation defines the magnitude of the curvature $1/r$ of the deflection curve and the third equation states that the y and z axes are the principal axes of inertia of the cross section (see Appendix A, Art. IV, p. 424) and that the planes xy and xz are the *principal planes* of the beam. This shows that in the general case of pure bending the plane of bending coincides with the plane of the acting couples only if the latter is one of the principal planes of the bar.

If the plane of the acting couples does not pass through one of the principal axes of the cross section of the bar, then the third of the equations of equilibrium (a) will not be satisfied. Thus, the direction of the neutral axis will not be perpendicular to the plane of the bending couples and must be found as follows: Assume that a couple M is acting in an axial plane of the bar which intersects the cross section along the line nm inclined to the principal axis y by an angle θ, Fig. 200. The

---

[1] The remaining three equations are always satisfied since the moments of all forces with respect to the x axis and the projections of the forces on the y and z axes vanish.

vector M representing the moment of the couple can be resolved into two components as shown in the figure. Since each component of the couple is acting in one of the principal planes of the beam, the corresponding stresses will be obtained

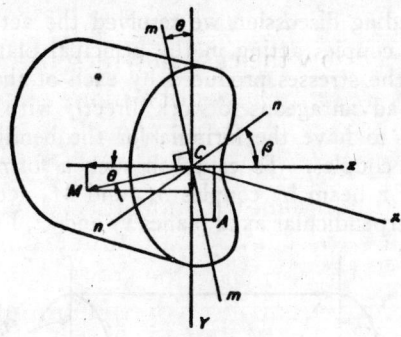

FIG. 200.

by applying the usual beam formulas and the total stress at any point A of the cross section will be given by the formula

$$\sigma = \frac{My \cos \theta}{I_z} + \frac{Mz \sin \theta}{I_y}. \qquad (b)$$

Equating this stress to zero, we obtain the equation of the neutral axis:

$$\frac{y \cos \theta}{I_z} + \frac{z \sin \theta}{I_y} = 0, \qquad (c)$$

and the tangent of the angle β defining the direction of the neutral axis nn will be

$$\tan \beta = -\frac{y}{z} = \frac{I_z}{I_y} \tan \theta. \qquad (d)$$

It is seen that the angle β is in general different from θ and the neutral axis is not perpendicular to the axial plane mm in which the bending couples are acting. The two angles are equal only if $\theta = 0$ or if $I_z = I_y$. In the first of these cases the bending couples are acting in the principal plane xy and the neutral axis coincides with the principal axis z. In the second case the two principal moments of inertia of the cross section are equal. The ellipse of inertia of the cross section (see Appendix A, Art. V, p. 427) in such a case becomes a circle and any pair of two perpendicular centroidal axes can be considered as the principal axes, and thus the neutral axis is always perpendicular to the plane of the bending couples.

In the preceding discussion we resolved the acting couples into two component couples acting in the principal planes of the beam and calculated the stresses produced by each of those components. Sometimes it is advantageous to work directly with the given bending couples and to have the formula for the bending stresses produced by those couples. To establish such a formula, let us consider bending of a beam by couples M_z and M_y acting in two arbitrarily chosen perpendicular axial planes xy and xz, Fig. 201. Assume

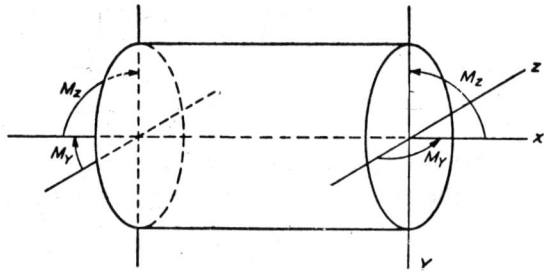

FIG. 201.

that the magnitudes of the couples are such that bending occurs i the xy plane, so that the neutral axis in each cross section is parallel to the z axis. Denoting by r_y the corresponding radius of curvature, the bending stresses will be $\sigma_x = Ey/r_y$ and we obtain the following values of the bending couples:

$$M_z = \int_A y\sigma_x dA = \frac{EI_z}{r_y}, \qquad M_y = -\int_A z\sigma_x dA = -\frac{EI_{yz}}{r_y}. \qquad (e)$$

Similarly, if the couples are such that bending in the xz plane is produced, then $\sigma_x = -Ez/r_z$, and we have

$$M_y = \frac{EI_y}{r_z}, \qquad M_z = -\frac{EI_{yz}}{r_z}. \qquad (f)$$

In the general case, when deflections in both planes are produced, the relations between the bending moments and the curvatures are obtained by combining eqs. (e) and (f), and we have

$$M_y = \frac{EI_y}{r_z} - \frac{EI_{yz}}{r_y}, \qquad M_z = \frac{EI_z}{r_y} - \frac{EI_{yz}}{r_z}. \qquad (g)$$

If the couples are acting only in the xy plane, then $M_y = 0$ and we obtain

$$\frac{EI_y}{r_z} - \frac{EI_{yz}}{r_y} = 0, \qquad \frac{1}{r_z} = \frac{1}{r_y} \cdot \frac{I_{yz}}{I_y}.$$

Substituting in the second of eqs. (g) we find

$$M_z = \frac{1}{r_y} \cdot \frac{E(I_z I_y - I_{yz}^2)}{I_y},$$

$$\frac{1}{r_y} = \frac{M_z I_y}{E(I_z I_y - I_{yz}^2)}, \qquad \frac{1}{r_z} = \frac{M_z I_{yz}}{E(I_z I_y - I_{yz}^2)}. \tag{131}$$

The bending stresses produced by the couple M_z will then be

$$\sigma_x = \frac{Ey}{r_y} - \frac{Ez}{r_z} = \frac{M_z}{I_z I_y - I_{yz}^2} (I_y y - I_{yz} z). \tag{132}$$

Similarly, if M_z vanishes, we obtain

$$\sigma_x = \frac{M_y}{I_z I_y - I_{yz}^2} (I_{yz} y - I_z z). \tag{133}$$

Eqs. (132) and (133) are especially useful for beams in which the web and flanges are parallel to the y and z axes.

Problems

1. A cantilever beam of rectangular cross section, Fig. 202, is bent by a couple M acting in the axial plane mm. What curve will be

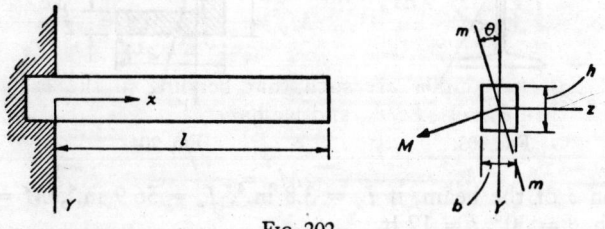

Fig. 202.

described by the end of the beam when the angle θ, defining the plane of the bending moment, varies from zero to 2π?

Solution. Resolving the bending moment M into two components $M \cos \theta$ and $M \sin \theta$ acting in the principal planes xy and xz, respectively, we find the two components u and v of the deflection of the end of the cantilever in the y and z directions, respectively, as:

$$u = -\frac{Ml^2 \cos\theta}{2EI_z}, \qquad v = -\frac{Ml^2 \sin\theta}{2EI_y},$$

from which

$$\frac{u^2}{(Ml^2/2EI_z)^2} + \frac{v^2}{(Ml^2/2EI_y)^2} = 1.$$

We see that the end of the cantilever describes an ellipse with the semi-axes $Ml^2/2EI_z$ and $Ml^2/2EI_y$.

2. Find for the preceding problem the numerical value of the ratio of the vertical and horizontal deflections of the end of the cantilever if $\theta = 45°$ and $h = 2b$.

Answer. $u/v = I_y/I_z = \frac{1}{4}$.

3. Find for the cantilever in Fig. 202 the angle of inclination β of the neutral axis to the horizontal and the magnitude of the maximum stress if $\theta = 45°$ and $h = 2b = 6$ in., $M = 1,200$ in. lb.

Answer. $\tan\beta = 4$, $\sigma_{max} = 141$ lb per sq in.

4. A standard 8 I 18.4 beam, simply supported, is bent by two equal and opposite couples M acting at the ends of the beam in the plane mm, Fig. 203. Find the maximum stress and the maximum

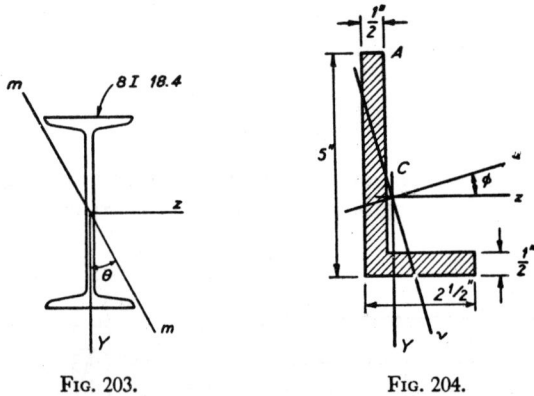

Fig. 203.   Fig. 204.

deflection δ of the beam, if $I_y = 3.8$ in.^{4}, $I_z = 56.9$ in.^{4}, $M = 5 \times 10^4$ in. lb, $\theta = 30°$, $l = 12$ ft.

Answer. $\sigma_{max} = \dfrac{4M\cos\theta}{I_z} + \dfrac{2M\sin\theta}{I_y} = 16,200$ lb per sq in.;

$$\delta_{max}{}^2 = \left(\frac{Ml^2\cos\theta}{8EI_z}\right)^2 + \left(\frac{Ml^2\sin\theta}{8EI_y}\right)^2, \text{ and } \delta_{max} = 0.573 \text{ in.}$$

5. A bar of angle cross section, Fig. 204, is bent by couples M applied at the ends and acting in the plane of the larger flange. Find the directions of the centroidal principal axes u and v, the magnitudes I_u and I_v of the principal moments of inertia and the magnitude of the maximum bending stress if $M = 10^4$ in. lb. Check the result by using formula (132).

Answer. $\varphi = 14°20'$, $I_u = 9.36$ in.^{4}, $I_v = 0.99$ in.^{4}, $|\sigma_{max}| = 4,850$ lb per sq in. at point A.

51. Bending of Beams Having Two Planes of Symmetry.

—If a beam has two planes of symmetry the problem of bending by transverse forces inclined to these planes and intersecting the axis of the beam can be readily solved by using the method of superposition. Each transverse force can be resolved into two components acting in the two planes of symmetry and, after solving the bending problem for each of those planes, the final stresses and deflections are obtained by superposition.

Let us consider, as an example, a cantilever of rectangular cross section, Fig. 205, with a transverse force P applied at

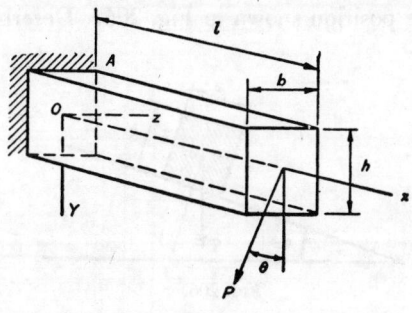

Fig. 205.

the end at an angle θ with the vertical plane of symmetry. Resolving the force into two components $P \cos \theta$ and $P \sin \theta$ and considering bending in the vertical and horizontal planes of symmetry, we find that the absolute values of the corresponding bending moments at the built-in end are $Pl \cos \theta$ and $Pl \sin \theta$. Taking into consideration the directions of these moments, the bending stress at any point of the built-in cross section will be

$$\sigma = -\frac{Ply \cos \theta}{I_z} + \frac{Plz \sin \theta}{I_y}.$$

The maximum tensile stress will be at point A and its magnitude is

$$\sigma_{max} = \frac{6Pl}{bh}\left(\frac{\cos\theta}{h} + \frac{\sin\theta}{b}\right).$$

The vertical and horizontal deflections of the loaded end will be

$$\delta_y = \frac{Pl^3\cos\theta}{3EI_z}, \qquad \delta_z = -\frac{Pl^3\sin\theta}{3EI_y},$$

and the total deflection will be obtained from the equation

$$\delta = \sqrt{\delta_y^2 + \delta_z^2}.$$

Problems

1. A horizontal wooden beam of rectangular cross section carries a vertical load uniformly distributed along the axis and is supported at the ends in the position shown in Fig. 206. Determine the maxi-

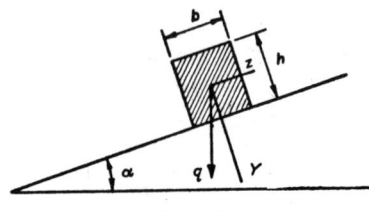

FIG. 206.

mum normal stress and the vertical deflection at the middle if the length of the beam $l = 10$ ft, the intensity of the load $q = 200$ lb per ft, $h = 8$ in., $b = 6$ in., $\tan\alpha = \frac{1}{3}$.
  Answer. $\sigma_{max} = 643$ lb per sq in., $\delta = 0.126$ in.
  2. Solve the preceding problem if the distance between the supports is 6 ft and the beam has two overhangs each 2 ft long.
  3. A horizontal circular bar of length l and with built-in ends carries a uniformly distributed vertical load of intensity q and a horizontal transverse load P concentrated at the middle. Find the maximum stress if $Pl = 24{,}000$ in. lb, $ql^2 = 48{,}000$ in. lb and the diameter of the bar $d = 4$ in.
  Answer. $\sigma_{max} = 796$ lb per sq in.

4. A horizontal square beam (Fig. 207) with simply supported ends is loaded at the third points by two equal forces P, one of which

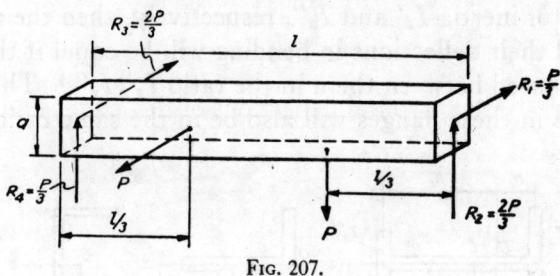

FIG. 207.

is vertical and the other horizontal. Find $\sigma_{\max}$ if $l = 12$ ft, $a = 12$ in., $P = 6{,}000$ lb.

Answer. $\sigma_{\max} = \dfrac{2Pl}{a^3} = 1{,}000$ lb per sq in.

52. Bending of Beams in a Principal Plane Which Is Not a Plane of Symmetry.—*Shear Center.*—In the discussion of pure bending (see p. 228) it was shown that the plane of the deflection curve coincides with the plane of the bending couples provided these couples act in one of the two *principal planes of bending.* In the case of bending of a beam by a coplanar system of transverse forces, the problem is more complicated. If the principal plane in which the forces are acting is not a plane of symmetry of the beam, such bending is usually accompanied by torsion of the beam. The following discussion will show how this torsion can be eliminated and *simple bending* established by a proper displacement of the plane of the acting forces parallel to itself.

We begin with simple examples in which the cross section of the beam has one axis of symmetry (z axis) and the forces are acting in a plane perpendicular to this axis, Fig. 208. Let us consider the case of a thin-walled beam shown in Fig. 208*a* and determine the position of the vertical plane in which the transverse loads must act in order to produce simple bending of the beam in a vertical plane. From our previous discussion of distribution of vertical shearing stresses τ_{zy} (see p. 122) we may conclude that practically the whole of the shearing force

V will be taken by the flanges alone. If we consider the flanges as two separate beams whose cross sections have moments of inertia $I_z{}'$ and $I_z{}''$, respectively, then their curvatures and their deflections in bending will be equal if the loads are distributed between them in the ratio $I_z{}':I_z{}''$.[2] The shearing forces in these flanges will also be in the same ratio. This

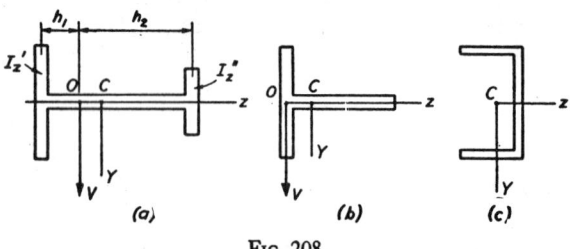

Fig. 208.

condition will be satisfied if the transverse loads act in the vertical plane passing through the point O (Fig. 208a), such that

$$\frac{h_1}{h_2} = \frac{I_z{}''}{I_z{}'},$$

where h_1 and h_2 are the distances of O from the centroids of the cross sections of the flanges. In this manner we find that the point O is displaced from the centroid C of the cross section towards the flange whose cross section has the larger moment of inertia. In the limiting case, shown in Fig. 208b, in which one of the flanges vanishes, it can be assumed with sufficient accuracy that the point O coincides with the centroid of the flange and that the transverse loads should act in the vertical plane through this point in order to have simple bending. The point O, through which the plane of loading must pass to eliminate torsion, is called the *shear center*.

Let us now consider the channel section (Fig. 208c) and determine the position of the plane in which the vertical loads must act to produce simple bending with the z axis as the neutral axis. For this purpose it is necessary to consider the dis-

---

[2] The effect of shearing force on deflection of flanges is neglected in this consideration.

tribution of the shearing stresses over the cross section in simple bending. To calculate the vertical shearing stresses τ_{xy} for the cross section of the web, the same method is used as in the case of an I beam (p. 122) and it can be assumed with sufficient accuracy that the vertical shearing force V is taken by the web alone. In the flanges there will be horizontal shearing stresses which we shall denote by τ_{xz}. To find the magnitude of these stresses, let us consider an element cut from the flange by two adjacent cross sections dx apart and by a vertical plane mnm_1n_1 parallel to the web (Fig. 209). If the beam is

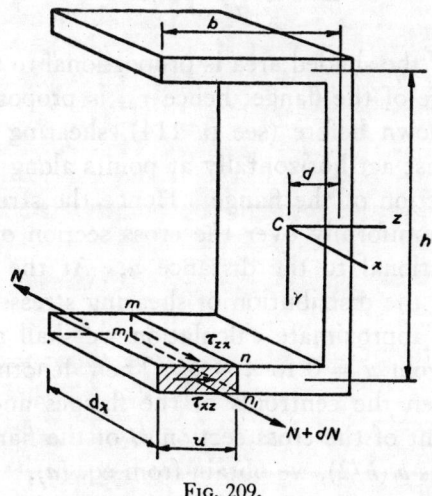

Fig. 209.

bent convex downward, the lower flange will be in tension and the tensile forces N and $N + dN$ acting on the above element will be equal to

$$N = \frac{M}{I_z} \int y \, dA$$

and

$$N + dN = \frac{M + \dfrac{dM}{dx} dx}{I_z} \int y \, dA,$$

where the integration must be extended over the shaded portion of the cross section of the flange. The integral represents the moment of the shaded area with respect to the z axis. The difference of the tensile forces N and $N + dN$ must be equal to the sum of the shearing stresses τ_{zx} acting over the side mnm_1n_1 of the element. Assuming that these stresses are uniformly distributed over this side and denoting by t the thickness of the flange, we obtain the following equation for calculating τ_{zx}:

$$\tau_{zx} t\, dx = -dN = -\frac{dM}{dx} \cdot \frac{dx}{I_z} \int y\, dA,$$

from which

$$\tau_{zx} = -\frac{V}{tI_z} \int y\, dA. \tag{a}$$

The moment of the shaded area is proportional to the distance u from the edge of the flange; hence τ_{zx} is proportional to u. As we have shown before (see p. 114), shearing stresses τ_{xz} equal to τ_{zx} must act horizontally at points along the line nn_1 in the cross section of the flange. Hence the stresses τ_{xz} are distributed nonuniformly over the cross section of the flange but are proportional to the distance u. At the junction of flange and web the distribution of shearing stresses is complicated. In our approximate calculation we shall assume that eq. (a) holds from $u = 0$ to $u = b$. Then denoting by h the distance between the centroids of the flanges and observing that the moment of the cross section bt of the flange with respect to axis z is $bt(h/2)$, we obtain from eq. (a),

$$(\tau_{zx})_{\max} = (\tau_{xz})_{\max} = -\frac{Vbh}{2I_z}. \tag{b}$$

The resultant R (Fig. 210) of the shearing stresses τ_{xz} distributed over the cross-sectional area bt of the flange is

$$R = \frac{Vbh}{2I_z} \cdot \frac{bt}{2} = \frac{Vb^2ht}{4I_z}. \tag{c}$$

The sum of the shearing stresses τ_{xz} over the cross section of the upper flange will evidently be an equal and opposite force. Thus the shearing stresses over a channel section reduce to the forces shown in Fig. 210. This system of forces is statically equivalent to a force V applied at a point O at a distance from the center of the web:

$$e = \frac{Rh}{V} = \frac{b^2h^2t}{4I_z}. \qquad (d)$$

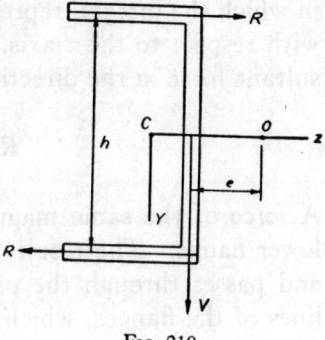

From this it is seen that in order to obtain simple bending with the z axis representing the neutral axis, the vertical plane in which the transverse loads act

Fig. 210.

must pass through the point O, which is called the *shear center*. At any other position of this plane, bending of the beam will be accompanied by twist, and the stresses will no longer follow the simple law in which σ_x is proportional to y and independent of the coordinate z.

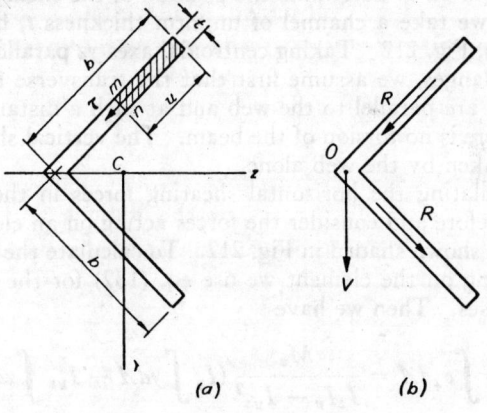

(a)   (b)

Fig. 211.

In the case of bending in a vertical plane of an angle section (Fig. 211), the shearing stress τ at points along mn will be in the direction shown and will be equal to [*]

$$\tau = \frac{V}{tI_z} \int y \, dA,$$

[*] The same method is used in calculating these stresses as in the case of channel sections.

in which the integral represents the moment of the shaded area with respect to the z axis. These shearing stresses yield a resultant force in the direction shown in Fig. 211*b* equal to

$$R = \frac{Vb^3t}{3I_z\sqrt{2}}.$$

A force of the same magnitude will also be obtained for the lower flange. The resultant of these two forces is equal to V and passes through the point of intersection O of the middle lines of the flanges, which is therefore the shear center in this case.

In the preceding cases we considered beams with one plane of symmetry which were deflected perpendicularly to that plane. In such a case the shear center is on the symmetry axis of the cross section and to determine its position we need only one coordinate. Let us consider now a nonsymmetrical beam for which two coordinates are required to determine the position of the shear center.[4] As an example we take a channel of uniform thickness t, but with unequal flanges, Fig. 212. Taking centroidal axes yz parallel to the web and to the flanges, we assume first that the transverse forces acting on the beam are parallel to the web and at such a distance from the web that there is no torsion of the beam. The vertical shearing force V_y will be taken by the web alone.

For calculating the horizontal shearing forces in the flanges we proceed as before and consider the forces acting on an element of the lower flange, shown shaded in Fig. 212. To calculate the longitudinal force N acting on the element we use eq. (132) for the longitudinal normal stresses. Then we have

$$N = \int \sigma_x dA = \frac{M_z}{I_z I_y - I_{yz}^2}\left(I_y \int y dA - I_{yz}\int z dA\right),$$

in which the integrals on the right-hand side represent the moments

---

[4] The problem of determining the shear center has been discussed by several authors. See, e.g., A. A. Griffith and G. I. Taylor, *Advisory Comm. Aeronaut.* (England), *Tech. Repts.*, Vol. 3, p. 950, 1917; R. Maillart, *Schweiz. Bauzeitung*, Vol. 77, p. 197, Vol. 79, p. 254 and Vol. 83, pp. 111 and 176; C. Weber, *Z. angew. Math. u. Mech.*, Vol. 4, p. 334, 1924; A. Eggenschwyler, *Proc. 2d Internat. Congr. Appl. Mech.*, Zürich, p. 434, 1926. See also writer's paper, *J. Franklin Inst.*, Vol. 239, p. 201, 1945. In recent times the problem has become of importance in airplane design. A review of the corresponding literature is given in a paper by P. Kuhn, *Nat. Advisory Comm. Aeronaut. Tech. Notes*, No. 691.

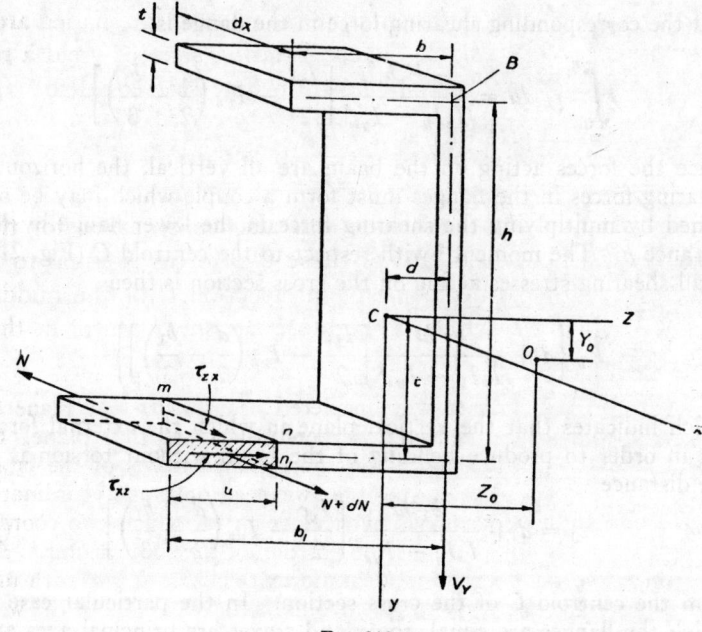

FIG. 212.

with respect to the z and y axes of the area shaded in Fig. 212, so that

$$\int y\,dA = ctu$$

and

$$\int z\,dA = -tu\left(b_1 - d - \frac{u}{2}\right) = t\left(\frac{u^2}{2} + du - b_1 u\right).$$

By differentiation we then obtain

$$\frac{dN}{dx}\,dx = \frac{V_y\,dx}{I_z I_y - I_{yz}^{2}}\left[I_y ctu - I_{yz}t\left(\frac{u^2}{2} + du - b_1 u\right)\right],$$

and the equation of equilibrium of the element under consideration is

$$t\tau_{zx}\,dx = \frac{-V_y\,dx}{I_z I_y - I_{yz}^{2}}\left[I_y ctu - I_{yz}t\left(\frac{u^2}{2} + du - b_1 u\right)\right].$$

The horizontal shearing stress on the cross section of the flange is then

$$\tau_{xz} = \tau_{zx} = \frac{-V_y}{I_z I_y - I_{yz}^{2}}\left[I_y cu - I_{yz}\left(\frac{u^2}{2} + du - b_1 u\right)\right].$$

and the corresponding shearing force in the flange is

$$t \int_0^{b_1} \tau_{xz} du = \frac{-V_y b_1{}^2 t}{I_z I_y - I_{yz}{}^2} \left[\frac{I_y c}{2} - I_{yz} \left(\frac{d}{2} - \frac{b_1}{3} \right) \right]. \qquad (e)$$

Since the forces acting on the beam are all vertical, the horizontal shearing forces in the flanges must form a couple which may be obtained by multiplying the shearing force in the lower flange by the distance h. The moment [5] with respect to the centroid C (Fig. 212) of all shearing stresses acting on the cross section is then

$$V_y \left\{ d + \frac{b_1{}^2 h t}{I_z I_y - I_{yz}{}^2} \left[\frac{I_y c}{2} - I_{yz} \left(\frac{d}{2} - \frac{b_1}{3} \right) \right] \right\},$$

which indicates that the vertical plane in which the external forces act in order to produce bending of the beam without torsion is at the distance

$$z_0 = d + \frac{b_1{}^2 h t}{I_z I_y - I_{yz}{}^2} \left[\frac{I_y c}{2} - I_{yz} \left(\frac{d}{2} - \frac{b_1}{3} \right) \right] \qquad (f)$$

from the centroid C of the cross section. In the particular case in which the flanges are equal, the y and z axes are principal axes and I_{yz} vanishes, $c = h/2$ and we obtain

$$z_0 = d + \frac{b_1{}^2 h^2 t}{4 I_z},$$

which is in agreement with the previously obtained result (d) (see p. 239).

The distance z_0 represents the horizontal coordinate of the shear center O. To calculate the coordinate y_0, we assume the external transverse forces to be acting in a plane parallel to the xz plane. For the calculations of the normal stresses σ_x we now use eq. (133). Considering again the element shown shaded in Fig. 212 and proceeding as before, we obtain

$$t \tau_{zx} = \frac{-V_z}{I_z I_y - I_{yz}{}^2} \left[I_{yz} c t u - I_z t \left(\frac{u^2}{2} + d u - b_1 u \right) \right].$$

The horizontal shearing force in the lower flange is

$$t \int_0^{b_1} \tau_{xz} du = \frac{-V_z b_1{}^2 t}{I_z I_y - I_{yz}{}^2} \left[\frac{I_{yz} c}{2} - I_z \left(\frac{d}{2} - \frac{b_1}{3} \right) \right].$$

[5] Clockwise moment is considered as positive.

Taking the moment of this force with respect to point B (Fig. 212) and dividing by the horizontal shearing force V_z, contributed by the external forces, we will obtain the distance f of the plane of the acting forces from point B. The required coordinate of the shear center will then be

$$y_0 = f + c - h = c - h + \frac{b_1^2 ht}{I_z I_y - I_{yz}^2}\left[\frac{I_{yz}c}{2} - I_z\left(\frac{d}{2} - \frac{b_1}{3}\right)\right].$$

(g)

For the particular case of equal flanges we have $I_{yz} = 0$, $c = h/2$, and by evaluating d and I_y it can be shown that eq. (g) vanishes and the shear center lies on the z axis. The coordinates y_0 and z_0 given by eqs. (f) and (g) completely define the position of the shear center for the channel shown in Fig. 212.

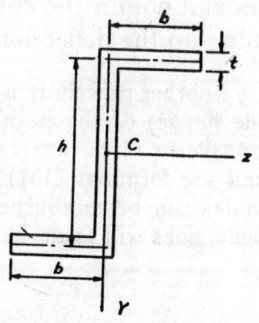

In the case of a Z section, Fig. 213, considering first the action of transverse forces in a plane parallel to the web and deriving an equation similar to eq. (e), it is readily proved that the horizontal shearing forces in the flanges vanish. Hence the horizontal coordinate z_0 of the shear center also vanishes. Next, considering transverse forces acting in a horizontal plane we will find that the horizontal shearing forces in the flanges

FIG. 213.

are equal to $\frac{1}{2}V_z$, indicating that the plane of horizontal loading must pass through the centroid C. Thus $y_0 = 0$ also and the shear center in this case coincides with the centroid C.

In all cases in which the flanges of a thin-walled beam intersect along one axis O, as in the examples shown in Fig.

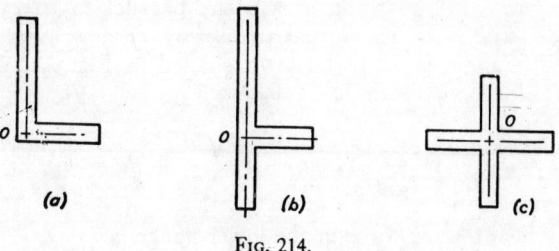

(a)   (b)   (c)

FIG. 214.

214, it will be found that the resultant shearing force passes through the same axis and this axis is evidently the shear center axis.

Returning now to the general case of bending of nonsymmetrical beams, we conclude from the preceding discussion that in order to have simple bending of a beam (bending without torsion) the external forces must be distributed along the shear center axis. For the calculation of deflections produced by these forces we can use the same methods as in the case of pure bending (see Art. 50). We can resolve each force into two components parallel to the principal centroidal axes of the cross section of the beam, investigate bending of the beam in each of the principal planes by applying the usual beam formulas and obtain the complete deflections by geometrically summing up the deflections found for the two principal planes.

Another procedure is to select y and z axes parallel to the web and the flanges of the beam as shown in Fig. 212, resolve each acting transverse force into two components parallel to the y and z axes and use formulas (131) for the forces in the xy plane. Similar formulas can be established for the forces in the xz plane. The final deflections will again be obtained by geometrical summation.

CHAPTER IX

COMBINED BENDING AND AXIAL LOAD.
THEORY OF COLUMNS

53. Bending Accompanied by Tension or Compression.—
It is assumed here that a prismatic bar is loaded by forces in
one of its planes of symmetry, but, whereas in the previous
discussion these forces were all transverse, here they may have
components along the axis of the bar. A simple case of this
kind is shown in Fig. 215, which represents a column loaded by
an inclined force P. This force is resolved into a transverse
component N and a longitudinal component T and it is assumed
that the column is comparatively stiff with a deflection so small
that it can be neglected in discussing the stresses produced by
the force T. Then the resultant stress at any point is obtained
by superposing the compressive stress due to the force T on
the bending stress produced by the
transverse load N. The case of a
flexible column in which the thrust,
due to deflection of the column (Fig.
215b), has a considerable effect on
the bending will be discussed later
(see Art. 56). The stress due to
force T is constant for all cross sections of the column and equal to
T/A, where A is the cross-sectional
area. The bending stress depends
upon the moment, which increases from zero at the top to a
maximum Nl at the bottom. Hence the dangerous section is
at the built-in end and the stress there, for a point at distance
y from the z axis, is

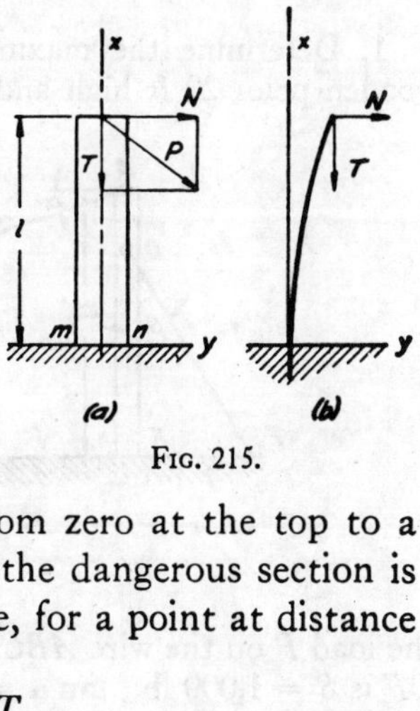

Fig. 215.

$$\sigma_x = -\frac{Nly}{I_z} - \frac{T}{A}. \qquad (a)$$

245

Assuming, for example, that the cross section of the column in Fig. 215a is a rectangle of dimensions $b \times h$ with the side h parallel to the plane of bending, we have $A = bh$ and $I_z = bh^3/12$. Then the maximum compressive stress will be at point n and is equal to

$$(\sigma_x)_{\min} = -\frac{6Nl}{bh^2} - \frac{T}{bh}. \qquad (b)$$

This stress is numerically the largest. At point m we obtain

$$(\sigma_x)_{\max} = \frac{6Nl}{bh^2} - \frac{T}{bh}.$$

When the force P is not parallel to one of the two planes of symmetry, the bending stresses produced by its transverse component N are found by resolving N into components parallel to those planes (see the discussion in Art. 51). The resultant stress at any point is obtained by superposing these bending stresses with the compressive stress produced by the longitudinal force.

Problems

1. Determine the maximum compressive stress in the circular wooden poles 20 ft high and 8 in. in diameter shown in Fig. 216, if

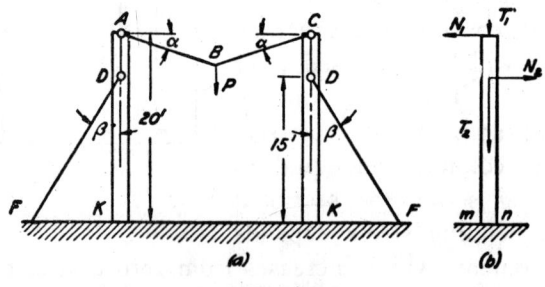

Fig. 216.

the load P on the wire ABC is 60 lb. The tensile force in each cable DF is $S = 1,000$ lb.; $\tan \alpha = \frac{1}{10}$; $\sin \beta = \frac{1}{5}$ and $\overline{DK} = 15$ ft.

Solution. The components of the force in the wire BC (Fig. 216b) are $N_1 = 300$ lb, $T_1 = 30$ lb. The components of the force in the cable DF are $N_2 = 200$ lb, $T_2 = 980$ lb. The maximum bending moment is found to be at the built-in end, where $M_{max} = 36,000$ in. lb. The thrust at the same cross section is $T_1 + T_2 = 1,010$ lb. The maximum compressive stress at the point m is

$$\sigma = \frac{4 \times 1,010}{\pi d^2} + \frac{32 \times 36,000}{\pi d^3} = 21 + 715 = 736 \text{ lb per sq in.}$$

2. Determine the maximum tensile stress in the rectangular wooden beam shown in Fig. 217 if $S = 4,000$ lb, $b = 8$ in., $h = 10$ in.

Answer. $(\sigma_x)_{max} = \dfrac{6 \times 72 \times 1,000}{8 \times 100} + \dfrac{4,000}{80} = 590$ lb per sq in.

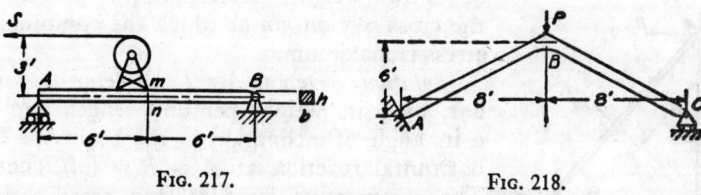

FIG. 217.   FIG. 218.

3. Determine the maximum compressive stress in the structure ABC, which supports a load $P = 2,000$ lb (Fig. 218), has a rigid connection between the bars at B, an immovable hinge at A and a movable support at C. The cross section of the bars AB and BC is a square 10 × 10 in.

Answer. $\sigma = \dfrac{6 \times 1,000 \times 8 \times 12}{10^3} + \dfrac{600}{10^2}$

$= 582$ lb per sq in.

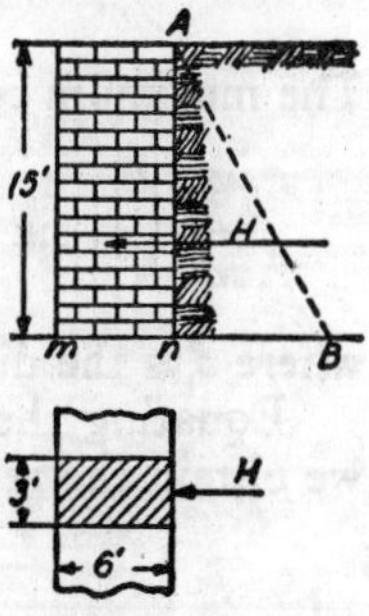

4. A brick wall 6 ft thick and 15 ft high supports sand pressure (Fig. 219). Determine the maximum tensile and compressive stresses at the bottom of the wall if its weight is $\gamma = 150$ lb per cu ft and the lateral pressure of the sand is 10,000 lb per yd of wall. The distribution of the sand pressure along the height of the wall follows a linear law, given by the line AB.

FIG. 219.

Answer. The stress at $m = -\dfrac{150 \times 15}{144} - \dfrac{10,000 \times 60 \times 6}{36 \times 72^2}$

$= -15.6 - 19.3 = -34.9$ lb per sq in. The stress at $n =$

$-\dfrac{150 \times 15}{144} + \dfrac{10,000 \times 60 \times 6}{36 \times 72^2} = 3.7$ lb per sq in.

5. Determine the required thickness of the wall in the previous problem to give zero stress at n.

Answer.   80 in.

6. A circular column 6 ft high, Fig. 215, is acted upon by a force P which has components N and T equal to 1,000 lb each. Find the diameter of the column if the maximum compressive stress is 1,000 lb per sq in.

7. Find σ_{max} and σ_{min} at the cross section at the middle of the bar BC (Fig. 218) if, instead of the concentrated load P, a uniform vertical load $q = 400$ lb per ft is distributed along the axis ABC.

8. A circular bar AB (Fig. 220) hinged at B and supported by a smooth vertical surface (no friction) at A is submitted to the action of its own weight. Determine the position of the cross section mn at which the compressive stress is maximum.

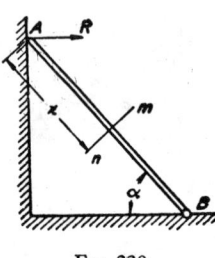

Solution. Denote by l the length of the bar, by q its weight per unit length and by α its angle of inclination to the horizon. The horizontal reaction at A is $R = (ql/2) \cot \alpha$. The compressive force at any cross section mn, distance x from A, is $qx \sin \alpha + (ql/2) \times (\cos^2 \alpha / \sin \alpha)$. The bending moment at the same cross section is

FIG. 220.

$$M = \frac{ql}{2} x \cos \alpha - \frac{qx^2 \cos \alpha}{2}.$$

The maximum compressive stress at the cross section mn is

$$\frac{4}{\pi d^2}\left(q_x \sin \alpha + \frac{ql}{2}\frac{\cos^2 \alpha}{\sin \alpha}\right) + \frac{32}{\pi d^3}\left(\frac{ql}{2} x \cos \alpha - \frac{qx^2 \cos \alpha}{2}\right),$$

where d is the diameter of the bar.

Equating the derivative of this stress with respect to x to zero, we obtain the required distance

$$x = \frac{l}{2} + \frac{d}{8} \tan \alpha.$$

9. The bar shown in Fig. 215 is 6 ft long and has a circular cross section of 12 in. diameter. Determine the magnitude of the force P if its components N and T are equal and the maximum compressive stress at n is 1,000 lb per sq in.

Answer. $P = 3,260$ lb.

10. A force P produces bending of the bar ABC built in at A (Fig. 221). Determine the angle of rotation of the end C, during

bending, if the bending moments at A and at B are numerically equal.

Solution. From the equality of the bending moments at A and B, it follows that the force P passes through the mid-point D of the bar AB. Then $P_x = P_y l/2a$ and the components P_x and P_y may now be calculated. The rotation of the cross section B due to bending of the portion AB by the component P_y is $P_y l^2/2EI$ in a clockwise direction. The rotation of the same cross section due to the component P_x is $P_x al/EI$ in a counter-clockwise direction. The rotation of the cross section C with

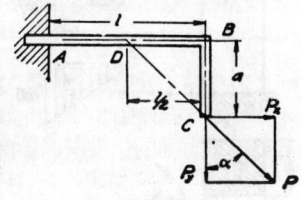

Fig. 221.

respect to the cross section B, due to bending of the portion BC of the bar, is $P_x a^2/2EI$ in a counter-clockwise direction. The total angle of rotation of the end C in a clockwise direction is

$$\frac{P_y l^2}{2EI} - \frac{P_x al}{EI} - \frac{P_x a^2}{2EI} = -\frac{P_x a^2}{2EI}.$$

11. A three-hinged frame ABC (Fig. 222) supports a vertical load P. Find the numerically largest bending moment $M_{\max}$ in the frame and the compressive force N in the horizontal bars.

Answer. $M_{\max} = \dfrac{Pl}{4}, N = \dfrac{Pl}{4h}.$

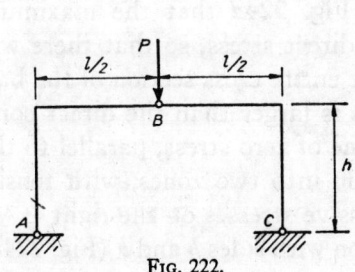

Fig. 222.

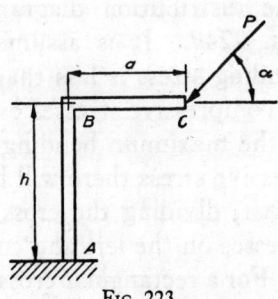

Fig. 223.

12. Find the angle of inclination α of the force P acting on the bar ABC (Fig. 223) if it is known that the deflection at B is zero.

Answer. $\tan \alpha = \dfrac{2h}{3a}.$

54. Eccentric Loading of a Short Strut.

Eccentric loading is a particular case of the combination of direct and bending stresses. When the length of the bar is not very large in com-

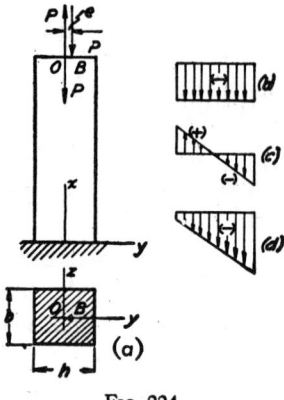

FIG. 224.

parison with its lateral dimensions, its deflection is so small that it can be neglected in comparison with the initial eccentricity e, and the method of superposition may be used.[1] Take, for example, the case of compression by a longitudinal force P applied with an eccentricity e (Fig. 224) on one of the two principal axes of the cross section. If we put two equal and opposite forces P at the centroid O of the cross section, the condition is not changed because they are equivalent to zero. We then obtain an axial compression by the force P producing direct compressive stresses $-P/A$, as shown in Fig. 224b, and bending in one of the principal planes by the couple Pe producing bending stresses $-(Pey/I_z)$, as shown in Fig. 224c. The total stress is then

$$\sigma_x = -\frac{P}{A} - \frac{Pey}{I_z}. \qquad (a)$$

The distribution diagram of this total stress is shown in Fig. 224d. It is assumed in Fig. 224d that the maximum bending stress is less than the direct stress, so that there will be compressive stresses over the entire cross section of the bar. If the maximum bending stress is larger than the direct compressive stress there will be a line of zero stress, parallel to the z axis, dividing the cross section into two zones, with tensile stresses on the left and compressive stresses on the right.

For a rectangular cross section with sides h and b (Fig. 224a) eq. (a) becomes

$$\sigma_x = -\frac{P}{bh} - \frac{12Pey}{bh^3} \qquad (a')$$

and we obtain, by putting $y = -(h/2)$,

$$(\sigma_x)_{\max} = -\frac{P}{bh} + \frac{6Pe}{bh^2} = \frac{P}{bh}\left(-1 + \frac{6e}{h}\right). \qquad (b)$$

---

[1] For the case of eccentric loading of long bars see Art. 56.

By putting $y = h/2$, we obtain

$$(\sigma_x)_{\min} = -\frac{P}{bh} - \frac{6Pe}{bh^2} = -\frac{P}{bh}\left(1 + \frac{6e}{h}\right). \qquad (c)$$

It may be seen that when $e < h/6$ there is no reversal of sign of the stresses over the cross section. When $e = h/6$, the maximum compressive stress, from eq. (c), is $2P/bh$ and the stress on the opposite side of the rectangular cross section is zero. When $e > h/6$, there is a reversal of sign of the stress and the position of the line of zero stress is obtained by equating to zero the general expression (a') for σ_x, giving

$$v = -\frac{h^2}{12e}, \qquad (d)$$

or, using the notation k_z for the *radius of gyration* with respect to the z axis (see Appendix),

$$v = -\frac{k_z^2}{e}. \qquad (134)$$

It will be seen that the distance of the line of zero stress from the centroid O diminishes as the eccentricity e increases. The same discussion applies as well to the case of eccentric loading in tension. Eq. (134) may also be used for other shapes of cross sections if the point of application of the load is on one of the principal axes of inertia.

Let us consider now the case in which B, the point of application of the eccentric compressive force P, is not on one of the two principal axes of the cross section, taken as the y and the z axes in Fig. 225. Using m and n as the coordinates of this point, the moments of P with respect to the y and z axes are Pn and Pm, respectively. By superposition, the stress at any point F of the cross section is

$$\sigma_x = -\frac{P}{A} - \frac{Pmy}{I_z} - \frac{Pnz}{I_y}, \qquad (e)$$

in which the first term on the right side represents the direct stress and the two other terms are the bending stresses pro-

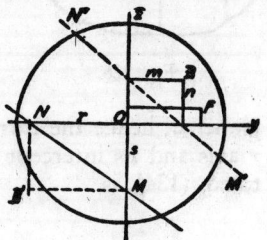

Fig. 225.

duced by the moments Pm and Pn, respectively. It may be seen that the stress distribution follows a linear law. The equation of the line of zero stress is obtained by equating the right side of eq. (e) to zero. Using the notation $I_z/A = k_z{}^2$ and $I_y/A = k_y{}^2$, where k_z and k_y are the radii of gyration with respect to the z and y axes, respectively, this gives

$$\frac{my}{k_z{}^2} + \frac{nz}{k_y{}^2} + 1 = 0. \qquad (f)$$

By substituting in this equation first $y = 0$ and then $z = 0$, we obtain the points M and N of intersection of the line of zero stress with the axes of coordinates z and y (Fig. 225). The coordinates of these points, s and r, are

$$s = -\frac{k_y{}^2}{n}, \qquad r = -\frac{k_z{}^2}{m}. \qquad (g)$$

From these equations we obtain

$$n = -\frac{k_y{}^2}{s}, \qquad m = -\frac{k_z{}^2}{r}.$$

These equations have the same form as eqs. (g) and it can be concluded that when the load is put at the point B' with the coordinates s and r, the corresponding line of zero stress will be the line $N'M'$, indicated in the figure by the dotted line, and cutting off from the y and z axes the lengths m and n.

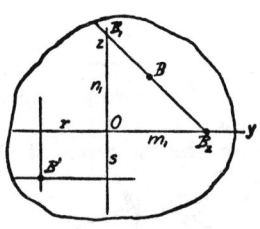

FIG. 226.

There is another important relation between the point of application B of the load and the position of the corresponding line of zero stress, namely, as B moves along a line B_1B_2 (Fig. 226), the corresponding line of zero stress turns about a certain constant point B'. This is proved as follows: Resolve the load at B into two parallel components, one at B_1 and the other at B_2. The component at B_1 acts in the principal plane xz; hence the corresponding line of zero stress is parallel to the y axis and its intercept on Oz, as found from an equation analogous to eq. (134), is

$$s = -\frac{k_y{}^2}{n_1}. \qquad (h)$$

Similarly the line of zero stress for the component B_2 is parallel to the z axis and its distance from this axis is

$$r = -\frac{k_z{}^2}{m_1}. \qquad (i)$$

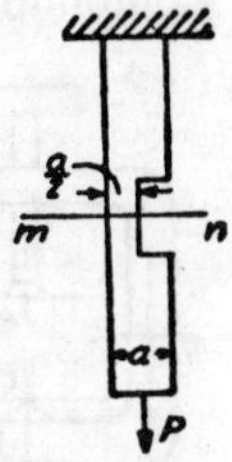

For any position of the load on the line B_1B_2 there will be zero stress at B'. Hence, as the point of application of the load moves along the straight line B_1B_2, the corresponding line of zero stress turns about the point B', the coordinates of which are determined by eqs. (h) and (i).

Fig. 227.

Problems

1. The cross-sectional area of a square bar is reduced by one half at mn (Fig. 227). Determine the maximum tensile stress at cross section mn produced by an axial load P.

Answer. $(\sigma_x)_{\max} = \dfrac{2P}{a^2} + \dfrac{Pa}{4}\dfrac{24}{a^3} = \dfrac{8P}{a^2}.$

2. Solve the above problem, assuming the bar to have a circular cross section.

3. A bar of $\perp$ section is eccentrically loaded by the forces P (Fig. 228). Determine the maximum tensile and compressive stresses in this bar if $d = 1$ in., $h = 5$ in., the width of the flange $b = 5$ in., $P = 4,000$ lb.

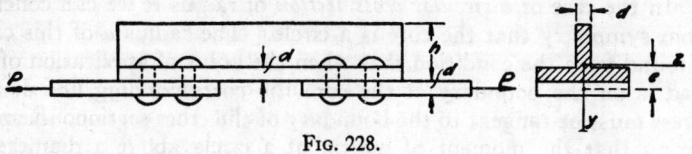

Fig. 228.

Solution. The distances of the centroid of the $\perp$ section from the bottom and the top are, respectively, $h_1 = \frac{29}{18}$ in. and $h_2 = \frac{61}{18}$ in. The eccentricity of the force P is $e = \frac{1}{2} + \frac{29}{18} = 2\frac{1}{9}$ in. The moment of inertia $I_z = 19.64$ in.[4] The bending stresses are

$$(\sigma_x)_{\max} = \frac{Peh_1}{I_z} = \frac{4,000 \times 2\frac{1}{9} \times 29}{19.64 \times 18} = 693 \text{ lb per sq in.},$$

$$(\sigma_x)_{\min} = -\frac{Peh_2}{I_z} = -\frac{4,000 \times 2\frac{1}{9} \times 61}{19.64 \times 18} = -1,458 \text{ lb per sq in.}$$

Combining with the direct stress $P/A = 4,000/9 = 444$ lb per sq in., we obtain the maximum tensile stress $693 + 444 = 1,137$ lb per sq in., maximum compressive stress $1,458 - 444 = 1,014$ lb per sq in.

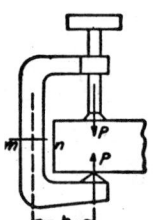

FIG. 229.

4. Determine the maximum tensile stress at the section mn of the clamp shown in Fig. 229 if $P = 300$ lb, $b = 3$ in. and the cross section is a rectangle with the dimensions $1 \times \frac{1}{4}$ in.

Answer. $\sigma_{max} = 22,800$ lb per sq in.

5. Determine the width of the cross section mn in the previous problem to make $\sigma_{max} = 20,000$ lb per sq in.

6. Find the maximum and the minimum stress at the built-in cross section of the rectangular column shown in Fig. 224, if $b = 10$ in., $h = 12$ in., $P = 5,000$ lb and the coordinates of the point B of application of the load (Fig. 225) are $m = n = 2$ in. Find the position of the neutral axis.

55. The Core of a Section.—In the previous article it was shown that for a small eccentricity e the normal stresses have the same sign over all of the cross section of an eccentrically loaded bar. For larger values of e the line of zero stress cuts the cross section and there is a reversal of sign of the stress. In the case of a material very weak in tension, such as concrete or brick work, the question arises of determining the region in which the compressive load may be applied without producing any tensile stress on the cross section. This region is called the *core of the cross section*. The method of determining the core is illustrated in the following simple examples.

In the case of a *circular cross section* of radius R we can conclude from symmetry that the core is a circle. The radius a of this circle is found from the condition that when the point of application of the load is on the boundary of the core the corresponding line of zero stress must be tangent to the boundary of the cross section. Remembering that the moment of inertia of a circle about a diameter is $\pi R^4/4$ (see Appendix), and hence the radius of gyration is $k = \sqrt{I/A} = R/2$, we find from eq. (134) (p. 251), by substituting a for e and R for $-y$, that

$$a = \frac{k^2}{R} = \frac{R}{4},$$ (135)

i.e., the radius of the core is one quarter of the radius of the cross section.

For the case of a *circular ring section* with outer radius R_o and inner radius R_i we have

$$I = \frac{\pi}{4}(K_o^4 - R_i^4), \qquad k^2 = \frac{I}{A} = \frac{R_o^2 + R_i^2}{4},$$

and the radius of the core, from eq. (134), becomes

$$a = \frac{k^2}{R_o} = \frac{R_o{}^2 + R_i{}^2}{4R_o}. \qquad (136)$$

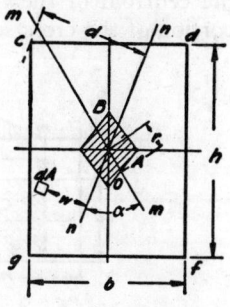

FIG. 230.

For $R_i = 0$, eq. (136) coincides with eq. (135). For a very narrow ring, when R_i approaches R_o the radius a of the core approaches the value $R_o/2$.

In the case of a *rectangular cross section* (Fig. 230), the line of zero stress coincides with the side cg when the load is applied at point A, a distance $b/6$ from the centroid (see p. 251). In the same manner the line of zero stress coincides with the side gf when the load is at the point B, a distance $h/6$ from the centroid. As the load moves along the line AB, the neutral axis rotates about the point g (see p. 252) without cutting the cross section. Hence AB is one of the sides of the core. The other sides follow from symmetry. The core is therefore a rhombus with diagonals equal to $h/3$ and $b/3$. As long as the point of application of the load remains within this rhombus the line of zero stress does not cut the cross section and there will be no reversal in the sign of the stress.

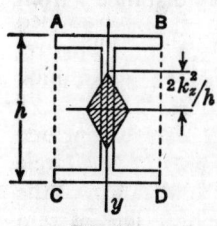

FIG. 231.

For an I section (Fig. 231) the extreme positions of the lines of zero stress, in which they do not cut the cross section, are given by the sides AB and CD and by the dotted lines AC and BD. The corresponding positions of the point of application of the load may be determined from eq. (134). From symmetry it may be concluded that these points will be the corners of a rhombus, shaded in Fig. 231.

If the point of application of the eccentric load is outside the core of a cross section, the corresponding line of zero stress crosses the section and the load produces not only compressive but also tensile stresses. If the material does not resist tensile stresses at all, part of the cross section will be inactive and the rest will carry compressive stresses only. Take, for example, a rectangular cross section (Fig. 232) with the point of application A of the load on the principal axis y and at a distance c from the edge of the section. If c is less than $h/3$, part of the cross section will be inactive. The working portion may be found from the condition that the distribution of the compressive forces over the cross section follows a linear law, represented in the figure by the line mn, and that the resultant of these forces is P. Since this resultant must pass through the centroid of the triangle *mns*, the dimension *ms* of the working portion of the cross section must be equal to $3c$.

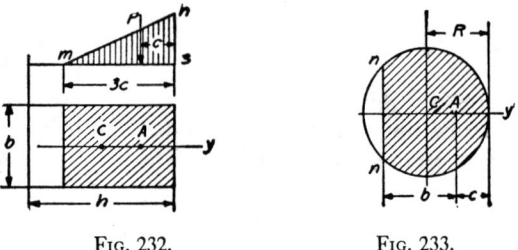

FIG. 232.   FIG. 233.

In the case of a circular cross section (Fig. 233), if the eccentricity CA of the load is larger than $R/4$ and the material does not resist tensile stresses, only a portion of the cross section will work. Let the line *nn*, perpendicular to AC, be the limit of this portion. Its distance b from the point A may be found from the conditions that (1) the compressive stresses are proportional to the distance y from *nn*, (2) the sum of the compressive forces over the working portion of the cross section is equal to the load P and (3) the moment of these forces with respect to *nn* is equal to the moment Pb of the load P with respect to the same axis. Denoting the maximum compressive stress by $\sigma_{\max}$, the compressive stress at any distance y from *nn* is

$$\sigma = \frac{y\sigma_{\max}}{b + c}$$

and the equations for determining b become

$$\int \frac{y\sigma_{\max}}{b + c} dA = P, \qquad \int \frac{y^2\sigma_{\max}}{b + c} dA = Pb,$$

from which

$$b = \frac{I_{nn}}{Q_{nn}}, \tag{a}$$

in which $I_{nn} = \int y^2 dA$ is the moment of inertia of the working portion of the cross section with respect to the *nn* axis and $Q_{nn} = \int y dA$ is the moment of the working portion of the cross section with respect to the same axis. By using eq. (a) the position of A for any given position of *nn* may easily be found. The same equation may also be used for other shapes of cross sections, provided A is on one of the principal axes.[2] If the load is not on a principal axis, the problem of determining the working portion of the cross section becomes more complicated.[3]

By using the notion of the core of a section, the calculation of maximum bending stresses when the bending is not in a principal plane may be greatly simplified. For example, in Fig. 230 let mm be the axial plane of the beam in which a bending moment M acts and nn, the corresponding neutral axis which makes an angle α with the plane mm (see p. 229). Denoting by σ_{max} the maximum stress in the most remote point c and by d its distance from the neutral axis nn, the stress at any other point, distance w from nn, is $\sigma = \sigma_{max}w/d$, and the moment of all forces distributed over the cross section with respect to the axis nn is

$$\frac{\sigma_{max}w^2}{d}\,dA = \frac{\sigma_{max}}{d}I_{nn}, \qquad (b)$$

in which I_{nn} is the moment of inertia of the cross section with respect to the nn axis. The moment of the external forces with respect to the same axis is $M \sin \alpha$. Equating this to (b), we have

$$\sigma_{max} = \frac{Md \sin \alpha}{I_{nn}}. \qquad (c)$$

This equation may be greatly simplified by using the property of the core of the cross section.[4] Let O be the point of intersection of the plane mm with the core and r its distance from the centroid of the cross section. From the property of the core it follows that a compressive force P at O produces zero stress at the corner c. Hence the tensile stress produced at c by the bending moment Pr, acting in the plane mm, is numerically equal to the direct compressive stress P/A; or, substituting Pr for M in eq. (c), we obtain

$$\frac{P}{A} = \frac{Prd \sin \alpha}{I_{nn}},$$

---

[2] For the cases of circular cross sections and circular ring sections, which are of importance in calculating stresses in chimneys, tables have been published which simplify these calculations. See Keck, *Z. Architekt. u. Ing.- Ver. (Hannover)*, p. 627, 1882; see also *Z. Ver. deut. Ing.*, p. 1321, 1902, and paper by G. Dreyer, in *Bautechnik*, 1925.

[3] Some calculations for a rectangular cross section will be found in the following papers: F. Engesser, *Zentr. Bauverwalt*, p. 429, 1919; O. Henkel, *ibid.*, p. 447, 1918; K. Pohl, *Der Eisenbau*, p. 211, 1918; F. K. Esling, *Proc. Inst. Civil Engrs.*, (London), Part 3, 1905–6.

[4] See R. Land, *Z. Architekt. u. Ingenieurw.*, p. 291, 1897.

from which

$$\frac{d \sin \alpha}{I_{nn}} = \frac{1}{Ar}. \qquad (d)$$

Substituting this into eq. (c), we obtain

$$\sigma_{max} = \frac{M}{Ar}. \qquad (137)$$

The product Ar is called the *section modulus* of the cross section in the plane *mm*. This definition agrees with the definition which we had previously (see p. 96), and for bending in a principal plane Ar becomes equal to Z.

Problems

1. Determine the core of a standard I beam of 24 in. depth, for which $A = 23.33$ sq in., $I_z = 2,087$ in.^{4}, $k_z = 9.46$ in., $I_y = 42.9$ in.^{4}, $k_y = 1.36$ in. The width of the flanges $b = 7$ in.

   Answer. The core is a rhombus with diagonals equal to 14.9 in. and 1.06 in.

2. Determine the radius of the core of a circular ring section if $R_o = 10$ in. and $R_i = 8$ in.

   Answer. The radius of the core $a = 4.10$ in.

3. Determine the core of a cross section in the form of an equilateral triangle.

4. Determine the core of the cross section of a square thin tube.

   Solution. If h is the thickness of the tube and b the side of the square cross section, we have

$$I_z = I_y \approx \frac{2}{3} hb^3, \qquad k_z{}^2 = k_u{}^2 = \frac{b^2}{6}.$$

The core is a square with diagonal

$$d = 2 \frac{k^2}{\frac{1}{2}b} = \frac{2b}{3}.$$

56. Eccentric Compression of a Slender Symmetrical Column.—In discussing the bending of a slender column under the action of an eccentric load, Fig. 234, the deflection δ can no longer be neglected as being small in comparison with the eccentricity e. Assuming that the eccentricity is in the plane of symmetry, the deflection occurs in the same axial plane xy

in which the load P acts, and the bending
moment at any cross section mn is

$$M = -P(\delta + e - y). \qquad (a)$$

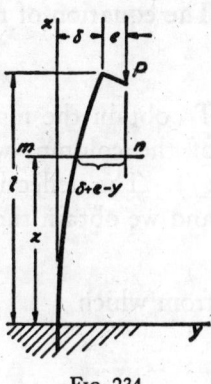

In determining the sign of the moment it
should be noted that by rotating Fig. 234 in
the clockwise direction by an angle $\pi/2$, the
same directions of the coordinate axes are
obtained as those used in deriving eq. (79).
Hence, to follow the rule shown in Fig. 63b,
the moment (a) is taken with a minus sign
since the deflection curve is concave in the

Fig. 234.

positive direction of the y axis. The differential equation of
the deflection curve obtained by substituting (a) in eq. (79) is

$$EI_z \frac{d^2y}{dx^2} = P(\delta + e - v). \qquad (b)$$

Using the notation

$$\frac{P}{EI_z} = p^2, \qquad (138)$$

we obtain from eq. (b)

$$\frac{d^2y}{dx^2} + p^2y = p^2(\delta + e). \qquad (c)$$

By substitution, it can be readily proved that

$$y = C_1 \sin px + C_2 \cos px + \delta + e \qquad (d)$$

is the solution of eq. (c). This solution contains two constants
of integration, C_1 and C_2, whose magnitudes must be adjusted
so as to satisfy the conditions at the ends of the column if we
are to obtain the true deflection curve of the column. At the
lower end, which is built in, the conditions are

$$(y)_{x=0} = 0, \qquad \left(\frac{dy}{dx}\right)_{x=0} = 0. \qquad (e)$$

Using these conditions together with expression (d) and its
first derivative, we obtain

$$C_1 = 0, \qquad C_2 = -(\delta + e).$$

The equation of the deflection curve (d) thus becomes

$$y = (\delta + e)(1 - \cos px). \tag{f}$$

To obtain the magnitude of the deflection δ at the upper end of the column, we substitute $x = l$ in the right side of eq. (f). The deflection y on the left side must then be equal to δ and we obtain the equation

$$\delta = (\delta + e)(1 - \cos pl),$$

from which

$$\delta = \frac{e(1 - \cos pl)}{\cos pl}. \tag{139}$$

Substituting this into eq. (f), we obtain the deflection curve

$$y = \frac{e(1 - \cos px)}{\cos pl}. \tag{140}$$

By using this equation the deflection at any cross section of the bar can readily be calculated.

In the case of short columns, which were considered in Art. 54, the quantity pl is small in comparison with unity and it is sufficiently accurate to take

$$\cos pl \approx 1 - \tfrac{1}{2}p^2 l^2. \tag{g}$$

Using this value of $\cos pl$ and neglecting the quantity $p^2 l^2 / 2$ in the denominator of expression (139), as being small in comparison with unity, we obtain

$$\delta = \frac{ep^2 l^2}{2} = \frac{Pel^2}{2EI_z}. \tag{h}$$

This represents the magnitude of the deflection at the end of a cantilever bent by a couple Pe applied at the end. Hence the use of the approximate expression (g) is equivalent to neglecting the effect of the deflections upon the magnitude of the bending moment and taking instead a constant moment equal to Pe.

If pl is not small, as is usually the case when the column is slender, we must use expression (139) in calculating δ. In this way we find that the deflection is no longer proportional to the load P. Instead, it increases more rapidly than P, as is seen from the values of this deflection as given in the second line of Table 2.

TABLE 2: DEFLECTIONS PRODUCED BY AN ECCENTRIC LONGITUDINAL LOAD

pl...............	0.1	0.5	1.0	1.5	$\pi/2$
δ................	$0.005e$	$0.139e$	$0.851e$	$13.1e$	∞
Approximate δ......	$0.005e$	$0.139e$	$0.840e$	$12.8e$	∞
sec pl	1.005	1.140	1.867	13.2	∞
P/P_{cr}.............	0.004	0.101	0.405	0.911	1

The maximum bending moment occurs at the built-in end of the column and has a magnitude

$$M_{\max} = P(e + \delta) = Pe \sec pl. \qquad (141)$$

A series of values of sec pl is given in the fourth line of the above table. These values show how rapidly the moment increases as pl approaches the value $\pi/2$. This phenomenon will be discussed in the next article.

Here, however, it should be repeated that in the case under discussion there is no proportionality between the magnitude of the compressive force and the deflection δ which it produces. Hence the method of superposition (p. 162) cannot be used here. An axially applied force P produces only compression of the bar, but when the same force acts in conjunction with a bending couple, Pe, it produces not only compression but also additional bending, so that the resulting deformation cannot be obtained by simple superposition of an axial compression due to the force P and a bending due to the couple Pe. The reason why in this case the method of superposition is not applicable can readily be seen if we compare this problem with the bending of a beam by transverse loads. In the latter case, it can be assumed that small deflections of the beam do not change the distances between the forces and the bending moments can be calculated without considering the deflection of the beam. In the case of eccentric compression of a column, the deflections produced by the couple Pe entirely change the character of the action of the axial load by causing it to have a bending action as well as a compressive action. In each case in which the deformation produced by one load changes the action of the other load, it will be found that the final deformation cannot be obtained by the method of superposition.

In the previous discussion bending in a plane of symmetry of the column was considered. If the column has two planes of symmetry and the eccentricity e is not in the direction of one of the principal axes of the cross section, it is necessary to resolve the bending couple Pe into two component couples each

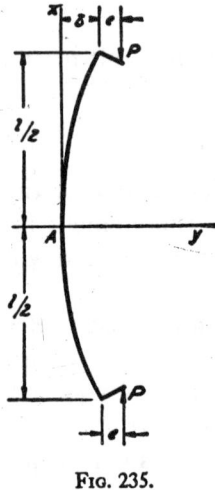

acting in a plane of symmetry of the column. The deflection in each of the two planes of symmetry can then be investigated in the same manner as discussed above.

The preceding discussion of bending of a column built in at one end can also be applied to the case of a strut eccentrically compressed by two equal and opposite forces P, Fig. 235. From symmetry it can be concluded that the middle cross section A does not rotate during bending and each half of the strut in Fig. 235 is in exactly the same condition as the strut in

FIG. 235.

Fig. 234. Hence the deflection and the maximum bending moment are obtained by substituting $l/2$ for l in eqs. (139) and (141). In this way we obtain

$$\delta = \frac{e\left(1 - \cos\dfrac{pl}{2}\right)}{\cos\dfrac{pl}{2}}, \qquad (142)$$

$$M_{max} = Pe \sec \frac{pl}{2}, \qquad (143)$$

and the equation for the maximum compressive stress becomes

$$\sigma_{max} = \frac{P}{A} + \frac{Pe}{Z} \sec \frac{pl}{2},$$

where Z denotes the section modulus.

Problems

1. Find the deflection at the middle and the maximum tensile and compressive stresses in an eccentrically compressed steel strut 10 ft long with hinged ends if the cross section is an 8 ⌐ 11.5 channel with depth 8 in., $I_z = 1.3$ in.^{4}, $I_y = 32.3$ in.^{4}, $A = 3.36$ in.^{2} and the width of the flanges 2.26 in. The distance between the centroid and the back of the channel is 0.58 in., and the compressive force $P = 4,000$ lb acts in the plane of the back of the channel and in the symmetry plane of the channel.

Answer. $\delta = 0.126$ in., $\sigma_{max} = 2,460$ lb per sq in. tension, $\sigma_{min} = 2,450$ lb per sq in. compression.

2. A square steel bar 2×2 in. and 6 ft long is eccentrically compressed by forces $P = 1,000$ lb. The eccentricity e is directed along a diagonal of the square and is equal to 1 in. Find the maximum compressive stress, assuming that the ends of the bar are hinged.

Answer. $\sigma = 1,330$ lb per sq in.

3. A steel bar 4 ft long and having a rectangular cross section 1×2 in. is compressed by two forces $P = 1,000$ lb applied at the corners of the end cross sections so that the eccentricity is in the direction of a diagonal of the cross section and is equal to one-half the length of the diagonal. Considering the ends to be hinged, find the maximum compressive stress.

Answer. $\sigma = 3,610$ lb per sq in.

57. Critical Load.—It was indicated in the preceding article that the deflection of an eccentrically compressed column increases very rapidly as the quantity pl in eq. (139) approaches the value $\pi/2$. When pl becomes equal to $\pi/2$, the formulas (139) for the deflections and (141) for the maximum bending moment both give infinite values. To find the corresponding value of the load we use formula (138). Substituting $p = \pi/2l$ in this expression, we find that the value of the load at which expressions (139) and (141) become infinitely large is

$$P_{cr} = \frac{\pi^2 E I_z}{4l^2}. \tag{144}$$

This value, which depends only on the dimensions of the column and on the modulus of the material, is called the *critical load* or Euler's load, since Euler was the first to derive its value in his famous study of elastic curves.[5] To see more clearly the

---

[5] *Loc. cit.*, p. 139.

physical significance of this load let us plot curves representing the relation between the load P and the deflection δ as given by eq. (139). Several curves of this kind, for various values of the eccentricity ratio e/k_z, are shown in Fig. 236. The abscissas of these curves are the values of the ratio δ/k_z while the ordinates are the ratio P/P_{cr}, i.e., the values of the ratio of the acting load to its critical value defined by eq. (144).

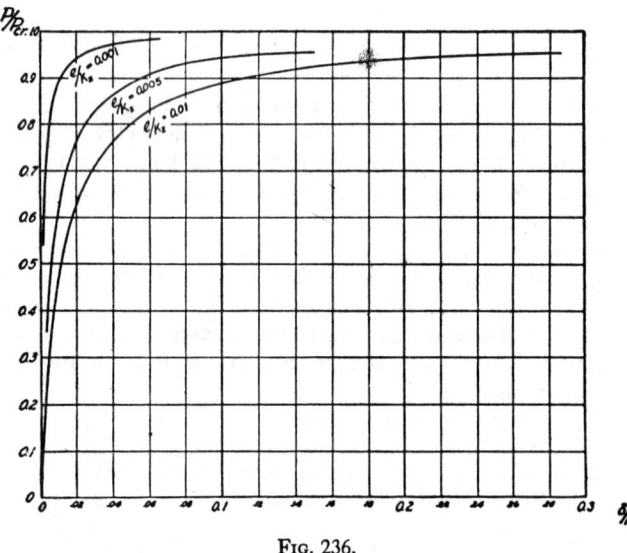

Fig. 236.

It is seen from the curves that the deflections δ become smaller and smaller and the curves approach closer and closer to the vertical axis as the eccentricity e decreases. At the same time the deflections increase rapidly as the load P approaches its critical value (144), and all the curves have as their asymptote the horizontal line $P/P_{cr} = 1$.

The differential equation of the deflection curve (79), which was used in the discussion of the preceding article, was derived on the assumption that the deflections are small in comparison with the length l of the column. Hence formula (139) for the deflection δ cannot give us an accurate result when P is very close to P_{cr}. However, the curves in Fig. 236 indicate that, irrespective of how small the eccentricity e may be, very large deflections are produced if the load P is sufficiently close to its critical value. If the deflection becomes large, the bending moment at the built-in end and the stresses are also large.

Experiments dealing with the compression of columns show that even when all practicable precautions are taken to apply the load centrally there are always some unavoidable small eccentricities. Consequently in such experiments the load P produces not only compression but also bending. The

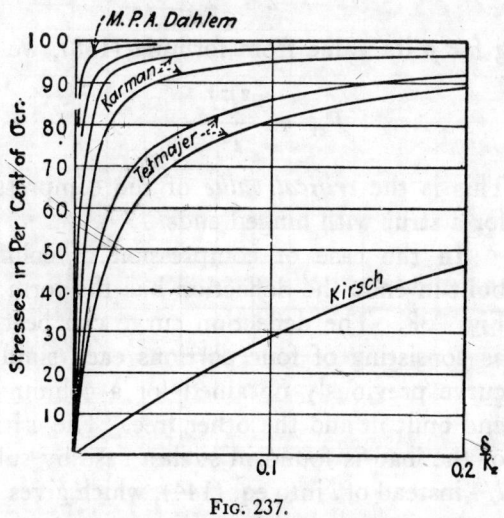

FIG. 237.

curves in Fig. 237 show the results of such experiments as obtained by several experimenters. It may be seen that, with increasing accuracy in the application of the load, the curves come closer and closer to the vertical axis and the rapid increase in the deflection as the load approaches its critical value becomes more and more pronounced. The loads P which are close to their critical values always produce large deformations which usually go beyond the elastic limit of the material, so that after such a loading the column loses its practical usefulness. This indicates that the critical value of the load, as given by eq. (144), must be considered as an *ultimate load* which will produce complete failure of the column. In practical applications the allowable load should be smaller than the critical load and is obtained by dividing the critical value of the load by a certain factor of safety. Further discussion of this question is given in the next two articles.

In the preceding discussion a column with one end built in and the other end free was considered. Similar conclusions can be made in the case of a strut with hinged ends, Fig. 235. Eqs. (142) and (143) give infinite values when

$$\frac{pl}{2} = \frac{\pi}{2}.$$

Substituting for p its value from formula (138), we obtain in this case

$$P_{cr} = \frac{\pi^2 EI_z}{l^2}. \tag{145}$$

Fig. 238.

This is the *critical value* of the compressive force for a strut with hinged ends.

In the case of compression of columns with built-in ends the deflection has the form shown in Fig. 238. The deflection curve can be considered as consisting of four portions each similar to the curve previously obtained for a column with one end built in and the other free. The critical value of the load is found in such a case by substituting $l/4$ instead of l into eq. (144), which gives

$$P_{cr} = \frac{4\pi^2 EI_z}{l^2}. \tag{146}$$

This is the *critical load* for a column with built-in ends.

It should be noted that in the derivation of eq. (139) it was assumed that the eccentricity was in the direction of the y axis and that this axis was an axis of symmetry. If the column has two planes of symmetry, similar formulas will be obtained if the initial eccentricity is in the direction of the z axis. The bending then occurs in the xz plane and, to calculate the deflections, I_y must be substituted in place of I_z in eq. (139). If an attempt is made to apply the load centrally and bending occurs as a result of small unavoidable eccentricities, we must consider deflections in both principal planes xy

and xz, and in calculating the critical value of the load, we must use the smaller of the two principal moments of inertia in eqs. (144), (145) and (146). In the following discussion it is assumed that I_z is the smaller principal moment of inertia and k_z is the corresponding radius of gyration.

In calculating deflections it is sometimes advantageous to use approximate formulas instead of the accurate formulas (139) and (142). It was shown in the preceding article (see p. 260) that for small loads, that is when pl is a small fraction, say less than $\frac{1}{10}$, the deflection is given with sufficient accuracy by the equation

$$\delta = \frac{Pel^2}{2EI_z}, \qquad (a)$$

in which the influence of the longitudinal force on the bending is neglected and a constant bending moment Pe is assumed. For larger loads eq. (a) is not accurate enough and the influence of the compressive force on bending should be considered. This influence depends principally on the ratio P/P_{cr} and the deflection can be obtained with very satisfactory accuracy from the approximate formula

$$\delta = \frac{Pel^2}{2EI_z} \cdot \frac{1}{1 - \dfrac{P}{P_{cr}}}. \qquad (b)$$

The deflections calculated from this formula are given in the third line of Table 2, p. 261. Comparison of these figures with those of the second line of the same table shows that formula (b) is sufficiently accurate almost up to the critical value of the load.

A similar approximate formula for the deflection of a strut with hinged ends is [6]

$$\delta = \frac{Pel^2}{8EI_z} \cdot \frac{1}{1 - \dfrac{P}{P_{cr}}}. \qquad (c)$$

---

[6] This approximate solution was given by Thomas Young in his famous book, *A Course of Lectures on Natural Philosophy and the Mechanical Arts*, London, 1807.

The first factor on the right side is the deflection produced by the two couples Pe applied at the ends. The second factor represents the effect on the deflection of the longitudinal compressive force P.

Eq. (c) is very useful for determining the critical load from an experiment with a compressed strut. If the results of such an experiment are represented in the form of a curve, such as those shown in Fig. 237, the horizontal asymptote to that curve must be drawn to determine P_{cr}. This operation cannot be done with sufficient accuracy, especially if the unavoidable eccentricities are not very small and the curve does not turn very sharply in approaching the horizontal asymptote. A more satisfactory determination of P_{cr} is obtained by using eq. (c). Dividing this equation by P/P_{cr} we obtain

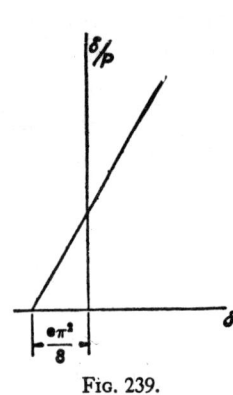

$$\frac{\delta}{P} \cdot P_{cr} = \frac{e\pi^2}{8} \frac{1}{1 - \dfrac{P}{P_{cr}}}$$

and

$$\frac{\delta}{P} \cdot P_{cr} - \delta = \frac{e\pi^2}{8}.$$

This equation shows that if we plot the ratio δ/P against the deflection δ measured during experiment, the points will fall on a straight line, Fig. 239. This line will cut the horizontal axis ($\delta/P = 0$)

Fig. 239.

at the distance $e\pi^2/8$ from the origin, and the inverse slope of the line gives the critical load.[7]

58. Critical Stress.—*Design of Columns.*—Considering the case of a strut with hinged ends, the critical stress is obtained by dividing the critical load given by eq. (145) by the cross-sectional area A. In this way we find

$$\sigma_{cr} = \frac{P_{cr}}{A} = \frac{\pi^2 E}{(l/k_z)^2}. \tag{147}$$

---

[7] This method, suggested by R. V. Southwell, *Proc. Roy. Soc. (London)*, A, Vol. 135, p. 601, 1932, has proved a very useful one and is now widely used in column tests.

It is seen that for a given material the value of the critical stress depends on the magnitude of the ratio l/k_z, which is called the *slenderness ratio*. In Fig. 240 the curve ACB represents [8] the relation between σ_{cr} and l/k_z for the case of steel having $E = 30 \times 10^6$ lb per sq in. It will be appreciated that the curve is entirely defined by the magnitude of the modulus of the material and is independent of its ultimate strength.

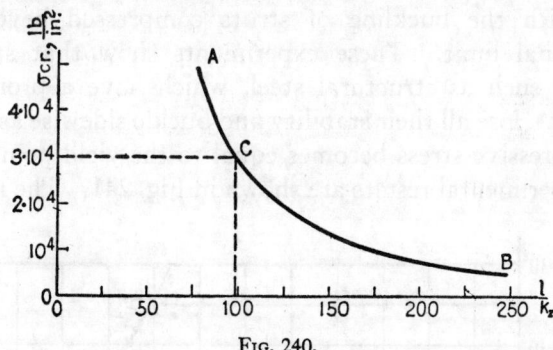

Fig. 240.

For large values of the slenderness ratio l/k_z the critical stress becomes small, which indicates that a very slender strut buckles sidewise and loses its strength at a very small compressive stress. This condition cannot be improved by taking a steel of higher strength, since the modulus of steel does not vary much with alloy and heat treatment and remains practically constant. The strut can be made stronger by increasing the moment of inertia I_z and the radius of gyration k_z, which can very often be accomplished without any increase in the cross-sectional area by placing the material of the strut as far as possible from the neutral axis. Thus tubular sections are more economical as columns than solid sections. As the slenderness ratio diminishes the critical stress increases and the curve ACB approaches the vertical axis asymptotically. However, there must be a certain limitation to the use of the Euler curve for shorter struts. The derivation of the expression for the critical load is based on the use of the differential eq. (79)

---

[8] This curve is sometimes called the *Euler curve*, since it is derived from Euler's formula for the critical load.

for the deflection curve, which equation assumes that the material is perfectly elastic and follows Hooke's law (see Art. 31). Hence the curve ACB in Fig. 240 gives satisfactory results only for comparatively slender bars for which σ_{cr} remains within the elastic region of the material. For shorter struts, for which σ_{cr} as obtained from eq. (147) is higher than the proportional limit of the material, the Euler curve does not give a satisfactory result and recourse must be had to experiments with the buckling of struts compressed beyond the proportional limit. These experiments show that struts of materials such as structural steel, which have a pronounced yield point, lose all their stability and buckle sidewise as soon as the compressive stress becomes equal to the yield point stress. Some experimental results are shown in Fig. 241. The material

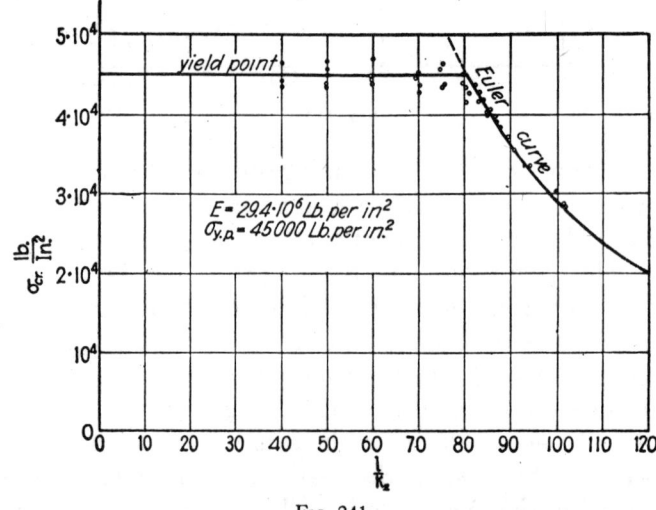

FIG. 241.

is structural steel having a very pronounced yield point at $\sigma_{Y.P.}$ = 45,000 lb per sq in. It is seen that for struts of relatively large slenderness ($l/k_z > 80$) the experimental values of the critical stresses coincide satisfactorily with the Euler curve, while for shorter struts the critical stress remains practically independent of the slenderness ratio l/k_z and is equal to the yield point stress.

In the case of an ordinary low-carbon structural steel the yield point is not as pronounced as in the preceding example and occurs at a much lower stress. For such steel we may take $\sigma_{Y.P.} = 34,000$ lb per sq in. The proportional limit is also much lower, so that the Euler curve is satisfactory only for slenderness ratios above $l/k_z = 100$, which corresponds to the compressive stress $\sigma_{cr} = 30,000$ lb per sq in. For higher stresses, i.e., for $l/k_z < 100$, the material does not follow Hooke's law and the Euler curve cannot be used. It is usually replaced in the inelastic region by the two straight lines AB and BC as shown in Fig. 242. The horizontal line AB corre-

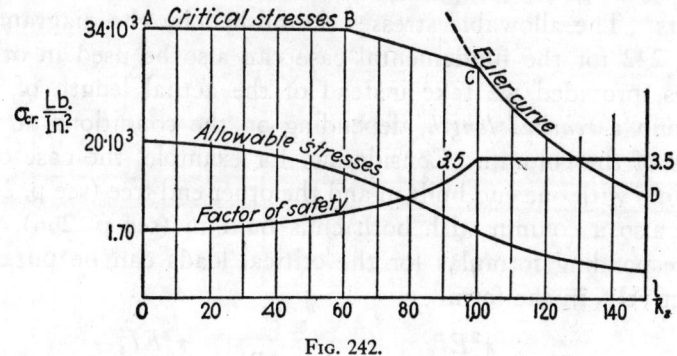

FIG. 242.

sponds to the yield point stress and the inclined line BC is taken for the stresses between the proportional limit and the yield point of the material.

Having such a diagram as the line $ABCD$ in Fig. 242, constructed for ordinary structural steel, the critical stress for a steel strut of any dimensions can readily be obtained. It is only necessary to calculate in each particular case the value of the slenderness ratio l/k_z and take the corresponding ordinate from the curve. To obtain the safe stress on the strut the critical stress must then be divided by a proper factor of safety. In selecting this factor it must be considered that as the slenderness ratio increases various imperfections, such as an initial crookedness of the column, are likely to increase. It appears logical therefore to introduce a variable factor of safety which increases with the slenderness ratio.

In some specifications the factor of safety increases from 1 7 for $l/k_z = 0$ to 3.5 for $l/k_z = 100$. It varies in such a way that the allowable stress in the inelastic range follows a parabolic law. For $l/k_z > 100$, the factor of safety is taken as constant at 3.5, and the allowable stresses are calculated from the Euler curve. In Fig. 242 curves are given which represent the allowable stress and the factor of safety as functions of the slenderness ratio for ordinary structural steel.

In the preceding discussion a strut with hinged ends was considered. This case is sometimes called the *fundamental case* of buckling of struts, since it is encountered very often in the design of compressed members of trusses with hinged joints. The allowable stresses established by the diagram in Fig. 242 for the fundamental case can also be used in other cases, provided we take instead of the actual length of the column a *reduced length*, depending on the conditions at the ends of the column. Considering, for example, the case of a column with one end built in and the other end free (see p. 259) and also a column with both ends built in (see p. 266), the corresponding formulas for the critical loads can be put, respectively, in the form

$$P_{cr} = \frac{\pi^2 E I_z}{(2l)^2} \quad \text{and} \quad P_{cr} = \frac{\pi^2 E I_z}{(\frac{1}{2}l)^2}.$$

Comparing these formulas with formula (145) for the fundamental case it can be concluded that in the design of a column with one end built in and the other free we must take a length two times larger than the actual length of the column when using the diagram of Fig. 242. In the case of a column with both ends built in the reduced length is equal to half of the actual length.

The selection of proper cross-sectional dimensions of a column is usually made by trial and error. Knowing the load P which acts on the column, we assume certain cross-sectional dimensions and calculate k_z and l/k_z for these dimensions. Then the safe value of the compressive stress is obtained from the diagram of Fig. 242. Multiplying this value by the area of the assumed cross section, the safe load on the column is obtained. If this load is neither smaller nor considerably larger than P, the assumed cross section is satisfactory. Otherwise the calculations must be repeated. In the case of built-up columns the gross cross section is used in calculating k_z, since the rivet holes do not appreciably affect the magnitude of the critical load. However, in calculating the safe load on the column the safe stress is multiplied by the net cross-sectional area in order to insure against excessive stresses in the column.

Problems

1. A steel bar of rectangular cross section 1×2 in. and with hinged ends is compressed axially. Determine the minimum length at which eq. (147) for the critical stress can be applied if the limit of proportionality of the material is 30,000 lb per sq in, and $E = 30 \times 10^6$ lb per sq in. Determine the magnitude of the critical stress if the length of the bar is 5 ft.

Answer. Minimum length = 28.9 in. The critical stress for $l = 5$ ft is 6,850 lb per sq in.

2. Solve the preceding problem, assuming a bar with circular cross section 1 in. in diameter and built-in ends.

3. Determine the critical compressive stress for a standard 6 I 12.5 section, which is 6 ft long and has hinged ends. $I_z = 1.8$ in.^{4}, $I_y = 21.8$ in.^{4} and $A = 3.61$ sq in. Determine the safe load from the curve of Fig. 242.

Answer. $\sigma_{cr} = 28,500$ lb per sq in., $l/k_z = 102$, safe load = 29,400 lb.

4. Solve the preceding problem, assuming that the ends of the column are built in. Use Fig. 242.

Answer. $\sigma_{cr} = 34,000$ lb per sq in., factor of safety = 2, safe load = 61,000 lb.

5. Calculate by the use of Fig. 242 the safe load on a member (Fig. 243) built up of two I beams of the same cross section as those in Prob. 3 above. The length of the member is 10 ft and the ends are hinged. Assume that the connecting details are so rigid that both I beams work together as a single bar.

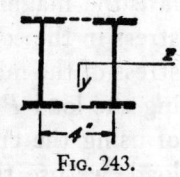

Fig. 243.

Solution. $I_z = 2 \times 21.8 = 43.6$ in.^{4}, $I_y = 2[1.8 + 3.61(2)^2] = 32.48$ in.^{4} Therefore the larger value of the slenderness ratio is

$$\frac{l}{k_y} = \frac{10 \times 12}{\sqrt{\dfrac{32.48}{2 \times 3.61}}} = 56.6,$$

and from Fig. 242 the allowable compressive stress is 16,000 lb per sq in.

$$\text{Safe load} = 16{,}000 \times 2 \times 3.61 = 116{,}000 \text{ lb.}$$

6. Solve the preceding problem, assuming that the ends of the member are built in.

7. A column 10 ft long with hinged ends is built up of two channels 8 in. deep having $I_z = 1.3$ in.^{4}, $I_y = 32.3$ in.^{4}, $A = 3.36$ in.^{2} and a distance of $c = 0.58$ in. between the centroid and the back of the channel. Find the safe load on the column if the back to back distance between the channels is 4 in.

8. Determine the necessary cross-sectional area of a square steel strut 6 ft long if the load $P = 40{,}000$ lb and the ends are hinged. Use Fig. 242.

9. Solve the preceding problem, assuming that the ends of the strut are built in.

59. Design of Columns on the Basis of Assumed Inaccuracies.

—In the preceding article the safe load on a column was obtained by dividing the critical load for the column by a proper factor of safety. The weakness of that method lies in a certain arbitrariness in the selection of the factor of safety which, as we have seen, varies with the slenderness ratio. To make the procedure of column design more rational, another method based on assumed inaccuracies has been developed.[9] On the basis of existing experimental data we can assume certain values for the magnitude of the unavoidable eccentricity e in the application of the compressive force P. Then, by using these values in the formulas of Art. 56, we can calculate the magnitude $P_{Y.P.}$ of the load at which the maximum stress in the compressed strut becomes equal to the yield point stress of the material. The safe load is then obtained by dividing the load $P_{Y.P.}$ by a proper factor of safety. Thus instead of using the critical load, which is equivalent to the ultimate load, we use the load at which yielding begins as a basis for calculating the safe load.

This method of column design can be simplified by the use of diagrams for which the calculations will now be explained. Taking the case of a strut with hinged ends, Fig. 235, the

---

[9] See D. H. Young, *Proc. Am. Soc. Civil Engrs.*, Dec., 1934; also H. K. Stephenson and K. Cloninger, Jr., *Texas Eng. Exp. Sta., Bull. No. 129*, 1953.

maximum bending moment is obtained from eq. (143) and the maximum compressive stress is

$$\sigma_{\max} = \frac{P}{A} + \frac{Pe}{Z} \sec \sqrt{\frac{P}{EI_z}} \frac{l}{2}. \qquad (a)$$

The first term on the right side is the direct stress and the second is the maximum compressive bending stress. The load at which yielding begins is obtained by substituting $\sigma_{Y.P.}$ for $\sigma_{\max}$ in this equation, which gives

$$\sigma_{Y.P.} = \frac{P_{Y.P.}}{A}\left(1 + \frac{e}{r} \sec \frac{l}{2k_z} \sqrt{\frac{P_{Y.P.}}{EA}}\right), \qquad (b)$$

in which we use the notation $r = Z/A$ for the radius of the core of the cross section (see p. 254) and $k_z = \sqrt{I_z/A}$ for the smaller principal radius of gyration. The quantity $P_{Y.P.}/A$ is the average compressive stress or *centroidal compressive stress* at which yielding begins. Denoting this stress by σ_c, we obtain

$$\sigma_{Y.P.} = \sigma_c\left(1 + \frac{e}{r} \sec \frac{l}{2k_z} \sqrt{\frac{\sigma_c}{E}}\right). \qquad (c)$$

From this equation, for a given value of the eccentricity ratio e/r, the value of σ_c can be obtained for any value of the slenderness ratio l/k_z. The results of such calculations for a structural steel having $\sigma_{Y.P.} = 36{,}000$ lb per sq in. are represented by curves in Fig. 244. By the use of these curves the average compressive stress σ_c and the compressive load $P_{Y.P.} = A\sigma_c$ at which yielding begins can readily be calculated if e/r and l/k_z are given. The safe load is then obtained by dividing $P_{Y.P.}$ by the factor of safety.

We assumed in the foregoing discussion that the unavoidable inaccuracies in the column could be represented by an eccentricity of the load. In a similar manner we can also consider the inaccuracies to be equivalent to an initial crookedness of the column. Denoting the maximum initial deviation of the axis of the column from a straight line [10] by a, curves

---

[10] A half wave of a sine curve is usually taken as representing the initial crookedness of a column.

similar to those shown in Fig. 244 and representing σ_c as a function of the ratio a/r and the slenderness ratio l/k_s can be obtained.

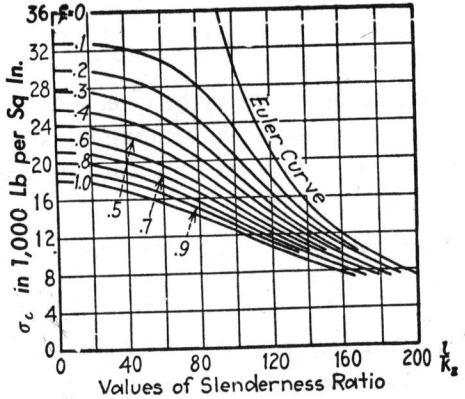

FIG. 244.

In practical design it is usually assumed that the initial deflection a is in a certain ratio to the length l of the column. Taking a certain magnitude for that ratio,[11] the magnitude

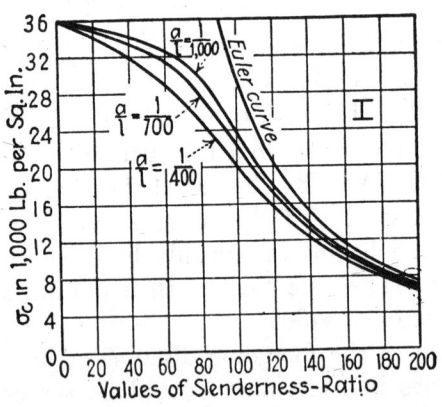

FIG. 245.

of a is calculated and the value of σ_c is then obtained from the above-mentioned curves. The results obtained in this way for

---

[11] It is usually taken within the limits $\dfrac{l}{400} \geqq a \geqq \dfrac{l}{1,000}$.

three different values of the ratio a/l and for $\sigma_{Y.P.} = 36,000$ lb per sq in. are shown for an I section in Fig. 245. For very short columns all three curves give $\sigma_c = 36,000$ lb per sq in. For very slender columns the values given by the curves approach those obtained from the Euler curve. Using one of the curves and dividing the value σ_c from the curve by a proper factor of safety, say 2, the safe value of the average compressive stress is obtained. The advantage of this method is that it employs a constant factor of safety, since the increase of inaccuracies with the length l of the column has already been taken into consideration by assuming that the eccentricity is proportional to the span. However, the magnitude of the inaccuracies which should be taken remains to a certain extent indefinite and dependent upon existing experimental data.

60. Empirical Formulas for Column Design.—In both of the methods of column design developed in the last two articles on the basis of theoretical considerations there occur some uncertainties, such as a variable factor of safety in the design procedure illustrated by Fig. 242 or the assumed inaccuracies as used in making the curves in Fig. 245. These quantities can be properly selected only on the basis of experiments with actual columns. Under such circumstances it is natural that many practical engineers prefer to use directly the results of experiments as represented by empirical formulas. Such a procedure is entirely legitimate so long as the application of these formulas remains within the limits for which they were established and for which there is sufficient experimental information. However, as soon as it is necessary to go beyond those limits, the formulas must be modified to conform with the new conditions. In this work the theoretical considerations become of primary importance.

One of the oldest empirical formulas was originated by Tredgold.[12] It was adapted by Gordon to represent the results of Hodgkinson's experiments and was given in final form by Rankine. The allowable average compressive stress as given

---

[12] Regarding the history of the formula see E. H. Salmon, *Columns*, London, 1921. See also Todhunter and Pearson, *History of the Theory of Elasticity*, Cambridge, Vol. 1, p. 105, 1886.

by the *Gordon-Rankine formula* is

$$\sigma_W = \frac{a}{1 + b\left(\dfrac{l}{k_z}\right)^2},\qquad (a)$$

in which a is a stress and b is a numerical factor, both of which are constant for a given material. By a proper selection of these constants the formula can be made to agree satisfactorily with the results of experiments within certain limits.

The *straight-line formula*, developed principally on the basis of experimental work by L. v. Tetmajer and formerly used by the American Railway Engineering Association as well as incorporated in the Chicago Building Code of 1924, gives the working stress (lb per sq in.) in the form

$$\sigma_W = 16{,}000 - 70l/k_z. \qquad (b)$$

This formula is to be used for $30 < l/k_z < 120$ for main members and as high as $l/k_z = 150$ for secondary members. For values of $l/k_z < 30$, $\sigma_W = 14{,}000$ lb per sq in. is used.

The *parabolic formula* proposed by A. Ostenfeld [13] is also sometimes used. It gives for the critical compressive stress

$$\sigma_{cr} = a - b\left(\frac{l}{k_z}\right)^2, \qquad (c)$$

in which the constants a and b depend upon the mechanical properties of the material. For structural steel, eq. (c) is sometimes taken in the form

$$\sigma_{cr} = 40{,}000 - 1.33\left(\frac{l}{k_z}\right)^2. \qquad (d)$$

This gives a parabola tangent to the Euler curve at $l/k_z = 122.5$ and makes $\sigma_{cr} = 40{,}000$ lb per sq in. for short columns. A suitable factor of safety varying from $2\frac{1}{2}$ to 3 should be used with this formula to obtain the working stress.

The American Institute of Steel Construction (AISC) specifications of 1948 specify a parabolic formula for the work-

---

[13] *Z. ver. deut. Ing.*, Vol. 42, p. 1462, 1898. See also C. E. Fuller and W. A. Johnston, *Applied Mechanics*, Vol. 2, p. 359, 1919.

ing stress in compression:

$$\sigma_W = 17,000 - 0.485 \left(\frac{l}{k_z}\right)^2, \qquad (e)$$

for $l/k_z < 120$ and for either main or secondary members. For secondary members (bracing, etc.) with $120 < l/k_z < 200$ the allowable compressive stress is given by a formula of the Gordon-Rankine type:

$$\sigma_W = \frac{18,000}{1 + \dfrac{l^2}{18,000 k_z^2}}. \qquad (f)$$

For main members with $120 < l/k_z < 200$ the allowable compressive stress is obtained by multiplying eq. (f) by the following fraction:

$$1.6 - \frac{l}{200 k_z}. \qquad (g)$$

The specifications of the American Railway Engineering Association (AREA) for 1946 and the specifications of the American Association of State Highway Officials (AASHO) for 1949 use a parabolic formula for compression members:

$$\sigma_W = 15,000 - \tfrac{1}{4}(l/k_z)^2, \quad \text{for } l/k_z < 140. \qquad (h)$$

The building code of New York City (1947) specifies formula (f) for members with $l/k_z < 120$, with a maximum value of 15,000 lb per sq in.

Problems

1. A structural steel column with hinged ends has $I_z = 12.2$ in.[4] and $A = 6.61$ sq in. Three lengths of column are to be considered, $l = 5$ ft, 10 ft and 13 ft 4 in. What are the safe loads on the column in pounds using (1) the AISC specifications, (2) the building code of New York City, (3) formula (b) and (4) the AREA specifications?

Answer.

	$l = 5$ ft	$l = 10$ ft	$l = 13$ ft 4 in.
(1)	106,000	87,500	68,100
(2)	99,000	83,000	67,400
(3)	85,400	65,000	51,400
(4)	96,000	86,200	76,400

2. Select a WF beam section to serve as a column 25 ft long with fixed ends and carrying a load of 200,000 lb. Use the AISC specifications.

Solution. Taking the reduced length $l = \frac{1}{2}(25)$ ft $= 150$ in., eq. (*e*) gives

$$\frac{200,000}{A} = 17,000 - 0.485 \left(\frac{150}{k_z}\right)^2, \tag{i}$$

assuming $l/k_z < 120$. The minimum area may be found by taking $\sigma_W = 17,000$ lb per sq in. as for a short column. This gives $A = 200,000/17,000 = 11.75$ sq in. We therefore need not try any section which has an area less than 11.75 sq in.

We try first an 8 WF 40 section for which $A = 11.76$ sq in., least radius of gyration $k_z = 2.04$ in. and $l/k_z = 73.5$ which is <120. The actual stress given by the left-hand side of eq. (*i*) is 17,000 lb per sq in., while the allowable stress given by the right-hand side of the equation is 14,400 lb per sq in. Thus the trial section is unsafe and we try next an 8 WF 48 section for which $A = 14.11$ sq in., least $k_z = 2.08$ in. and $l/k_z = 72.1$. The actual stress is 14,180 lb per sq in. and the allowable stress is 14,480 lb per sq in., so the section chosen is satisfactory.

CHAPTER X

TORSION AND COMBINED BENDING AND TORSION

61. Torsion of a Circular Shaft.—Let us consider a circular shaft built in at the upper end and twisted by a couple applied to the lower end (Fig. 246). It can be shown by measurements

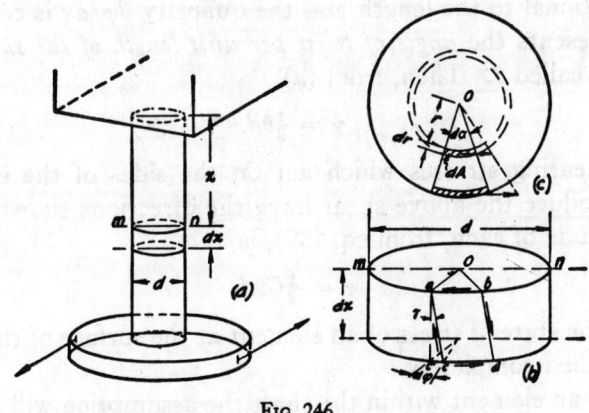

Fig. 246.

at the surface that circular cross sections of the shaft remain circular during twist and that their diameters and the distances between them do not change, provided the angle of twist is small.

A disc isolated as in Fig. 246b will be in the following state of strain: There will be a rotation of its lower cross section with respect to its top cross section through an angle $d\varphi$, where φ measures the rotation of the section mn with respect to the built-in end. A rectangular element $abcd$ of the lateral surface of the disc takes the form shown in Fig. 246b. The lengths of the sides remain essentially the same and only the angles at the corners change. The element is in a state of *pure shear* (see Art. 16) and the magnitude of the shearing strain γ is found from the small triangle cac':

$$\gamma = \frac{c'c}{ac'}.$$

Since $c'c$ is the small arc of radius $d/2$ corresponding to the difference $d\varphi$ in the angle of rotation of the two adjacent cross sections, $c'c = (d/2)d\varphi$ and we obtain

$$\gamma = \frac{1}{2}\frac{d\varphi}{dx}d. \qquad (a)$$

For a shaft twisted by a torque at the end the angle of twist is proportional to the length and the quantity $d\varphi/dx$ is constant. It represents the *angle of twist per unit length of the shaft* and will be called θ. Then, from (a),

$$\gamma = \tfrac{1}{2}\theta d. \qquad (148)$$

The shearing stresses which act on the sides of the element and produce the above shear have the directions shown. The magnitude of each, from eq. (39), is

$$\tau = \tfrac{1}{2}G\theta d. \qquad (149)$$

Thus the state of stress of an element at the surface of the shaft is specified completely.

For an element within the shaft the assumption will now be made that not only the circular boundaries of the cross sections of the shaft remain undistorted but also the cross sections themselves remain plane and rotate as if absolutely rigid, i.e., every diameter of the cross section remains straight and rotates through the same angle. Tests of circular shafts show that the theory developed on this assumption is in very good agreement with experimental results. This being the case, the discussion for the element $abcd$ at the surface of the shaft (Fig. 246b) will also hold for a similar element on the surface of an inner cylinder, whose radius r replaces $d/2$ (Fig. 246c). The thickness dr of the element in the radial direction is considered as very small. Such elements are then also in pure shear and the shearing stress on their sides is

$$\tau = Gr\theta. \qquad (b)$$

This states that the shearing stress varies directly as the distance r from the axis of the shaft. Fig. 247 pictures this stress distribution. The maximum stress occurs in the surface layer of the shaft. For a ductile material, plastic flow begins first in this outer layer. For a material which is weaker in shear longitudinally than transversely, e.g., a wooden shaft with the fibers parallel to the axis, the first cracks will be produced by shearing stresses acting in the axial sections and they will appear on the surface of the shaft in the longitudinal direction. In the case of a material which is weaker in tension than in shear,

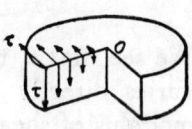

FIG. 247.

e.g., a circular shaft of cast iron or a cylindrical piece of chalk, a crack along a helix inclined at 45° to the axis of the shaft often occurs (Fig. 248). The explanation is simple. We recall that the state of pure shear is equivalent to one of tension in one direction and equal compression in the perpendicular direction (see Fig. 45). A rectangular element cut from the outer layer of a twisted shaft with sides at 45° to the axis will be submitted to the stresses shown in Fig. 248. The tensile stresses shown produce the helical crack mentioned.

FIG. 248.

We seek now the relationship between the applied twisting couple M_t and the stresses which it produces. From the equilibrium of the portion of the shaft between the bottom and the cross section mn (Fig. 246a) we conclude that the shearing stresses distributed over the cross section are statically equivalent to a couple equal and opposite to the torque M_t. For each element of area dA (Fig. 246c) the shearing force is τdA. The moment of this force about the axis of the shaft is $(\tau dA)r = G\theta r^2 dA$, from eq. (b). The torque M_t is the summation, taken over the entire cross-sectional area, of these moments, i.e.,

$$M_t = \int_A G\theta r^2 dA = G\theta \int_A r^2 dA = G\theta I_p, \qquad (c)$$

where I_p is the polar moment of inertia of the circular cross section. For a circle of diameter d we have (see Appendix A, p. 420) $I_p = \pi d^4/32$, and therefore

$$M_t = G\theta \frac{\pi d^4}{32}$$

and

$$\theta = \frac{M_t}{G} \frac{32}{\pi d^4} = \frac{M_t}{GI_p}. \tag{150}$$

We see that θ, the angle of twist per unit length of the shaft, varies directly as the applied torque and inversely as the modulus of shear G and the fourth power of the diameter. If the shaft is of length l, the total angle of twist will be

$$\varphi = \theta l = \frac{M_t l}{GI_p}. \tag{151}$$

This equation has been checked by numerous experiments which prove the accuracy of the assumptions made in deriving the theory.

It should be noted that experiments in torsion are commonly used for determining the shearing modulus of elasticity G for various materials. If the angle of twist produced in a given shaft by a given torque is measured, the magnitude of G can easily be obtained from eq. (151).

Substituting θ from eq. (150) into eq. (149), we obtain an equation for calculating the maximum shearing stress in torsion for a circular shaft:

$$\tau_{\max} = \frac{M_t d}{2I_p} = \frac{16M_t}{\pi d^3}. \tag{152}$$

The maximum shear stress is proportional to the torque M_t and inversely proportional to the cube of the diameter of the shaft.

In practical applications the required diameter of a shaft must frequently be calculated from the horsepower H which it transmits. Given H, the torque is obtained in units of inch-pounds from the well-known equation:

$$M_t \cdot \frac{2\pi n}{60} = 550 \times 12 \times H, \tag{153}$$

in which n denotes the number of revolutions of the shaft per minute. The quantity $2\pi n/60$ is the angle of rotation per second and the left side of eq. (153) represents the work done during one second by the torque M_t measured in inch-pounds. The right side of the equation represents the work done (inch-pounds per second) as calculated from the horsepower H. Taking M_t from eq. (153) and substituting into eq. (152), we obtain

$$d \text{ (in.)} = 68.5 \sqrt[3]{\frac{H}{n\tau_{\max}}}. \tag{154}$$

Taking, for example, the working stress for shear as $\tau_W = 9,000$ lb per sq in., we have

$$d \text{ (in.)} = 3.29 \sqrt[3]{\frac{H}{n}}.$$

Problems

1. Determine the shaft diameter d for a machine of 200 hp and speed $n = 120$ rpm, if the working stress $\tau_W = 3,000$ lb per sq in. *Answer.* $d = 5.63$ in.

2. Determine the horsepower transmitted by a shaft if $d = 6$ in., $n = 120$ rpm, $G = 12 \times 10^6$ lb per sq in. and the angle of twist, as measured between two cross sections 25 ft apart, is $\frac{1}{15}$ of a radian.

Solution. From eq. (151),

$$M_t = \frac{\pi d^4}{32} \cdot \frac{\varphi \cdot G}{l} = \frac{\pi \times 6^4}{32} \cdot \frac{12 \times 10^6}{15 \times 25 \times 12}.$$

The power transmitted, from eq. (153), is

$$H = \frac{M_t \cdot 2\pi n}{60 \times 550 \times 12}$$

$$= \frac{\pi \times 6^4 \times 12 \times 10^6 \times 2\pi \times 120}{32 \times 15 \times 25 \times 12 \times 60 \times 550 \times 12} = 646 \text{ hp}$$

3. A shaft of diameter $d = 3.5$ in. makes 45 rpm. Determine the power transmitted if the maximum shearing stress is 4,500 lb per sq in.

4. A steel wire ($G = 12 \times 10^6$ lb per sq in.) is to have proportions such that the maximum shearing stress is 13,500 lb per sq in.
for an angle of twist of 90°. Determine the ratio l/d.
  Answer. $l/d = 698$.

5. A steel shaft with built-in ends (Fig. 249) is submitted to the
action of a torque M_t applied at an intermediate cross section mn.
Determine the angle of twist if the working stress τ_W is known.

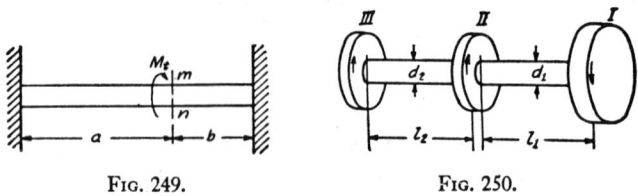

FIG. 249.   FIG. 250.

Solution. For both parts of the shaft the angles of twist are
equal. Therefore, from eq. (151), the twisting moments are inversely
proportional to the lengths of these parts. If $a > b$, the greater
twisting moment is in the right-hand part of the shaft and its magnitude is $M_t \cdot a/(a + b)$. Substituting this for the torque, and τ_W for
τ_{max} in eq. (152), the following equation for d is obtained:

$$d = \sqrt[3]{\frac{16aM_t}{(a + b)\pi\tau_W}}.$$

The angle of twist can now be obtained by using eq. (151).

6. 500 hp is transmitted from pulley I, 200 hp to pulley II and
300 hp to pulley III (Fig. 250). Find the ratio of the diameters d_1
and d_2 to give the same maximum stress in both parts of the shaft.
Find the ratio of the angles of twist for these two parts.
  Solution. The torques in the two parts of the shaft are in the
ratio 5:3. In order to have the same maximum stress from eq.
(152), we have

$$\frac{d_1}{d_2} = \sqrt[3]{\frac{5}{3}}$$

The angles of twist, from eqs. (151) and (152), will be in the ratio

$$\frac{\varphi_1}{\varphi_2} = \frac{l_1}{l_2}\sqrt[3]{\frac{3}{5}}.$$

7. Assuming that the shaft of the preceding problem has a
constant diameter and turns at 200 rpm, find the magnitude of

the diameter if $\tau_W = 6,000$ lb per sq in. Find the angle of twist for each portion of the shaft if $G = 12 \times 10^6$ lb per sq in. and $l_1 = l_2 = 4$ ft.

8. Determine the length of a steel shaft of 2 in. diameter ($G = 12 \times 10^6$ lb per sq in.) if the maximum stress is equal to 13,500 lb per sq in. when the angle of twist is 6°.

Answer. $l = 93$ in.

9. Determine the diameter at which the angle of twist of the shaft, and not the maximum stress, is the controlling factor in design, if $G = 12 \times 10^6$ lb per sq in., $\tau_W = 3,000$ lb per sq in. and the maximum allowable twist is $\frac{1}{4}°$ per yd.

Solution. Eliminating M_t from the equations

$$\frac{16M_t}{\pi d^3} = 3,000 \quad \text{and} \quad \frac{32M_t}{G \cdot \pi d^4} = \frac{\pi}{180 \times 4 \times 36},$$

we obtain $d = 4.12$ in., so that for $d < 4.12$ in. the angle of twist is the controlling factor in design.

10. Determine the torque in each portion of a shaft with built-in ends which is twisted by the moments M_t' and M_t'' applied at two intermediate sections (Fig. 251).

Solution. Determining the torques produced in each portion of the shaft

Fig. 251.

by each of the moments M_t' and M_t'' (see Prob. 5 above) and adding these moments for each portion, we obtain

$$\frac{M_t'(b + c) + M_t''c}{l}, \quad \frac{M_t'a - M_t''c}{l}, \quad \frac{M_t'a + M_t''(a + b)}{l}.$$

11. Determine the diameters and the angles of twist for the shaft of Prob. 6 if $n = 120$ rpm, $\tau_{\max} = 3,000$ lb per sq in., $l_1 = 6$ ft, $l_2 = 4$ ft.

62. Torsion of a Hollow Shaft.

—From the previous discussion of torsion of a solid shaft, it is seen (see Fig. 247) that only the material at the outer surface of the shaft will be stressed to the limit assigned as the working stress. The material within will work at a lower stress. Hence in cases in which reduction in weight is of great importance, e.g., propeller shafts of airplanes, it is advisable to use hollow shafts. In discussing the torsion of hollow shafts the same assumptions are made as in the case of solid shafts. The general expression for shearing stresses will then be the same as given by eq. (*b*) of the preceding article. In calculating the moment of the shearing stresses, however, the radius r varies from the radius of the inner hole, which we will denote by $\frac{1}{2}d_1$, to the outer radius of the shaft which, as before, will be $\frac{1}{2}d$. Then eq. (*c*) of the previous article must be replaced by the following equation:

$$G\theta \int_{\frac{1}{2}d_1}^{\frac{1}{2}d} r^2 dA = M_t = G\theta I_p,$$

where $I_p = (\pi/32)(d^4 - d_1^4)$ is the polar moment of inertia of the ring section. Then

$$\theta = \frac{32M_t}{\pi(d^4 - d_1^4)G} = \frac{M_t}{GI_p} \tag{155}$$

and the angle of twist will be

$$\varphi = \theta l = \frac{M_t l}{GI_p}. \tag{156}$$

Substituting eq. (155) in eq. (149), we obtain

$$\tau_{\max} = \frac{16M_t}{\pi d^3 \left(1 - \dfrac{d_1^4}{d^4}\right)} = \frac{M_t d}{2I_p}. \tag{157}$$

From eqs. (156) and (157) we see that by taking, for example, $d_1 = \frac{1}{2}d$ the angle of twist and the maximum stress, as compared with the same quantities for a solid shaft of diameter d, will increase about 6 per cent while the reduction in the weight of the shaft will be 25 per cent.

Problems

1. A hollow cylindrical steel shaft, 10 in. outside diameter and 6 in. inside diameter, turns at 1,000 rpm. What horsepower is being transmitted if $\tau_{\max} = 8,000$ lb per sq in.?
Answer. $H = 21,700$ hp.

2. Find the maximum torque that may be applied to a hollow circular shaft if $d = 6$ in., $d_1 = 4$ in. and $\tau_w = 8,000$ lb per sq in.

3. A hollow propeller shaft of a ship transmits 8,000 hp at 100 rpm with a working stress of 4,500 lb per sq in. If $d/d_1 = 2$, find d.

Solution.

$$M_t = \frac{8,000 \times 12 \times 33,000}{2\pi \times 100}.$$

Eq. (157) becomes

$$\tau_{max} = \frac{16}{15} \cdot \frac{16M_t}{\pi d^3},$$

from which

$$d = \sqrt[3]{\frac{16 \times 16 \times 8,000 \times 12 \times 33,000}{15 \times 2\pi \times 100 \times \pi \times 4,500}} = 18.2 \text{ in.}$$

Then $d_1 = 9.1$ in.

63. Shaft of Rectangular Cross Section.—The problem of the twist of a shaft of rectangular cross section is complicated, due to the warping of the cross section during twist. This warping can be shown experimentally with a rectangular bar of rubber on whose faces a system of small squares has been traced. It is seen from the photograph [1] (Fig. 252) that during twist the lines originally perpendicular to the axis of the bar become curved. This indicates that the distortion of the small squares, mentioned above, varies along the sides of this cross section, reaching a maximum value at the middle and becoming zero at the corners. We therefore expect that the shearing stress varies as this distortion, namely, it is maximum at the middle of the sides and zero at the corners of the cross section. Rigorous investigation of the problem [2] indicates that the maximum shearing stress occurs at the middle of the longer sides of the rectangular cross section and is given by the equation

Fig. 252.

$$\tau_{max} = \frac{M_t}{\alpha b c^2}, \qquad (158)$$

in which b is the longer and c the shorter side of the rectangular cross section and α is a numerical factor depending upon the ratio b/c.

---

[1] The photograph is taken from C. Bach, *Elasticität und Festigkeit*, 6th Ed., p. 312, 1911.

[2] The complete solution is due to St.-Venant, *Mém. sav. étrangers*, Vol. 14, 1855. An account of this work will be found in Todhunter and Pearson's *History of the Theory of Elasticity*, Cambridge, Vol. 2, p. 312, 1893.

Several values of α are given in Table 3 below. The magnitude of the maximum stress can be calculated with satisfactory accuracy from the following approximate equation:

$$\tau_{max} = \frac{M_t}{bc^2}\left(3 + 1.8\,\frac{c}{b}\right).$$

TABLE 3: DATA FOR THE TWIST OF A SHAFT OF RECTANGULAR CROSS SECTION

$\dfrac{b}{c} =$	1.00	1.50	1.75	2.00	2.50	3.00	4.00	6	8	10	∞
$\alpha =$	0.208	0.231	0.239	0.246	0.258	0.267	0.282	0.299	0.307	0.313	0.333
$\beta =$	0.141	0.196	0.214	0.229	0.249	0.263	0.281	0.299	0.307	0.313	0.333

The angle of twist per unit length in the case of a rectangular cross section is given by the equation:

$$\theta = \frac{M_t}{\beta bc^3 G}. \tag{159}$$

The values of the numerical factor β are given in the third line of Table 3.

In all cases considered the angle of twist per unit length is proportional to the torque and can be represented by the equation

$$\theta = \frac{M_t}{C}, \tag{a}$$

where C is a constant called the *torsional rigidity* of the shaft. In the case of a circular shaft (eq. 150), $C = GI_p$.

For a rectangular shaft (eq. 159), $C = \beta bc^3 G$.

64. Helical Spring.—*Close Coiled.*—Assume that a helical spring of circular cross section is submitted to the action of axial forces P (Fig. 253), and that any one coil lies nearly in a plane perpendicular to the axis of the helix. Considering the equilibrium of the upper portion of the spring bounded by an axial section such as mn (Fig. 253b), it can be concluded from the equations of statics that the stresses over the cross section mn of the coil reduce to a shearing force P through the center of the cross section and a couple acting in a counter-clockwise direction in the plane of the cross section of magnitude PR, where R is the radius of the cylindrical surface containing the center line of the spring. The couple PR twists the coil and causes a maximum shearing stress which, from eq. (152), is

$$\tau_1 = \frac{16PR}{\pi d^3}, \tag{a}$$

where d is the diameter of the cross section mn of the coil. Upon this stress due to twist must be superposed the stress due

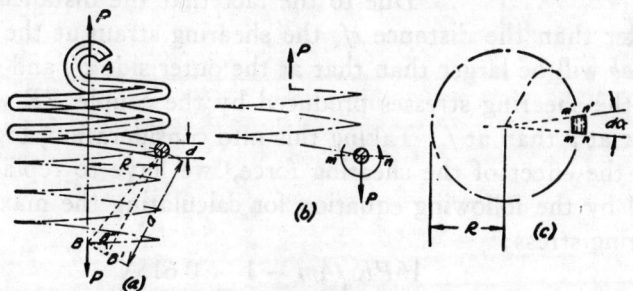

FIG. 253.

to the shearing force P. For a rough approximation this shearing force is assumed to be uniformly distributed over the cross section. The corresponding shearing stress will then be

$$\tau_2 = \frac{4P}{\pi d^2}. \tag{b}$$

At the point m (Fig. 253b) the directions of τ_1 and τ_2 coincide so that the maximum shearing stress occurs here and has the magnitude

$$\tau_{\max} = \tau_1 + \tau_2 = \frac{16PR}{\pi d^3}\left(1 + \frac{d}{4R}\right). \tag{160}$$

It can be seen that the second term in the parentheses, which represents the effect of the shearing force, increases with the ratio d/R. It becomes of practical importance in heavy helical springs, such as are used on railway cars. Points such as m on the inner side of a coil are in a more unfavorable condition than points such as n. Experience with heavy springs shows that cracks usually start on the inner side of the coil.

There is another reason to expect higher stresses at the inner side of the coil. In calculating the stresses due to twist we used eq. (*a*), which was derived for cylindrical shafts. In reality each element of the spring will be in the condition shown in Fig. 254. It is seen that if the cross section *bf* rotates with respect to *ac*, due to twist, the displacement of the point *b* with respect to *a* will be the same as that of the point *f* with respect to *c*.

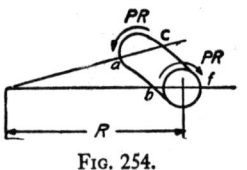

FIG. 254.

Due to the fact that the distance *ab* is smaller than the distance *cf*, the shearing strain at the inner side *ab* will be larger than that at the outer side *cf*, and therefore the shearing stresses produced by the couple *PR* will be larger at *b* than at *f*. Taking this into consideration, together with the effect of the shearing force,[3] we have to replace eq. (160) by the following equation for calculating the maximum shearing stress:

$$\tau_{\max} = \frac{16PR}{\pi d^3}\left(\frac{4m - 1}{4m - 4} + \frac{0.615}{m}\right),\qquad(161)$$

in which

$$m = \frac{2R}{d}.$$

It can be seen that the *correction factor* in the parentheses increases with a decrease of *m*. E.g., in case *m* = 4 this factor is about 1.40 and if *m* = 10 it is equal to 1.14.

In calculating the deflection of the spring usually only the effect of twist of the coils is taken into consideration. For the angle of twist of an element between the two adjacent cross sections *mn* and *m'n'* (Fig. 253*c*), using eq. (151) in which *Rdα* is used instead of *l*, we obtain

$$d\varphi = \frac{PR \cdot Rd\alpha}{GI_p}.$$

---

[3] Such investigations were made by V. Roever, *Z. Ver. deut. Ing.*, Vol. 57, p. 1906, 1913; also by A. M. Wahl, *Trans. A. S. M. E.*, 1928. The latter also determined the stresses experimentally by making measurements at the surface of the coil. A complete study of various kinds of springs is given by Wahl, *loc. cit.*, p. 212.

Due to this twist the lower portion of the spring rotates with respect to the center of mn (Fig. 253a), and the point of application B of the force P describes the small arc BB' equal to $ad\varphi$. The vertical component of this displacement is

$$B'B'' = BB'\frac{R}{a} = Rd\varphi = \frac{PR^3 da}{GI_p}. \qquad (c)$$

The complete deflection of the spring is obtained by summation of the deflections $B'B''$ due to each element $mnm'n'$, along the length of the spring. Then

$$\delta = \int_0^{2\pi n} \frac{PR^3}{GI_p} da = \frac{64nPR^3}{Gd^4}, \qquad (162)$$

in which n denotes the number of coils.

For a spring of other than circular cross section, the method given above can be used to calculate stresses and deflections if, instead of eqs. (151) and (152), we take the corresponding equations for this shape of cross section. For example, in the case of a rectangular cross section eqs. (158) and (159) should be used.

Problems

1. Determine the maximum stress and the extension of the helical spring (Fig. 253) if $P = 250$ lb, $R = 4$ in., $d = 0.8$ in., the number of coils is 20 and $G = 12 \times 10^6$ lb per sq in.

Answer. $\tau_{max} = 11,300$ lb per sq in., $\delta = 4.17$ in.

2. Solve the previous problem, assuming that the coil has a square cross section 0.8 in. on a side.

Solution. Assuming that the correction factor for the shearing force and the curvature of the coils (see eq. 161) in this case is the same as for a circular cross section, we obtain from eq. (158)

$$\tau_{max} = \frac{PR}{0.208 \times b^3} 1.14 = \frac{250 \times 4 \times 1.14}{0.208 \times 0.8^3} = 10,700 \text{ lb per sq in.}$$

In calculating the extension, $0.141d^4$ (see eq. 159) instead of $\pi d^4/32$ must be used in eq. (162). Then

$$\delta = \frac{4.17\pi}{32 \times 0.141} = 2.90 \text{ in.}$$

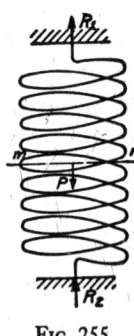

FIG. 255.

3. Compare the weights of two helical springs, one of circular and the other of square cross section, designed for the conditions stated in Prob. 1 and having the same maximum stress. Take the correction factor in both cases as 1.14. Compare the deflections of these two springs.

Solution. The length of the side of the square cross section is found from the equation $\pi d^3/16 = 0.208b^3$, from which $b = \sqrt[3]{0.944}\ d = 0.981d$. The weights of the springs are in the same ratio as the cross-sectional areas, i.e., in the ratio

$$\frac{\pi d^2}{4} : (0.981d)^2 = 0.816.$$

The deflections of the two springs are in the ratio

$$0.141b^4 : \frac{\pi d^4}{32} = 0.141 \times 0.926 : \frac{\pi}{32} = 1.33.$$

4. How will the load P be distributed between the two ends of the helical spring shown in Fig. 255 if the number of coils above the point of application of the load is 6 and below this point is 5?

Answer. $R_1 : R_2 = 5:6$.

5. Two helical springs of the same material and of equal circular cross sections and lengths, assembled as shown in Fig. 256, are compressed between two parallel p.anes. Determine the maximum stress in each spring if $d = 0.5$ and $P = 100$ lb.

Solution. From eq. (162) it follows that the load P is distributed between the two springs in inverse proportion to the cubes of the radii of the coils, i.e., the forces compressing the outer and the inner springs will be in the ratio 27:64. The maximum stresses in these springs are then (from eq. 161) 2,860 lb per sq in. and 5,380 lb per sq in., respectively.

6. What will be the limiting load for the spring of Prob. 1 if the working stress is $\tau_W = 20,000$ lb per sq in.? What will be the deflection of the spring at this limiting load?

Answer. 442 lb, $\delta = 7.38$ in.

7. A conical spring (Fig. 257) is submitted to the action of axial forces P. Determine the safe magnitude of P for a working stress $\tau_W = 45,000$ lb per sq in.; diameter of the cross section $d = 1$ in.; radius of the cone at the top of the spring $R_1 = 2$ in. and at the bottom, $R_2 = 8$ in. Determine the extension of the spring if the number of coils

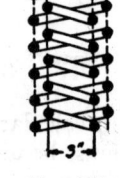

FIG. 256.

is n, and the horizontal projection of the center line of the spring is a spiral given by the equation

$$R = R_1 + \frac{(R_2 - R_1)\alpha}{2\pi n}.$$

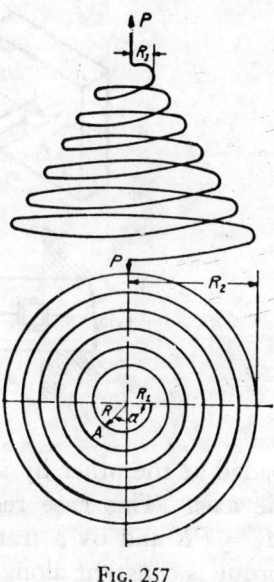

Solution. For any point A on the centerline of the spring, determined by the magnitude of the angle α, the distance from the axis of the spring is

$$R = R_1 + \frac{(R_2 - R_1)\alpha}{2\pi n}$$

and the corresponding torque is

$$M_t = P\left[R_1 + \frac{(R_2 - R_1)\alpha}{2\pi n}\right].$$

The maximum torque, at $\alpha = 2\pi n$, is $P \cdot R_2$. The safe limit for P, from eq. (161), will be

$$P = \frac{45,000 \times \pi}{16 \times 8 \times 1.09} = 1,010 \text{ lb.}$$

Fig. 257

The deflection of the spring will be obtained from eq. (c) (see p. 293) as follows:

$$\delta = \frac{32P}{\pi d^4 G}\int_0^{2\pi n}\left[R_1 + \frac{(R_2 - R_1)\alpha}{2\pi n}\right]^3 d\alpha$$

$$= \frac{16Pn}{d^4 G}(R_1{}^2 + R_2{}^2)(R_1 + R_2).$$

8. Determine the necessary cross-sectional area of the coils of a conical spring, designed for the same conditions as in the previous problem but of square cross section. Take 1.09 as the correction factor (see previous problem).

Answer. $b^2 = 0.960$ sq in.

65. Combined Bending and Torsion of Circular Shafts.—

In the previous discussion of torsion (see p. 281) it was assumed that the circular shaft was in simple torsion. In practical applications we often have cases where torque and bending moment are acting simultaneously. The forces transmitted to a shaft by a pulley, a gear or a flywheel can usually be reduced to a torque and a bending force. A simple case of this kind is shown in Fig. 258. A circular shaft is built in at one end and

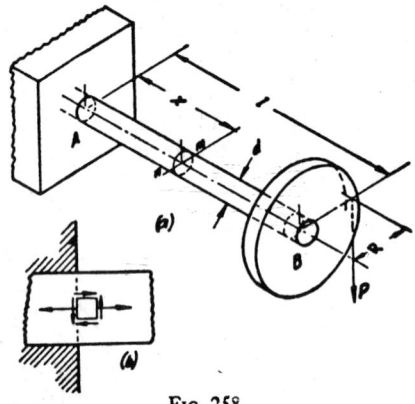

FIG. 258.

loaded at the other by a vertical force P at a distance R from the axis. This case reduces to one of loading by a torque $M_t = PR$ and by a transverse force P at the free end.[4] The torque is constant along the axis and the bending moment due to P, at any cross section, is

$$M = -P(l - x). \qquad (a)$$

In discussing the maximum stress produced in the shaft it is necessary to consider (1) shearing stresses due to the torque M_t, (2) normal stresses due to the bending moment (a) and (3) shearing stresses due to the shearing force P. The maximum torsional stress occurs at the circumference of the shaft and has the value

$$\tau_{max} = \frac{16M_t}{\pi d^3}. \qquad (b)$$

The maximum normal stress σ_x due to bending occurs in the fibers most remote from the neutral axis at the built-in end, where the bending moment is numerically a maximum, and it has the value

$$(\sigma_x)_{max} = \frac{M}{Z} = \frac{32M}{\pi d^3}. \qquad (c)$$

The weights of the shaft and of the pulley are neglected in this problem.

The stress due to the shearing force is usually of only secondary importance. Its maximum value occurs at the neutral axis where the normal stress due to bending is zero. Hence the maximum combined stress usually occurs at the point where stresses (1) and (2) are a maximum; in this case at the top and bottom surfaces at the built-in end.

Fig. 258*b* is a top view of the portion of the shaft at the built-in end, showing an element and the stresses acting on it. The principal stresses on this element are found from eqs. (72) and (73) (p. 126):

$$\sigma_{max} = \frac{\sigma_z}{2} + \frac{1}{2}\sqrt{\sigma_z^2 + 4\tau^2},$$

or, using eqs. (*b*) and (*c*),

$$\sigma_{max} = \frac{1}{2Z}(M + \sqrt{M^2 + M_t^2})$$

$$= \frac{16}{\pi d^3}(M + \sqrt{M^2 + M_t^2}). \tag{163}$$

In the same manner, using eq. (73),

$$\sigma_{min} = \frac{1}{2Z}(M - \sqrt{M^2 + M_t^2})$$

$$= \frac{16}{\pi d^3}(M - \sqrt{M^2 + M_t^2}). \tag{163'}$$

It should be noted that σ_{max} would have the same value for a case of simple bending in which the *equivalent bending moment* is

$$M_{equivalent} = \tfrac{1}{2}(M + \sqrt{M^2 + M_t^2}).$$

The maximum shearing stress at the same element (Fig. 258*b*), from eq. (34) (p. 51), is

$$\tau_{max} = \frac{\sigma_{max} - \sigma_{min}}{2} = \frac{16}{\pi d^3}\sqrt{M^2 + M_t^2}. \tag{164}$$

For ductile metals such as are used in shafts it is now common practice to use the maximum shearing stress to determine the safe diameter of the shaft. Denoting the working stress in shear by τ_W, and substituting it into eq. (164) for τ_{max}, the diameter must be

$$d = \sqrt[3]{\frac{16}{\pi\tau_W}} \sqrt{M^2 + M_t^2}. \tag{165}$$

The above discussion can also be used in the case of a hollow shaft of outer diameter d and inner diameter d_1. Then

$$Z = \frac{\pi(d^4 - d_1^4)}{32d} = \frac{\pi d^3}{32}\left[1 - \left(\frac{d_1}{d}\right)^4\right].$$

and setting $d_1/d = n$, eqs. (163) and (163′) for a hollow shaft become

$$\sigma_{max} = \frac{16}{\pi d^3(1 - n^4)}(M + \sqrt{M^2 + M_t^2}), \tag{166}$$

$$\sigma_{min} = \frac{16}{\pi d^3(1 - n^4)}(M - \sqrt{M^2 + M_t^2}). \tag{167}$$

The maximum shearing stress is

$$\tau_{max} = \frac{16}{\pi d^3(1 - n^4)}\sqrt{M^2 + M_t^2}, \tag{168}$$

and d becomes

$$d = \sqrt[3]{\frac{16}{\pi\tau_W(1 - n^4)}}\sqrt{M^2 + M_t^2}. \tag{169}$$

If several parallel transverse forces act on the shaft, the total bending moment M and the total torque M_t at each cross section must be used in calculating the necessary diameter at that point, from eq. (165) or (169). If the transverse forces acting on the shaft are not parallel, the bending moments due to them must be added vectorially to get the resultant bending moment M. An example of such a calculation is discussed in Prob. 3 below.

Problems

1. A $2\frac{1}{2}$-in. circular shaft carries a 30-in. diameter pulley weighing 500 lb. (Fig. 259). Determine the maximum shearing stress at cross

section *mn* if the horizontal tensions in the upper and lower portions of the belt are 1,750 lb and 250 lb, respectively.

Solution. At cross section *mn*,

$$M_t = (1,750 - 250)15 = 22,500 \text{ in. lb},$$

$$M = 6\sqrt{500^2 + 2,000^2} = 12,370 \text{ in. lb.}$$

Then, from eq. (164),

$$\tau_{max} = 8,370 \text{ lb per sq in.}$$

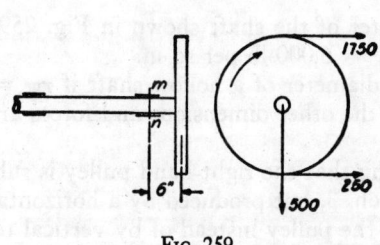

FIG. 259.   FIG. 260.

2. A vertical tube, shown in Fig. 260, is submitted to the action of a horizontal force $P = 250$ lb acting 3 ft from the axis of the tube. Determine σ_{max} and τ_{max} if the length of the tube is $l = 25$ ft and the section modulus $Z = 10$ in.[3]

Answer. $\sigma_{max} = 7,530$ lb per sq in., $\tau_{max} = 3,780$ lb per sq in.

3. Determine the necessary diameter for a uniform shaft (Fig. 261) carrying two equal pulleys 30 in. in diameter and weighing 500 lb each. The horizontal forces in the belt for one pulley and the vertical forces for the other are shown in the figure. $\tau_W = 6,000$ lb per sq in.

Solution. The weakest cross sections are *mn* and m_1n_1, which carry the full torque and the highest bending moments. The torque at both cross sections is $M_t = (1,500 - 500)15 = 15,000$ in. lb. The bending moment at *mn* is $(1,500 + 500 + 500)6 = 15,000$ in. lb. The bending moment at m_1n_1 in the horizontal plane is

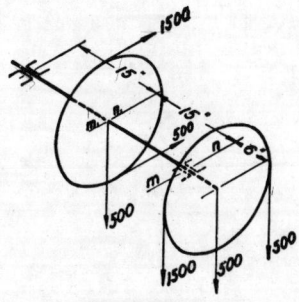

FIG. 261.

$$\tfrac{1}{4}(1,500 + 500) \times 30 = 15,000 \text{ in. lb.}$$

The bending moment at the same cross section in the vertical plane is

$$\frac{500 \times 30}{4} - \frac{2,500 \times 6 \times 15}{30} = -3,750 \text{ in. lb.}$$

The combined bending moment at cross section $m_1 n_1$ is

$$M = \sqrt{15,000^2 + 3,750^2} = 15,460 \text{ in. lb.}$$

This is larger than the moment at cross section mn and should there-fore be used together with the above calculated M_t in eq. (165), from which

$$d = 2.63 \text{ in.}$$

4. Determine the diameter of the shaft shown in Fig. 259 if the working stress in shear is $\tau_W = 6,000$ lb per sq in.

5. Determine the outer diameter of a hollow shaft if $\tau_W = 6,000$ lb per sq in., $d_1/d = \frac{1}{3}$ and the other dimensions and forces are as in Fig. 261.

6. Solve Prob. 3 assuming that the right-hand pulley is subjected to the same torque as in Prob. 3, but produced by a horizontal force tangent to the periphery of the pulley instead of by vertical tensions of 1,500 lb and 500 lb in the belt.

CHAPTER XI

STRAIN ENERGY AND IMPACT

66. Elastic Strain Energy in Tension.—In the discussion of a bar in simple tension (see Fig. 1), we saw that during elongation under a gradually increasing load, work was done on the bar, and that this work was transformed either partially or completely into potential energy of strain. If the strain remains within the elastic limit, the work done will be completely transformed into potential energy and can be recovered during a gradual unloading of the strained bar.

If the final magnitude of the load is P and the corresponding elongation is δ, the tensile test diagram will be as shown in Fig. 262, in which the abscissas are the elongations and the ordinates are the corresponding loads. P_1 represents an intermediate value of the load and δ_1 the corresponding elongation. An increment dP_1 in the load causes an increment $d\delta_1$ in the elongation. The work done by P_1 during this elongation is $P_1 d\delta_1$, represented in the figure by the shaded area. If

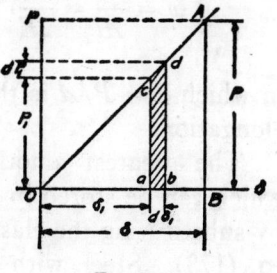

FIG. 262.

allowance is made for the increase of P_1 during the elongation, the work done will be represented by the area of the trapezoid *abcd*. The total work done in the process of loading, when the load is increasing from O to P, is the summation of such elemental areas and is given by the area of the triangle OAB. This represents the total energy U stored up in the bar during loading. Then

$$U = \frac{P\delta}{2}. \tag{170}$$

301

By the use of eq. (1), we obtain the following two expressions for the strain energy in a prismatic bar:

$$U = \frac{P^2 l}{2AE},$$ (171)

$$U = \frac{AE\delta^2}{2l}.$$ (172)

The first of these equations gives the strain energy as a function of the load P and the second gives the same energy as a function of the elongation δ. For a bar of given dimensions and a given modulus of elasticity the strain energy is completely determined by the value of the force P or the value of the elongation δ.

In practical applications the strain energy per unit volume w is often of importance. This is, from eqs. (171) and (172),

$$w = \frac{U}{Al} = \frac{\sigma^2}{2E},$$ (173)   or $$w = \frac{E\epsilon^2}{2},$$ (174)

in which $\sigma = P/A$ is the tensile stress and $\epsilon = \delta/l$ is the unit elongation.

The greatest amount of strain energy per unit volume which can be stored in a bar without permanent set [1] is found by substituting the elastic limit of the material in place of σ in eq. (173). Steel, with an elastic limit of 30,000 lb per sq in. and $E = 30 \times 10^6$ lb per sq in., gives $w = 15$ in. lb per cu in.; rubber, with a modulus of elasticity $E = 150$ lb per sq in. and an elastic limit of 300 lb per sq in., gives $w = 300$ in. lb per cu in. It is also sometimes of interest to know the greatest amount of strain energy per unit weight w_1 which can be stored without producing permanent set. This quantity is calculated from eq. (173) by substituting the elastic limit for σ and dividing w by the weight of one cubic inch of the material. Several numerical values calculated in this manner are given in Table 4.

---

[1] This quantity is sometimes called the *modulus of resilience*.

TABLE 4

Material	Specific Gravity	E lb per sq in.	Elastic Limit lb per sq in.	w in. lb per cu in.	w_1 in. lb per lb
Structural steel....	7.8	30×10^6	30,000	15.0	53
Tool steel.........	7.8	30×10^6	120,000	240	850
Copper...........	8.5	16×10^6	4,000	0.5	1.6
Oak..............	1.0	1.5×10^6	4,000	5.3	146
Rubber.......!...	0.93	150	300	300	8,900

This indicates that the quantity of energy which can be stored in a given weight of rubber is about 10 times larger than for tool steel and about 170 times larger than for structural steel.

Problems

1. A prismatic steel bar 10 in. long and 4 sq in. in cross-sectional area is compressed by a force $P = 4,000$ lb. Determine the amount of strain energy.

Answer. $U = \frac{2}{3}$ in. lb.

2. Determine the amount of strain energy in the previous problem if the cross-sectional area is 2 sq in. instead of 4 sq in.

Answer. $U = 1\frac{1}{3}$ in. lb.

3. Determine the amount of strain energy in a vertical uniform steel bar strained by its own weight if the length of the bar is 100 ft and its cross-sectional area 1 sq in., the weight of steel being 490 lb per cu ft.

Answer. $U = 0.772$ in. lb.

4. Determine the amount of strain energy in the previous problem if in addition to its own weight the bar carries an axial load $P = 1,000$ lb applied at the end.

Answer. $U = 27.58$ in. lb.

5. Check the solution of the problem shown in Fig. 18, p. 20, for the case in which all bars have the same cross section and the same modulus by equating the strain energy of the system to the work done by the load P.

Solution. If X is the force in the vertical bar, its elongation is Xl/AE and the work done by P is $\frac{1}{2}P(Xl/AE)$. Equating this to the strain energy, we obtain

$$\frac{1}{2} P \frac{Xl}{AE} = \frac{X^2 l}{2AE} + 2\frac{(X \cos^2 \alpha)^2 l}{2AE \cos \alpha},$$

from which

$$X = \frac{P}{1 + 2\cos^3 \alpha},$$

which checks the previous solution.

6. Check Prob. 2, p. 10, by showing that the work done by the load is equal to the strain energy of the two bars.

7. A steel bar 30 in. long and of 1 sq in. cross-sectional area is stretched 0.02 in. Find the amount of strain energy.

Answer. From eq. (172),

Fig. 263.

$$U = \frac{(0.02)^2 \times 30 \times 10^6}{2 \times 30} = 200 \text{ in. lb.}$$

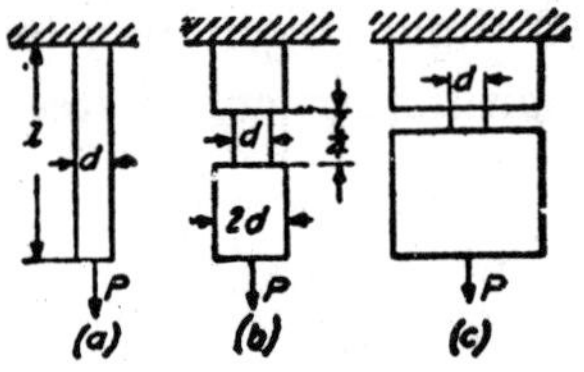

8. Compare the amounts of strain energy in the two circular bars shown in Figs. 263a and 263b, assuming a uniform distribution of stresses over cross sections of the bars.

Solution. The strain energy of the prismatic bar is

$$U = \frac{P^2 l}{2AE}.$$

The strain energy of the grooved bar is

$$U_1 = \frac{P^2 \frac{1}{4} l}{2AE} + \frac{P^2 \frac{3}{4} l}{8AE} = \frac{7}{16}\frac{P^2 l}{2AE}.$$

Hence

$$\frac{U_1}{U} = \frac{7}{16}.$$

For a given maximum stress the quantity of energy stored in a grooved bar is less than that in a bar of uniform thickness. It takes only a very small amount of work to bring the tensile stress to a dangerous limit in a bar such as shown in Fig. 263c, having a very narrow groove and a large outer diameter, although its diameter at the weakest place is equal to that of the cylindrical bar.

67. Tension Produced by Impact.—A simple arrangement

for producing tension by impact is shown in Fig. 264. A weight W falls from a height h onto the flange mn and during the impact produces an extension of the vertical bar AB, which is fixed at the upper end. If the masses of the bar and flange are small in comparison with the mass of the falling body, a satisfactory approximate solution is obtained by neglecting the mass of the bar and assuming that there are no losses of energy during impact. After striking the flange mn the body

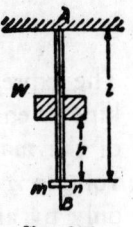

Fig. 264.

W continues to move downward, causing an extension of the bar. Due to the resistance of the bar the velocity of the moving body diminishes until it becomes zero. At this moment the elongation of the bar and the corresponding tensile stresses are a maximum and their magnitudes are calculated on the assumption that the total work done by the weight W is transformed into strain energy of the bar.[2] If δ denotes the maximum elongation, the work done by W is $W(h + \delta)$. The strain energy of the bar is given by eq. (172). Then the equation for calculating δ is

$$W(h + \delta) = \frac{AE}{2l}\delta^2, \qquad (a)$$

from which

$$\delta = \delta_{st} + \sqrt{\delta_{st}^2 + \frac{1}{g}\delta_{st}v^2}, \qquad (175)$$

where

$$\delta_{st} = \frac{Wl}{AE}$$

is the static elongation of the bar by the load W and $v = \sqrt{2gh}$ is the velocity of the falling body at the moment of striking the flange mn. If the height h is large in comparison with δ_{st}, this reduces to approximately

$$\delta = \sqrt{\frac{1}{g}\delta_{st}v^2}.$$

---

[2] In actual cases part of the energy will be dissipated and the actual elongation will always be less than that calculated on the above assumption.

The corresponding tensile stress in the bar is

$$\sigma = \frac{\delta E}{l} = \frac{E}{l}\sqrt{\frac{1}{g}\delta_{st}v^2} = \sqrt{\frac{2E}{Al}\cdot\frac{Wv^2}{2g}}. \tag{176}$$

The expression under the radical is directly proportional to the kinetic energy of the falling body, to the modulus of elasticity of the material of the bar, and inversely proportional to the volume Al of the bar. Hence the stress can be diminished not only by an increase in the cross-sectional area but also by an increase in the length of the bar or by a decrease in the modulus E. This is quite different from static tension of a bar where the stress is independent of the length l and the modulus E.

By substituting the working stress for σ in eq. (176) we obtain the following equation for proportioning a bar submitted to an axial impact:

$$Al = \frac{2E}{\sigma_W{}^2}\cdot\frac{Wv^2}{2g} = \frac{2EWh}{\sigma_W{}^2}, \tag{177}$$

i.e., for a given material the volume of the bar must be proportional to the kinetic energy of the falling body in order to keep the maximum stress constant.

Let us consider now another extreme case in which h is equal to zero, i.e., the body W is suddenly put on the support mn (Fig. 264) without an initial velocity. Although in this case we have no kinetic energy at the beginning of extension of the bar, the problem is quite different from that of static loading of the bar. In the case of static tension we assume a gradual application of the load and consequently there is always equilibrium between the acting load and the resisting forces of elasticity in the bar. The question of the kinetic energy of the load does not enter into the problem at all under such conditions. In the case of a sudden application of the load, the elongation of the bar and the stress in the bar are zero at the beginning and the suddenly applied load begins to fall under the action of its own weight. During this motion the resisting force of the bar gradually increases until it just equals W

when the vertical displacement of the weight is δ_{st}. But at this moment the load has a certain kinetic energy, acquired during the displacement δ_{st}, hence it continues to move downward until its velocity is brought to zero by the resisting force in the bar. The maximum elongation for this condition is obtained from eq. (175) by setting $v = 0$. Then

$$\delta = 2\delta_{st}, \qquad (178)$$

i.e., a suddenly applied load, due to dynamic conditions, produces a deflection which is twice as great as that which is obtained when the load is applied gradually.

This may also be shown graphically as in Fig. 265. The inclined line OA is the tensile test diagram for the bar shown in Fig. 264. Then for any elongation such as OC the

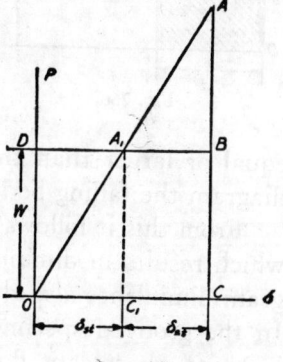

FIG. 265.

area AOC gives the corresponding strain energy in the bar. The horizontal line DB is at distance W from the δ axis and the area $ODBC$ gives the work done by the load W during the displacement OC. When δ is equal to δ_{st}, the work done by W is represented in the figure by the area of the rectangle ODA_1C_1. At the same time the energy stored in the bar is given by the area of the triangle OA_1C_1, which is only half the area of the above rectangle. The other half of the work done is transformed into the kinetic energy of the moving body. Due to its acquired velocity the body continues to move and comes to rest only at the distance $\delta = 2\delta_{st}$ from the origin. At this point the total work done by the load W, represented by the rectangle $ODBC$, equals the amount of energy stored in the bar and represented by the triangle OAC.

The above discussion of impact is based on the assumption that the stress in the bar remains within the elastic limit. Beyond this limit the problem becomes more involved because the elongation of the bar is no longer proportional to the tensile force. Assuming that the tensile test diagram does not depend upon the speed of straining the bar,[3] elongation

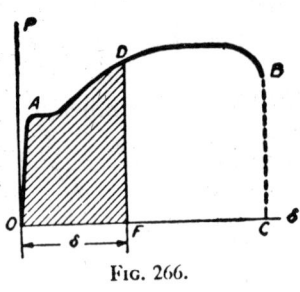

Fig. 266.

beyond the elastic limit during impact can be determined from an ordinary tensile test diagram such as shown in Fig. 266. For any assumed maximum elongation δ the corresponding area $OADF$ gives the work necessary to produce such an elongation; this must equal the work $W(h + \delta)$ produced by the weight W. When $W(h + \delta)$ is equal or larger than the total area $OABC$ of the tensile test diagram the falling body will fracture the bar.

From this it follows that any change in the form of the bar which results in diminishing the total area $OABC$ of the diagram diminishes also the resisting power of the bar to impact. In the grooved specimens shown in Figs. 263b and 263c, for instance, the plastic flow of metal will be concentrated at the groove and the total elongation and the work necessary to produce fracture will be much smaller than in the case of the cylindrical bar shown in the same figure. Such grooved specimens are very weak in impact. A slight shock may produce fracture, although the material itself is ductile. Members having rivet holes or any sharp variation in cross section are similarly weak against impact.[4]

In the previous discussion we neglected the mass of the bar in comparison with the mass of the falling body W. Only then may we assume that the total energy of the falling body is transformed into strain energy of the bar. The actual conditions of impact are more complicated and when the bar has an appreciable mass a part of the energy will be lost during impact. It is well known that when a mass W/g moving with a velocity v strikes cen-

[3] Experiments with ductile steel show that with a high velocity of straining the yield point is higher and the amount of work necessary to produce fracture is greater than in a static test. See N. N. Davidenkoff, *Bull. Polytech. Inst. (St. Petersburg)*, 1913; Welter, *Z. Metallkunde*, 1924; and M. J. Manjoine, *J. Appl. Mech.*, Vol. 11, p. 211, 1944.

[4] See Hackstroh, *Baumaterialienkunde*, p. 321, 1905; and H. Zimmermann, *Zentr. Bauverwalt.*, p. 265, 1899.

trally a stationary mass W_1/g and the deformation at the point of
contact is plastic, the final common velocity v_a of the two bodies is

$$v_a = \frac{W}{W + W_1} v. \qquad (b)$$

In the case of the bar shown in Fig. 264 the conditions are more
complicated. During impact the upper end A is at rest while ..e
lower end B acquires the velocity of the moving body W. Hence,
to calculate the final velocity v_a from eq. (b) we use a *reduced mass*
in place of the actual mass of the bar. Assuming that the velocity
of the bar varies linearly along its length, it can be shown that the
reduced mass in such a case is equal to one third of the mass of the
bar.[5] For a bar of weight q per unit length, eq. (b) becomes

$$v_a = \frac{W}{W + \dfrac{ql}{3}} v.$$

This is the common velocity of the load W and the lower end of the
bar which is established at the first moment of impact. Assuming
plastic deformation at the surface of contact between the falling
load and the support mn (Fig. 264) so that there will be no question
of rebounding, the corresponding kinetic energy is

$$\frac{v_a^2}{2g}(W + ql/3) = \frac{Wv^2}{2g} \cdot \frac{1}{1 + \dfrac{ql}{3W}}.$$

This quantity must be substituted for

$$\frac{Wv^2}{2g} = Wh$$

in eq. (a) in order to take into account the loss of energy at the
first moment of impact. Then, instead of eq. (175), we obtain

$$\delta = \delta_{st} + \sqrt{\delta_{st}^2 + \frac{1}{g}\delta_{st}v^2 \frac{1}{1 + \dfrac{ql}{3W}}}. \qquad (179)$$

The method described gives satisfactory results as long as the mass
of the bar is small in comparison with the mass of the falling body.

---

[5] This solution was obtained by H. Cox, *Trans. Cambridge Phil. Soc.*, p.
73, 1849. See also Todhunter and Pearson, *History of the Theory of Elasticity*, Cambridge, Vol. 1, p. 895, 1886.

Otherwise a consideration of longitudinal vibrations of the bar becomes necessary.[6] The local deformation at the point of contact during impact has been discussed by J. E. Sears [7] and J. E. P. Wagstaff.[8]

Problems

1. A weight of 10 lb attached to a steel wire $\frac{1}{8}$ in. in diameter (Fig. 267) falls from A with the acceleration g. Determine the stress produced in the wire when its upper end A is suddenly stopped. Neglect the mass of the wire.

FIG. 267.

Solution. If the acceleration of the weight W is equal to g, there is no tensile stress in the wire. The stress after stopping the wire at A is obtained from eq. (176), in which δ_{st} is neglected. Substituting $v^2 = 2gh$ and $l = h$, we obtain

$$\sigma = \sqrt{\frac{2EW}{A}} = \sqrt{\frac{2 \times 30 \times 10^6 \times 10 \times 8^2}{0.785}} = 221 \times 10^3 \text{ lb per sq in.}$$

It may be seen that the stress does not depend upon the height h through which the load falls, because the kinetic energy of the body increases in the same proportion as the volume of the wire.

2. A weight $W = 1,000$ lb falls from a height $h = 3$ ft upon a vertical wooden pole 20 ft long and 12 in. in diameter, fixed at the lower end. Determine the maximum compressive stress in the pole, assuming that for wood $E = 1.5 \times 10^6$ lb per sq in. and neglecting the mass of the pole and the quantity δ_{st}.

Answer. $\sigma = 2,000$ lb per sq in.

3. A weight $W = 10,000$ lb attached to the end of a steel wire rope (Fig. 267) moves downwards with a constant velocity $v = 3$ ft per sec. What stresses are produced in the rope when its upper end is suddenly stopped? The free length of the rope at the moment of impact is $l = 60$ ft, its net cross-sectional area is $A = 2.5$ sq in. and $E = 15 \times 10^6$ lb per sq in.

Solution. Neglecting the mass of the rope and assuming that the kinetic energy of the moving body is completely transformed into the potential energy of strain of the rope, the equation for de-

---

[6] The longitudinal vibrations of a prismatic bar during impact were considered by Navier. A more comprehensive solution was developed by St.-Venant; see his translation of Clebsch, *Theorie der Elasticität fester Körper*, 1883, note on par. 61. See also J. Boussinesq, *Application des Potentiels*, p. 508, 1885; and C. Ramsauer, *Ann. Physik*, Vol. 30, 1909.

[7] *Trans. Cambridge Phil. Soc.*, Vol. 21, p. 49, 1908.

[8] *Proc. Roy. Soc. (London), A.*, Vol. 105, p. 544, 1924.

termining the maximum elongation δ of the rope is

$$\frac{AE\delta^2}{2l} - \frac{AE\delta_{st}^2}{2l} = \frac{W}{2g}v^2 + W(\delta - \delta_{st}), \qquad (d)$$

in which δ_{st} denotes the statical elongation of the rope. Noting that $W = AE\delta_{st}/l$ we obtain, from eq. (d),

$$\frac{AE}{2l}(\delta - \delta_{st})^2 = \frac{Wv^2}{2g},$$

from which

$$\delta = \delta_{st} + \sqrt{\frac{Wv^2l}{AEg}}.$$

Hence, upon sudden stopping of the motion, the tensile stress in the rope increases in the ratio

$$\frac{\delta}{\delta_{st}} = 1 + \frac{v}{\delta_{st}}\sqrt{\frac{Wl}{AEg}} = 1 + \frac{v}{\sqrt{g\delta_{st}}}. \qquad (e)$$

For the above numerical data

$$\delta_{st} = \frac{Wl}{AE} = \frac{10,000 \times 60 \times 12}{2.5 \times 15 \times 10^6} = 0.192 \text{ in.,}$$

$$\frac{\delta}{\delta_{st}} = 1 + \frac{3 \times 12}{\sqrt{386 \times 0.192}} = 5.18.$$

Hence

$$\sigma = 5.18 \frac{W}{A} = 20,700 \text{ lb per sq in.}$$

4. Solve the previous problem if a spring which elongates $\frac{1}{2}$ in. per 1,000-lb load is put between the rope and the load.

Solution. $\delta_{st} = 0.192 + 0.5 \times 10 = 5.192$ in. Substituting into eq. (e),

$$\frac{\delta}{\delta_{st}} = 1 + 0.80 = 1.80, \qquad \sigma = 1.80 \frac{W}{A} = 7,200 \text{ lb per sq in.}$$

Comparison with the solution of the preceding problem shows that a spring incorporated between the rope and the load has a great effect in reducing the magnitude of σ_{max} at impact.

5. For the case shown in Fig. 264 determine the height h for which the maximum stress in the bar during impact is 30.000 lb per

sq in. Assume $W = 25$ lb, $l = 6$ ft, $A = \frac{1}{2}$ sq in., $E = 30 \times 10^6$ lb per sq in. Neglect the mass of the bar.

Answer. $h = 21.6$ in.

68. Elastic Strain Energy in Shear and Torsion.—The strain energy stored in an element submitted to pure shearing stress (Fig. 268) may be calculated by the method used in the case of simple tension. If the lower side *ad* of the element is taken as fixed, only the work done during strain by the force P at the upper side *bc* need be considered. Assuming that the material follows Hooke's law, the shearing strain is proportional to the shearing stress and the diagram showing this relationship is analogous to that shown in Fig. 262. The work done by the force P and stored in the form of elastic strain energy is then (see eq. 170, p. 301)

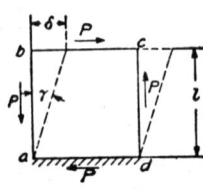

Fig. 268.

$$U = \frac{P\delta}{2}. \tag{170'}$$

Remembering that

$$\frac{\delta}{l} = \gamma = \frac{\tau}{G} = \frac{P}{AG},$$

we obtain the following two equations from (170'):

$$U = \frac{P^2 l}{2AG}, \quad (180) \qquad U = \frac{AG\delta^2}{2l}. \quad (181)$$

We obtain two expressions for the shearing strain energy per unit volume by dividing these equations by the volume Al of the block:

$$w = \frac{\tau^2}{2G}, \quad (182) \qquad w = \frac{\gamma^2 G}{2}, \quad (183)$$

in which $\tau = P/A$ is the shearing stress and $\gamma = \delta/l$ is the shearing strain. The amount of shear energy per unit volume, which can be stored in the block without permanent set, is obtained by substituting the elastic limit for τ in eq. (182).

The energy stored in a twisted circular shaft is easily calculated by use of eq. (182). If τ_{max} is the maximum shearing stress at the surface of the shaft, then $\tau_{max}(2r/d)$ is the shearing stress at a point a distance r from the axis, where d is the diameter of the shaft. The energy per unit volume at this point is, from eq. (182),

$$w = \frac{2\tau_{max}^2 r^2}{Gd^2}. \qquad (a)$$

The energy stored in the material included between two cylindrical surfaces of radii r and $r + dr$ is

$$\frac{2\tau_{max}^2 r^2}{Gd^2} 2\pi l r dr,$$

where l is the length of the shaft. Then the total energy stored in the shaft is

$$U = \int_0^{d/2} \frac{2\tau_{max}^2 r^2}{Gd^2} 2\pi l r dr = \frac{1}{2} \frac{\pi d^2 l}{4} \frac{\tau_{max}^2}{2G}. \qquad (184)$$

This shows that the total energy is only half what it would be if all elements of the shaft were stressed to the maximum shearing stress τ_{max}.

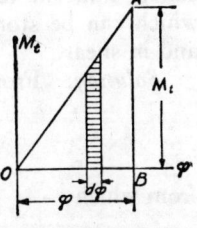

Fig. 269.

The energy of torsion may also be calculated from a diagram of torsion (Fig. 269) in which the torque is represented by the ordinates and the angle of twist by the abscissas. Within the elastic limit, the angle of twist is proportional to the twisting moment, as represented by an inclined line OA. The small area shaded in the figure represents the work done by the torque during an increment $d\varphi$ in the angle of twist φ. The area $OAB = M_t\varphi/2$ represents the total energy stored in the shaft during twist. Recalling that $\varphi = M_t l / GI_p$, we obtain

$$U = \frac{M_t^2 l}{2GI_p} \quad \text{or} \quad U = \frac{\varphi^2 GI_p}{2l}. \qquad (185)$$

In the first of these two equations the energy is represented as a function of the torque; in the second, as a function of the angle of twist.

In the general case of any shape of cross section and a torque varying along the length of the shaft, the angle of twist between the two adjacent cross sections is given by the equation (see p. 290)

$$\frac{d\varphi}{dx} dx = \frac{M_t}{C} dx.$$

The strain energy of the portion of the shaft between two adjacent cross sections is

$$\frac{1}{2} M_t \frac{d\varphi}{dx} dx = \frac{C}{2} \left(\frac{d\varphi}{dx}\right)^2 dx,$$

and the total energy of twist is

$$U = \frac{C}{2} \int_0^l \left(\frac{d\varphi}{dx}\right)^2 dx = \frac{1}{2C} \int_0^l M_t^2 dx. \tag{186}$$

Problems

1. Determine the ratio between the elastic limit in shear and the elastic limit in tension if the amount of strain energy per cu in., which can be stored without permanent set, is the same in tension and in shear.

Solution. From eqs. (173) and (182),

$$\frac{\sigma^2}{2E} = \frac{\tau^2}{2G},$$

from which

$$\frac{\tau}{\sigma} = \sqrt{\frac{G}{E}}.$$

For steel

$$\tau = \sigma \sqrt{\frac{1}{2.6}} = 0.62\sigma.$$

2. Determine the deflection of a helical spring (Fig. 253) by using the expression for the strain energy of torsion.

Solution. Denote by P the force acting in the direction of the axis of the helix (Fig. 253), by R the radius of the coils and by n the number of coils. The energy of twist stored in the spring, from eq. (185), is

$$U = \frac{(PR)^2 2\pi Rn}{2GI_p}.$$

Equating this to the work done, $P\delta/2$, we obtain

$$\delta = \frac{2\pi n P R^3}{G I_p} = \frac{64 n P R^3}{G d^4},$$

which coincides with eq. (162).

3. The weight of a steel helical spring is 10 lb. Determine the amount of energy which can be stored in this spring without producing permanent set if the elastic limit in shear is 74,300 lb per sq in.

Solution. The amount of energy per cu in., from eq. (182), is

$$w = \frac{(74,300)^2}{2 \times 11.5 \times 10^6} = 240 \text{ in. lb.}$$

The energy per lb of material (see p. 302) is 850 in. lb. Then the total energy of twist [9] which can be stored in the spring is

$$\tfrac{1}{2} \times 40 \times 850 = 4,250 \text{ in. lb.}$$

4. A solid circular shaft and a thin tube of the same material and the same weight are submitted to twist. In what ratio are the amounts of energy in shaft and tube if the maximum stresses in both are equal?

Answer. $\tfrac{1}{2} : 1.$

5. A circular steel shaft with a flywheel at one end rotates at 120 rpm. It is suddenly stopped at the other end. Determine the maximum stress in the shaft during impact if the length of the shaft $l = 5$ ft, the diameter $d = 2$ in., the weight of the flywheel $W = 100$ lb, its radius of gyration $r = 10$ in.

Solution. Maximum stress in the shaft is produced when the total kinetic energy of the flywheel is transformed into strain energy of the twisted shaft. The kinetic energy of the flywheel is

$$\frac{W r^2 \omega^2}{2g} = \frac{100 \times 10^2 \times (4\pi)^2}{2 \times 386} = 2,050 \text{ in. lb.}$$

Substituting this for U in eq. (184),

$$\tau_{max} = \sqrt{\frac{16 \times 11.5 \times 10^6 \times 2,050}{\pi \times 4 \times 60}} = 22,400 \text{ lb per sq in.}$$

6. Two circular bars of the same material and the same length but different cross sections A and A_1, are twisted by the same torque.

---

[9] The stress distribution is assumed to be the same as that in a twisted circular bar.

In what ratio are the amounts of strain energy stored in these two bars?

Answer. Inversely proportional to the squares of the cross-sectional areas.

69. Elastic Strain Energy in Bending.—Let us begin with the case of pure bending. For a prismatic bar built in at one

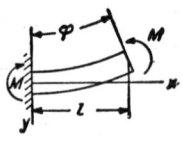

FIG. 270.

end, bent by a couple M applied at the other end (Fig. 270) and acting in one of the principal planes, the angular displacement at the free end is

$$\varphi = \frac{Ml}{EI_z}. \qquad (a)$$

This displacement is proportional to the bending moment M, and by using a diagram similar to that in Fig. 269 we may conclude from similar reasoning that the work done during deflection by the bending moment M, or the energy stored in the bar, is

$$U = \frac{M\varphi}{2}. \qquad (b)$$

By use of eq. (a) this energy may be expressed in either of these forms:

$$U = \frac{M^2 l}{2EI_z}, \qquad (187) \qquad\qquad U = \frac{\varphi^2 EI_z}{2l}. \qquad (188)$$

It is sometimes useful to have the potential energy expressed as a function of the maximum normal stress $\sigma_{max} = M_{max}/Z$. Thus, for a rectangular bar $\sigma_{max} = 6M/bh^2$ or $M = bh^2\sigma_{max}/6$, and eq. (187) becomes

$$U = \frac{1}{3} bhl \frac{\sigma_{max}{}^2}{2E}. \qquad (189)$$

In this case the total energy is evidently only one-third as much as it would be if all fibers carried the stress σ_{max}.

In discussing the bending of bars by transverse forces the strain energy of shear will be neglected for the present. The energy stored in an element of the beam between two adjacent cross sections dx apart is, from eqs. (187) and (188),

$$dU = \frac{M^2 dx}{2EI_z} \qquad \text{or} \qquad dU = \frac{EI_z(d\varphi)^2}{2dx}.$$

Here the bending moment M is variable with respect to x, and

$$d\varphi = \frac{dx}{r} = \left|\frac{d^2y}{dx^2}\right| dx$$

(see p. 138). The total energy stored in the beam is consequently

$$U = \int_0^l \frac{M^2 dx}{2EI_z} \quad (190) \qquad \text{or} \qquad U = \int_0^l \frac{EI_z}{2}\left(\frac{d^2y}{dx^2}\right)^2 dx. \quad (191)$$

Take, e.g., the cantilever AB (Fig. 271a). The bending moment at any cross section mn is $M = -Px$. Substitution into eq. (190) gives

$$U = \int_0^l \frac{P^2 x^2 dx}{2EI_z} = \frac{P^2 l^3}{6EI_z}. \quad (c)$$

For a rectangular bar, $\sigma_{max} = 6Pl/bh^2$, and eq. (c) may be put in the form

$$U = \frac{1}{9} bhl \frac{\sigma_{max}^2}{2E}. \quad (c')$$

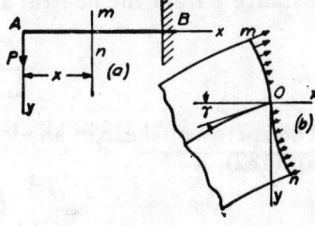

FIG. 271.

This shows that the quantity of energy which can be stored in a rectangular cantilever beam, loaded at the end, without producing permanent set, is one-third of that for pure bending of the same bar and one-ninth of that for the same bar in simple tension. This consideration is of importance in designing springs, which must absorb a given amount of energy without damage and yet have as small a weight as possible. The capacity of a cantilever to absorb energy may be increased by giving it a variable cross section. For example, a cantilever of *uniform strength* with a rectangular cross section of constant depth h (Fig. 188), and with the same values for P, h and σ_{max}, has a deflection and hence an amount of stored energy 50 per cent greater than for the prismatic bar. At the same time the bar of uniform strength has in this case half the weight of the prismatic bar, so that it can store three times as much energy per pound of material.

Returning to eq. (c) and equating the strain energy to the work done by the load P during deflection, we obtain

$$\frac{P\delta}{2} = \frac{P^2 l^3}{6EI_z}, \tag{d}$$

from which the deflection at the end is

$$\delta = \frac{Pl^3}{3EI_z},$$

which coincides with eq. (95).

The additional deflection due to shear may also be determined from the potential energy of strain. For the cantilever shown in Fig. 271, with a rectangular cross section, the shearing stress at a distance y from the neutral axis is (see eq. 65, p. 118)

$$\frac{P}{2I_z}\left(\frac{h^2}{4} - y^2\right).$$

The energy of shear in an elemental volume $bdxdy$ is, therefore, from eq. (182),

$$\frac{P^2}{8GI_z{}^2}\left(\frac{h^2}{4} - y^2\right)^2 bdxdy,$$

and

$$U = \int_0^l \int_{-h/2}^{+h/2} \frac{P^2}{8GI_z{}^2}\left(\frac{h^2}{4} - y^2\right)^2 bdxdy = \frac{P^2 l h^2}{20GI_z}. \tag{e}$$

This must be added [10] to the right-hand side of eq. (d) above to obtain the equation for determining the total deflection:

$$\frac{P\delta}{2} = \frac{P^2 l^3}{6EI_z} + \frac{P^2 l h^2}{20GI_z}. \tag{f}$$

Consequently

$$\delta = \frac{Pl^3}{3EI_z}\left(1 + \frac{3}{10}\frac{h^2}{l^2}\frac{E}{G}\right). \tag{g}$$

The second term in the parentheses represents the effect of the shearing stresses on the deflection of the beam.

[10] Such an addition of the energy of shear to the energy due to normal stresses is justified, because the shearing stresses acting on an element (Fig. 268) do not change the lengths of the sides of the element and if normal forces act on these sides, they do no work during shearing strain. Hence shearing stresses do not change the amount of energy due to tension or compression and the two kinds of energy may be simply added together.

By use of the method developed in Art. 39 under the assumption that the element of the cross section at the centroid of the built-in end remains vertical (Fig. 271b), the additional slope due to shear is

$$\gamma = \frac{\tau_{max}}{G} = \frac{3}{2} \frac{P}{bhG},$$

and the additional deflection is

$$\frac{3}{2} \frac{Pl}{bhG}.$$

Hence

$$\delta = \frac{Pl^3}{3EI_z} + \frac{3}{2} \frac{Pl}{bhG} = \frac{Pl^3}{3EI_z} \left(1 + \frac{3}{8} \frac{h^2}{l^2} \frac{E}{G}\right). \qquad (g')$$

We see that eqs. (g) and (g') do not coincide. The discrepancy is explained as follows: The derivation of Art. 39 was based on the assumption that the cross sections of the beam can warp freely under the action of shearing stresses. In such a case the built-in cross section will be distorted to a curved surface *mon* (Fig. 271b) and in calculating the total work done on the cantilever we must consider not only the work done by the force P, Fig. 271a, but also the work done by the stresses acting on the built-in cross section, Fig. 271b. If this latter work is taken into account, the deflection calculated from the consideration of the strain energy coincides with that obtained in Art. 39 and given in eq. (g') above

In the case of a simply supported beam loaded at the middle, the middle cross section does not warp, as can be concluded from considerations of symmetry. In such a case eq. (g), if applied to each half of the beam, will give a better approximation for the deflection than will eq. (g'). This can be seen by comparing the approximate eqs. (g) and (g') with the more rigorous solution given in Art. 39.

Problems

1. A wooden cantilever beam, 6 ft long, of rectangular cross section 8 × 5 in. carries a uniform load $q = 200$ lb per ft. Determine the amount of strain energy stored if $E = 1.5 \times 10^6$ lb per sq in.

Answer. $U = \dfrac{q^2 l^5}{40EI_z} = \dfrac{1,200^2 \times 72^3 \times 12}{40 \times 1.5 \times 10^6 \times 5 \times 8^3} = 42$ in. lb.

2. In what ratio does the amount of strain energy calculated in the preceding problem increase if the depth of the beam is 5 in. and the width 8 in.?

Answer. The strain energy increases in the ratio $8^2 : 5^2$.

3. Two identical bars, one simply supported, the other with built-in ends, are bent by equal loads applied at the middle. In what ratio are the amounts of strain energy stored?

Answer. 4:1.

4. Solve the preceding problem for a uniformly distributed load of the same intensity q for both bars.

5. Find the ratio of the amounts of strain energy stored in beams of rectangular section equally loaded, having the same length and the same width of cross sections but whose depths are in the ratio 2:1.

Solution. For a given load the strain energy is proportional to the deflection and this is inversely proportional to the moment of inertia of the cross section. By halving the depth the deflection is therefore increased 8 times and the amount of strain energy increases in the same proportion.

70. Deflection Produced by Impact.

—The dynamic deflection of a beam which is struck by a falling body W may be determined by the method used in the case of impact causing tension (Art. 67). Take, as an example, a simply supported beam struck at the middle (Fig. 272), and assume that the mass of the beam may be neglected in comparison with the mass of the falling body and that the beam is not stressed beyond the yield point. Then there will be no loss of energy during impact and the work done by the weight W during its fall is completely transformed into strain energy of bending of the beam.[11] Let δ denote the maximum deflection of the beam during impact. If we assume that the deflection curve during impact has the same shape as that during static deflection, the force which would cause such a deflection is, from eq. (90),

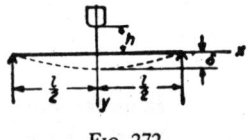

Fig. 272.

$$P = \delta \cdot \frac{48EI_z}{l^3}. \qquad (a)$$

The total energy stored in the beam is equal to the work done by the force P:

$$U = \frac{P\delta}{2} = \delta^2 \frac{24EI_z}{l^3}.$$

---

[11] Local deformation at the surface of contact of the load and the beam is neglected in this calculation.

If h denotes, as before, the distance fallen before impact, the equation for determining δ is

$$W(h + \delta) = \delta^2 \frac{24EI_z}{l^3}, \qquad (b)$$

from which

$$\delta = \delta_{st} + \sqrt{\delta_{st}^2 + \frac{1}{g}\delta_{st}v^2}, \qquad (192)$$

where

$$\delta_{st} = \frac{Wl^3}{48EI_z} \quad \text{and} \quad v = \sqrt{2gh}.$$

Eq. (192) is exactly the same as that for impact causing tension (eq. 175).

It should be noted that the form of the equation remains the same for any other case of impact, provided the deflection at the point of impact is proportional to the force P, exerted at this point. If we represent by α the factor of proportionality which depends upon the structure, we have

$$\alpha P = \delta \quad \text{and} \quad U = \frac{P\delta}{2} = \frac{\delta^2}{2\alpha}.$$

Then

$$W(h + \delta) = \frac{\delta^2}{2\alpha},$$

and since $\delta_{st} = W\alpha$, this reduces to eq. (192) above.

It should be noted also that the deflection δ calculated from (192) represents the upper limit which the maximum dynamic deflection approaches when there are no losses of energy during impact. Any such loss will reduce the dynamic deflection. When the dynamic deflection is found from eq. (192), the corresponding stresses can be found by multiplying by δ/δ_{st} the stresses obtained for the statical application of the load W.

When h is large in comparison with δ_{st} or if the impact is in the horizontal direction, eq. (192) takes the simpler form

$$\delta = \sqrt{\frac{1}{g}\delta_{st}v^2}. \qquad (c)$$

For the case of a beam supported at the ends and struck at the middle this equation gives

$$\delta = \sqrt{\frac{Wv^2}{2g} \frac{l^3}{24EI_z}}. \qquad (d)$$

The maximum bending moment in this case is

$$M_{max} = \frac{Pl}{4} = \frac{\delta \cdot 48EI_z}{l^3} \frac{l}{4}$$

and

$$\sigma_{max} = \frac{M_{max}}{Z} = \frac{\delta \cdot 48EI_z}{l^3} \frac{l}{4Z}.$$

For a rectangular cross section, using eq. (d),

$$\sigma_{max} = \sqrt{\frac{Wv^2}{2g} \frac{18\overline{E}}{iA}}. \qquad (e)$$

This indicates that the maximum stress depends upon the kinetic energy of the falling body and the volume Al of the beam.

In determining the effect of the mass of the beam on the maximum deflection we will assume that the deflection curve during impact has the same shape as during static deflection. Then it can be shown that the reduced mass of the beam [12] supported at the ends is $(17/35)(ql/g)$ and the common velocity which will be established at the first moment of impact is

$$v_a = \frac{W}{W + (17/35)ql} v.$$

The total kinetic energy after the establishment of the common velocity v_a is

$$\frac{v_a{}^2}{2g} [W + (17/35)ql] = \frac{Wv^2}{2g} \frac{1}{1 + \frac{17}{35}\frac{ql}{W}}.$$

Using this instead of

$$\frac{Wv^2}{2g} = Wh$$

---

[12] See H. Cox, loc. cit., p. 309.

in eq. (b), we obtain

$$\delta = \delta_{st} + \sqrt{\delta_{st}^2 + \frac{\delta_{st}v^2}{g}\,\frac{1}{1 + \frac{17}{35}\frac{ql}{W}}},\qquad(193)$$

which takes account of the effect of the mass of the beam on the deflection δ.[13]

In the case of a cantilever, if the weight W strikes the beam at the end, the magnitude of the reduced mass of the beam is $\frac{33}{140}(ql/g)$. When a beam simply supported at the ends is struck at a point whose distances from the supports are a and h, respectively, the reduced mass is

$$\frac{1}{105}\left[1 + 2\left(1 + \frac{l^2}{ab}\right)^2\right]\frac{ql}{g},$$

and eq. (193) must be changed accordingly.

Problems

1. A simply supported rectangular wooden beam 9 ft long is struck at the middle by a 40-lb weight falling from a height $h = 12$ in. Determine the necessary cross-sectional area if the working stress is $\sigma_W = 1{,}000$ lb per sq in., $E = 1.5 \times 10^6$ lb per sq in. and δ_{st} is neglected in comparison with h.

Solution. Using eq. (e), p. 322,[14]

$$A = \frac{Wv^2}{2g}\frac{18E}{l\sigma_W^2} = 40 \times 12 \frac{18 \times 1.5 \times 10^6}{9 \times 12 \times 1{,}000^2} = 120 \text{ sq in.}$$

---

[13] Several examples of the application of this equation will be found in the paper by Prof. Tschetsche, *Z. Ver. deut. Ing.*, p. 134, 1894. A more accurate theory of transverse impact on a beam is based on the investigation of its lateral vibration together with the local deformations at the point of impact. See St.-Venant's translation of Clebsch's book, p. 537, *loc. cit.*; also *Compt. rend.*, Vol. 45, p. 204, 1857; and writer's paper in *Z. Math. u. Phys.*, Vol. 62, p. 198, 1913. Experiments with beams subjected to impact have been made in Switzerland and are in satisfactory agreement with the above approximate theory, see M. Roš, *Tech Komm. Verband. Schweiz. Brückenbau-u. Eisenhochbaufabriken*, Mar. 1922. See also Z. Tuzi and M. Nisida, *Phil. Mag.* (Ser. 7), Vol. 21, p. 448; R. N. Arnold, *Proc. Inst. Mech. Engrs. (London)*, Vol. 137, p. 217, 1937; and E. H. Lee, *J. Appl. Mech.*, Vol. 7, p. 129, 1940.

[14] Local deformation at the surface of contact of the load and the beam is neglected in this calculation.

2. In what proportion should the area in the preceding problem be changed if (1) the span of the beam increases from 9 to 12 ft and (2) the weight W increases by 50 per cent?

Answer. (1) The area should be diminished in the ratio 3:4. (2) The area should be increased by 50 per cent.

3. A weight $W = 100$ lb drops 12 in. upon the middle of a simply supported I beam, 10 ft long. Find the safe dimensions if $\sigma_W = 30 \times 10^3$ lb per sq in.

Solution. Neglecting δ_{st} in comparison with h (see eq. *c*), the ratio between the dynamic and the static deflections is

$$\frac{\delta}{\delta_{st}} = \sqrt{\frac{v^2}{g\delta_{st}}} = \sqrt{\frac{2h}{\delta_{st}}}.$$

If the deflection curve during impact is of the same shape as for static deflection, the maximum bending stresses will be in the same ratio as the deflections. Hence

$$\sqrt{\frac{2h}{\delta_{st}}\frac{Wl}{4Z}} = \sigma_W, \text{ from which } \frac{Z}{c} = \frac{6EW}{\sigma_W^2}\frac{h}{l},$$

in which Z is the section modulus and c is the distance from the neutral axis of the most remote fiber, which is half the depth of the beam in this case. Substituting the numerical data,

$$\frac{Z}{c} = \frac{6 \times 30 \times 10^6 \times 100 \times 12}{30,000^2 \times 120} = 2 \text{ in.}^2$$

The necessary I beam is 6 I 12.5 (see Appendix).

4. What stress is produced in the beam of the preceding problem by a 200-lb weight falling onto the middle of the beam from a height of 6 in.?

Answer. $\sigma_{max} = 27,200$ lb per sq in.

5. A wooden cantilever beam 6 ft long and of square cross section 12 $\times$ 12 in. is struck at the end by a weight $W = 100$ lb falling from a height $h = 12$ in. Determine the maximum deflection, taking into account the loss in energy due to the mass of the beam.

Solution. Neglecting δ_{st} in comparison with h, the equation analogous to eq. (193) becomes

$$\delta = \sqrt{\frac{\delta_{st}v^2}{g}\frac{1}{1 + \dfrac{33}{140}\dfrac{ql}{W}}}.$$

For $ql = 40 \times 6 = 240$ lb,

$$\delta = \sqrt{\delta_{st} \cdot \frac{24}{1 + \dfrac{33 \times 240}{140 \times 100}}} = \sqrt{\frac{15.3 \times 100 \times 72^3}{3 \times 1.5 \times 10^6 \times 12^3}} = 0.271 \text{ in.}$$

6. A beam simply supported at the ends is struck at the middle by a weight W falling down from the height h. Neglecting δ_{st} in comparison with h, find the magnitude of the ratio ql/W at which the effect of the mass of the beam reduces the dynamical deflection by 10 per cent.

Answer. $\dfrac{ql}{W} = 0.483.$

71. The General Expression for Strain Energy.—In the discussion of problems in tension, compression, torsion and bending it has been shown that the energy of strain can be represented in each case by a function of the second degree in the external forces (eqs. 171, 180 and 187) or by a function of the second degree in the displacements (eqs. 172, 181 and 188). This is also true for the general case of deformation of an elastic body, with the following provisions: (1) the material follows Hooke's law and (2) the conditions are such that the small displacements, due to strain, do not affect the action of the external forces and are negligible in calculating the stresses.[15] With these two provisions the displacements of an elastic system are linear functions of the external loads. If these loads increase in a certain proportion, all the displacements increase in the same proportion. Consider a body submitted to the action of the external forces P_1, P_2, P_3, $\cdots$ (Fig. 273) and supported in such a manner that movement as a rigid body is impossible and displacements are due to elastic deformations only. Let δ_1, δ_2, δ_3, $\cdots$ denote the displacements of the points of application of the forces, each measured in the direction of the corresponding force.[16] If the external forces increase

---

[15] Such problems as the bending of bars by lateral forces with simultaneous axial tension or compression do not satisfy the above condition and are excluded from this discussion. Regarding these exceptional cases see Art. 76.

[16] The displacements of the same points in the directions perpendicular to the corresponding forces are not considered in the following discussion.

gradually so that they are always in equilibrium with the resisting internal elastic forces, the work which they do during
deformation will be equal to the strain
energy stored in the deformed body. The
amount of this energy does not depend
upon the order in which the forces are
applied and is completely determined by
their final magnitudes. Let us assume
that all external forces P_1, P_2, P_3, $\cdots$
increase simultaneously in the same ratio.

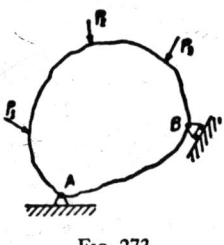

FIG. 273.

Then the relation between each force and
its corresponding displacement can be represented by a diagram
analogous to that shown in Fig. 262, and the work done by all
the forces P_1, P_2, P_3, $\cdots$, equal to the strain energy stored in
the body, is

$$U = \frac{P_1\delta_1}{2} + \frac{P_2\delta_2}{2} + \frac{P_3\delta_3}{2} + \cdots, \qquad (194)$$

i.e., the total energy of strain is equal to half the sum of the
products of each external force and its corresponding displacement.[17] On the assumptions made above, the displacements
δ_1, δ_2, δ_3, $\cdots$ are homogeneous linear functions of the forces
P_1, P_2, P_3, $\cdots$. The substitution of these functions into eq.
(194) gives a general expression for the strain energy in the
form of a homogeneous function of the second degree in the
external forces. If the forces be represented as linear functions
of the displacements and these functions be substituted into
eq. (194), an expression for the strain energy in the form of a
homogeneous function of the second degree in the displacements is obtained.

In the above discussion the reactions at the supports were
not taken into consideration. The work done by these reactions during the deformation is equal to zero since the displacement of an immovable support, such as A (Fig. 273),
is zero and the displacement of a movable support, such as B,
is perpendicular to the reaction, friction at the supports being

---

[17] This conclusion was obtained first by Clapeyron; see Lamé, *Leçons sur
la théorie mathématique de l'élasticité*, 2d Ed., p. 79, 1866.

neglected. Consequently, the reactions add nothing to the expression for the potential energy (194).

As an example of the application of eq. (194) let us consider the energy stored in a cubic element submitted to uniform tension in three perpendicular directions (Fig. 54). If the edge of the cube is of unit length, the tensile forces on its faces are numerically σ_x, σ_y and σ_z and the corresponding elongations are ϵ_x, ϵ_y and ϵ_z. Then the strain energy stored in one cubic inch from eq. (194), is

$$w = \frac{\sigma_x \epsilon_x}{2} + \frac{\sigma_y \epsilon_y}{2} + \frac{\sigma_z \epsilon_z}{2}.$$

Substituting for the elongations the values given by Eq. (43),[18] we obtain

$$w = \frac{1}{2E}(\sigma_x{}^2 + \sigma_y{}^2 + \sigma_z{}^2) - \frac{\mu}{E}(\sigma_x\sigma_y + \sigma_y\sigma_z + \sigma_z\sigma_x). \quad (195)$$

This expression can also be used when some of the normal stresses are compressive, in which case they must be given a negative sign.

If in addition to normal stresses there are shearing stresses acting on the faces of the element, the energy of shear can be added to the energy of tension or compression (see p. 318), and using eq. (182) the total energy stored in one cubic inch is

$$w = \frac{1}{2E}(\sigma_x{}^2 + \sigma_y{}^2 + \sigma_z{}^2) - \frac{\mu}{E}(\sigma_x\sigma_y + \sigma_y\sigma_z + \sigma_z\sigma_x)$$
$$+ \frac{1}{2G}(\tau_{xy}{}^2 + \tau_{yz}{}^2 + \tau_{zz}{}^2). \quad (196)$$

As a second example let us consider a beam simply supported at the ends, loaded at the middle by a force P and bent by a couple M applied at the end A. The deflection at the middle is, from eqs. (90) and (105),

$$\delta = \frac{Pl^3}{48EI} + \frac{Ml^2}{16EI}. \quad (a)$$

The slope at the end A is, from eqs. (88) and (104),

$$\theta = \frac{Pl^2}{16EI} + \frac{Ml}{3EI}. \quad (b)$$

[18] In this calculation the changes in temperature due to strain are considered of no practical importance. For further discussion see T. Weyrauch, *Theorie elastischer Körper*, Leipzig, p. 163, 1884. See also *Z. Architek. u. Ingenieurw.*, Vol. 54, pp. 91 and 277, 1908.

Then the strain energy of the beam, equal to the work done by the force P and by the couple M, is

$$U = \frac{P\delta}{2} + \frac{M\theta}{2} = \frac{1}{EI}\left(\frac{P^2 l^3}{96} + \frac{M^2 l}{6} + \frac{MP l^2}{16}\right). \qquad (c)$$

This expression is a homogeneous function of the second degree in the external force and the external couple. Solving eqs. (a) and (b) for M and P and substituting into eq. (c), an expression for the strain energy in the form of a homogeneous function of the second degree in the displacements may be obtained. It must be noted that when external couples are acting on the body the corresponding displacements are the angular displacements of the surface elements on which these couples are acting.

72. The Theorem of Castigliano.—Having the expressions for the strain energy in various cases, a very simple method for calculating the displacements of points of an elastic body during deformation may be established. For example, in the case of simple tension (Fig. 1), the strain energy as given by eq. (171) is

$$U = \frac{P^2 l}{2AE}.$$

By taking the derivative of this expression with respect to P, we obtain

$$\frac{dU}{dP} = \frac{Pl}{AE} = \delta.$$

Thus the derivative of the strain energy with respect to the load gives the displacement corresponding to the load, i.e., the displacement of the point of application of the load in the direction of the load. In the case of a cantilever loaded at the end, the strain energy is (eq. c, p. 317)

$$U = \frac{P^2 l^3}{6EI}.$$

The derivative of this expression with respect to the load P gives the known deflection $Pl^3/3EI$ at the free end.

In the torsion of a circular shaft the strain energy is (eq. 185)

$$U = \frac{M_t^2 l}{2GI_p}.$$

The derivative of this expression with respect to the torque gives

$$\frac{dU}{dM_t} = \frac{M_t l}{GI_p} = \varphi,$$

which is the angle of twist of the shaft, and represents the displacement corresponding to the torque.

When several loads act on an elastic body the same method of calculation of displacements may be used. For example, expression (c) of the preceding article gives the strain energy of a beam bent by a load P at the middle and by a couple M at the end. The partial derivative of this expression with respect to P gives the deflection under the load and the partial derivative with respect to M gives the angle of rotation of the end of the beam on which the couple M acts.

The theorem of Castigliano is a general statement of these results.[19] If the material of the system follows Hooke's law and the conditions are such that the small displacements due to deformation can be neglected in discussing the action of the forces, the strain energy of such a system may be given by a homogeneous function of the second degree in the acting forces (see Art. 71). Then the partial derivative of strain energy with respect to any such force gives the displacement corresponding to this force (for exceptional cases see Art. 76). The terms *force* and *displacement* here may have their generalized meanings, i.e., they include *couple* and *angular displacement*, respectively.

Let us consider a general case such as shown in Fig. 273. Assume that the strain energy is represented as a function of the forces $P_1, P_2, P_3, \cdots$, so that

$$U = f(P_1, P_2, P_3, \cdots). \tag{a}$$

[19] See the paper by Castigliano, "Nuova teoria intorno dell' equilibrio dei sistemi elastici," *Atti acc. sci. Torino*, 1875. See also his *Théorie de l'équilibre des systèmes élastiques*, Turin, 1879. For an English translation of Castigliano's work see E. S. Andrews, *Elastic Stresses in Structures*, London, 1919.

If a small increment dP_n is given to any external load P_n, the strain energy will increase and its new amount will be

$$U + \frac{\partial U}{\partial P_n} dP_n. \qquad (b)$$

But the magnitude of the strain energy does not depend upon the order in which the loads are applied to the body—it depends only upon their final values. It can be assumed, for example, that the infinitesimal load dP_n was applied first, and afterwards the loads P_1, P_2, P_3, $\cdots$. The final amount of strain energy remains the same, as given by eq. (b). The load dP_n, applied first, produces only an infinitesimal displacement, so that the corresponding work (equal to the product of the small force and the corresponding small displacement) is a small quantity of the second order and can be neglected. Applying now the loads P_1, P_2, P_3, $\cdots$, it must be noticed that their effect will not be modified by the load dP_n previously applied [20] and the work done by these loads will be equal to U (eq. a), as before. But during the application of these forces, however, the force dP_n is given some displacement δ_n in the direction of P_n and it does the work $(dP_n)\delta_n$. The two expressions for the work must be equal. Therefore

$$U + \frac{\partial U}{\partial P_n} (dP_n) = U + (dP_n)\delta_n,$$

$$\delta_n = \frac{\partial U}{\partial P_n}, \qquad (197)$$

and Castigliano's theorem is proved.

As an application of the theorem let us consider a cantilever beam carrying a load P and a couple M_a at the end, Fig. 274. The bending moment at a cross section mn is $M = -Px - M_a$ and the strain energy, from eq. (187), is

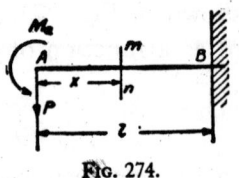

Fig. 274.

$$U = \int_0^l \frac{M^2 dx}{2EI}.$$

---

[20] This follows from the provisions made on p. 325 on the basis of which the strain energy was obtained as a homogeneous function of the second degree.

To obtain the deflection δ at the end of the cantilever we have only to take the partial derivative of U with respect to P,[21] which gives

$$\delta = \frac{\partial U}{\partial P} = \frac{1}{EI} \int_0^l M \frac{\partial M}{\partial P} dx.$$

Substituting for M its expression, in terms of P and M_a, we obtain

$$\delta = \frac{1}{EI} \int_0^l (Px + M_a)x dx = \frac{Pl^3}{3EI} + \frac{M_a l^2}{2EI}.$$

The same expression would have been obtained by applying one of the previously described methods, such as the area-moment method or the method of integration of the differential eq. (79) of the elastic curve.

To obtain the slope at the end we calculate the partial derivative of the strain energy with respect to the couple M_a, which gives

$$\theta = \frac{\partial U}{\partial M_a} = \frac{1}{EI} \int_0^l M \frac{\partial M}{\partial M_a} dx = \frac{1}{EI} \int_0^l (Px + M_a)dx$$

$$= \frac{Pl^2}{2EI} + \frac{M_a l}{EI}.$$

The positive signs obtained for δ and θ indicate that the deflection and rotation of the end have the same directions, respectively, as the force and the couple in Fig. 274.

It should be noted that the partial derivative $\partial M/\partial P$ is the rate of increase of the moment M with respect to the increase of the load P and can be visualized by the bending moment diagram for a load equal to unity, as shown in Fig. 275a. The partial derivative $\partial M/\partial M_a$ can be visualized in the same manner by the bending moment diagram in Fig. 275b. Using the notations

$$\frac{\partial M}{\partial P} = M_p \quad \text{and} \quad \frac{\partial M}{\partial M_a} = M_m',$$

---

[21] The simplest procedure is to first differentiate under the integral sign and then integrate, rather than to calculate first the integral and then differentiate.

we can represent our previous results in the following form:

$$\delta = \frac{1}{EI} \int_0^l MM_p'dx, \qquad \theta = \frac{1}{EI} \int_0^l MM_m'dx. \qquad (198)$$

These equations, derived for the particular case shown in Fig. 274, also hold for the general case of a beam with any kind of loading and any kind of support. They can also be used in the case of distributed loads.

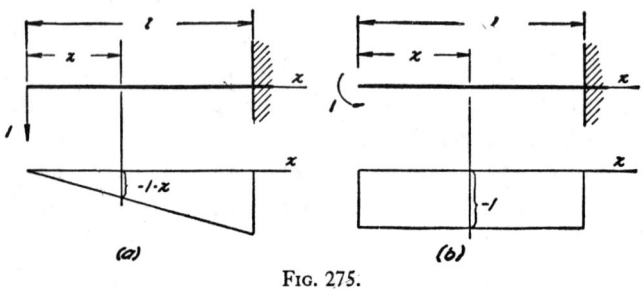

Fig. 275.

Let us consider, for example, the case of a uniformly loaded and simply supported beam, Fig. 276, and calculate the deflection at the middle of this beam by using the Castigliano theorem. In the preceding cases concentrated forces and couples were acting, and partial derivatives with respect to these forces and couples gave the corresponding displacements and

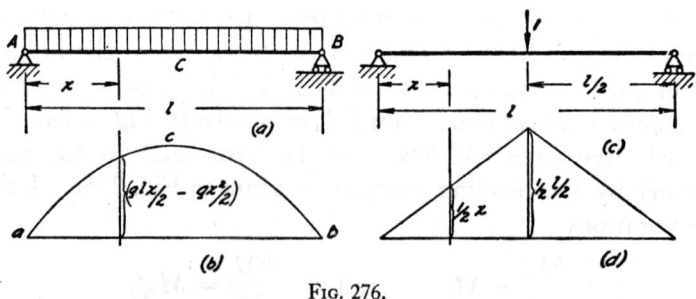

Fig. 276.

rotations. In the case of a uniform load, however, there is no vertical force acting at the middle of the beam which would correspond to the deflection at the middle which it is desired to calculate. Thus we cannot proceed as in the preceding problem. This difficulty can, however, be readily removed by assuming that there is a fictitious load P of infinitely small magnitude at the middle. Such a force evidently will not affect the deflection or the bending moment diagram shown in Fig. 276b. At the same time, the rate of increase of the bending moment due to the increase of P, represented by the partial derivative $\partial M/\partial P$, is as represented by Figs. 276c and 276d. With these values of M and $\partial M/\partial P$, the value of the deflection is

$$\delta = \frac{\partial U}{\partial P} = \frac{1}{EI_z} \int_0^l M \frac{\partial M}{\partial P} dx.$$

Observing that M and $\partial M/\partial P$ are both symmetrical with respect to the middle of the span, we obtain

$$\delta = \frac{2}{EI_z} \int_0^{l/2} M \frac{\partial M}{\partial P} dx = \frac{2}{EI_z} \int_0^{l/2} \left(\frac{qlx}{2} - \frac{qx^2}{2} \right) \frac{x}{2} dx$$

$$= \frac{5}{384} \frac{ql^4}{EI_z}.$$

If it is required to calculate the slope at the end B of the beam in Fig. 276a by using the Castigliano theorem, we have only to assume an infinitely small couple M_b applied at B. Such a couple does not change the bending moment diagram in Fig. 276b. The partial derivative $\partial M/\partial M_b$ is then represented in Figs. 277a and 277b. The required rotation of the end B of the beam is

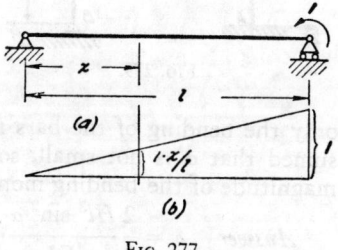

(a)

(b)

Fig. 277.

$$\theta = \frac{\partial U}{\partial M_b} = \frac{1}{EI_z} \int_0^l M \frac{\partial M}{\partial M_b} dx = \frac{1}{EI_z} \int_0^l \left(\frac{qlx}{2} - \frac{qx^2}{2} \right) \frac{x}{l} dx$$

$$= \frac{ql^3}{24EI_z}.$$

We see that the results obtained by the use of Castigliano's theorem coincide with those previously obtained (p. 141).

Problems

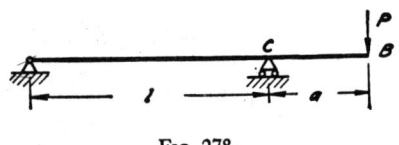

FIG. 278.

1. Determine by the use of Castigliano's theorem the deflection and the slope at the end of a uniformly loaded cantilever beam.

2. Determine the deflection at the end B of the overhang of the beam shown in Fig. 278.

3. What horizontal displacement of the support B of the frame shown in Fig. 279 is produced by the horizontal force H?

Answer. $\delta_h = \dfrac{2\,Hh^3}{3\,EI_1} + \dfrac{Hh^2 l}{EI}.$

4. Determine the increase in the distance AB produced by forces H (Fig. 280) if the bars AC and BC are of the same dimensions and

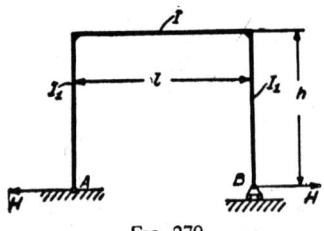

FIG. 279.

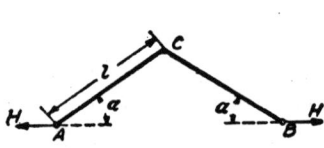

FIG. 280.

only the bending of the bars need be taken into account. It is assumed that α is not small, so that the effect of deflections on the magnitude of the bending moment can be neglected.

Answer. $\delta = \dfrac{2\,Hl^3 \sin^2 \alpha}{3\quad EI}.$

5. Determine the deflection at a distance a from the left end of the uniformly loaded beam shown in Fig. 276a.

Solution. Applying an infinitely small load P at a distance a from the left end, the partial derivative $\partial M / \partial P$ is as visualized in Figs. 281a and 281b. Using for M the parabolic diagram in Fig. 276b, the desired deflection is

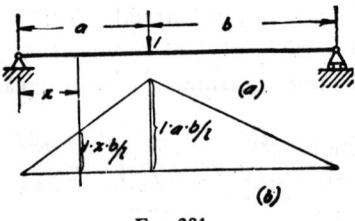

FIG. 281.

$$\delta = \frac{\partial U}{\partial P} = \frac{1}{EI}\int_0^l M\frac{\partial M}{\partial P}\,dx = \frac{1}{EI}\int_0^a \left(\frac{qlx}{2} - \frac{qx^2}{2}\right)\frac{xb}{l}\,dx$$

$$+ \frac{1}{EI}\int_a^l \left(\frac{qlx}{2} - \frac{qx^2}{2}\right)\frac{a(l-x)}{l}\,dx$$

$$= \frac{qab}{24EI}(a^2 + b^2 + 3ab).$$

Substituting x for a and $l - x$ for b, this result can be brought into agreement with the equation for the deflection curve previously obtained (p. 141).

73. Deflection of Trusses.—The Castigliano theorem is especially useful in the calculation of deflections in trusses. As an example let us consider the case shown in Fig. 282. All

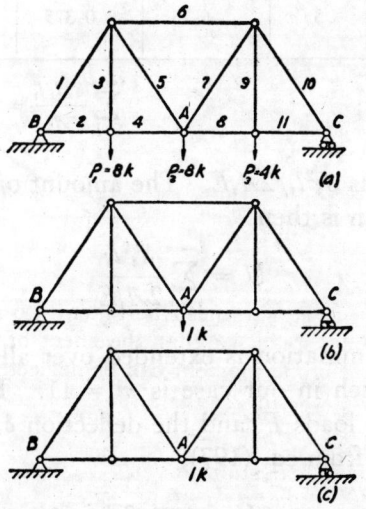

Fig. 282.

members of the system are numbered and their lengths and cross-sectional areas given in Table 5. The force S_i produced in any bar i of the system by the loads P_1, P_2, P_3 may be calculated from simple equations of statics. These forces are given in column 4 of Table 5. The strain energy of any bar i,

STRENGTH OF MATERIALS

TABLE 5: DATA FOR THE TRUSS IN FIG. 282

1	2	3	4	5	6	7
i	l_i in.	A_i in.^{2}	S_i kips [22]	S_i'	$\dfrac{S_i S_i' l_i}{A_i}$	S_i''
1	250	6	−13.75	−0.625	358	0
2	150	3	8.25	0.375	155	1
3	200	2	8.00	0	0	0
4	150	3	8.25	0.375	155	1
5	250	2	3.75	0.625	293	0
6	300	4	−10.50	−0.750	59	0
7	250	2	6.25	0.625	488	0
8	150	3	6.75	0.375	127	0
9	200	2	4.00	0	0	0
10	250	6	−11.25	−0.625	293	0
11	150	3	6.75	0.375	127	0

$$\sum_{i=1}^{i=m} \frac{S_i S_i' l_i}{A_i} = 2,055 \text{ kips per in.}$$

from eq. (171), is $S_i^2 l_i / 2 A_i E$. The amount of strain energy in the whole system is then

$$U = \sum_{i=1}^{i=m} \frac{S_i^2 l_i}{2 A_i E}, \tag{199}$$

in which the summation is extended over all the members of the system, which in our case is $m = 11$. The forces S_i are functions of the loads P, and the deflection δ_n under any load P_n is therefore, from eq. (197),

$$\delta_n = \frac{\partial U}{\partial P_n} = \sum_{i=1}^{i=m} \frac{S_i l_i}{A_i E} \frac{\partial S_i}{\partial P_n}. \tag{200}$$

The derivative $\partial S_i / \partial P_n$ is the rate of increase of the force S_i with increase of the load P_n. Numerically it is equal to the force produced in the bar i by a unit load applied in the posi-

---

[22] One kip (or kilo-pound) = 1,000 pounds.

tion of P_n, and we will use this fact in finding the above derivative. These derivatives will hereafter be denoted by S_i'. The equation for calculating the deflections then becomes

$$\delta_n = \sum_{i=1}^{i=m} \frac{S_i S_i' l_i}{A_i E}. \tag{201}$$

Consider for instance the deflection δ_2 corresponding to P_2 at A in Fig. 282a. The magnitudes S_i' tabulated in column 5 above are obtained by simple statics from the loading conditions shown in Fig. 282b, in which all actual loads are removed and a vertical load of one kip is applied at the hinge A. The values tabulated in column 6 are calculated from those entered in columns 2 through 5. Summation and division by the modulus $E = 30 \times 10^3$ kips per sq in. give the deflection at A, eq. (201),

$$\delta_2 = \frac{2{,}055}{30 \times 10^3} = 0.0685 \text{ in.}$$

The above discussion was concerned with the computation of displacements δ_1, δ_2, $\cdots$ corresponding to the given external forces P_1, P_2, $\cdots$. In investigating the deformation of an elastic system, it may be necessary to calculate the displacement of a point at which there is no load at all, or the displacement of a loaded point in a direction different from that of the load. The method of Castigliano may also be used here. We merely apply at that point an additional infinitely small *imaginary load* Q in the direction in which the displacement is wanted, and calculate the derivative $\partial U/\partial Q$. In this derivative the added load Q is put equal to zero and the desired displacement obtained. For example, in the truss shown in Fig. 282a, let us calculate the horizontal displacement of the point A. A horizontal force Q is applied at this point and the corresponding horizontal displacement is

$$\delta_h = \left(\frac{\partial U}{\partial Q}\right)_{Q=0} = \sum_{i=1}^{i=m} \frac{S_i l_i}{A_i E} \frac{\partial S_i}{\partial Q}, \tag{a}$$

in which the summation is extended over all the members of

the system. The forces S_i in eq. (a) have the same meaning
as before, because the added load Q is zero, and the derivatives
$\partial S_i/\partial Q = S_i''$ are obtained as the forces in the bars of the
truss produced by the loading shown in Fig. 282c. These
values are tabulated in column 7. Substituting these forces
into eq. (a), we find that the horizontal displacement of A is
equal to the sum of the elongations of the bars 2 and 4, namely,

$$\delta_h = \frac{1}{E}\left(\frac{S_2 l_2}{A_2} + \frac{S_4 l_4}{A_4}\right) = \frac{150}{3 \times 30 \times 10^3}(8.25 + 8.25)$$

$$= 0.0275 \text{ in.}$$

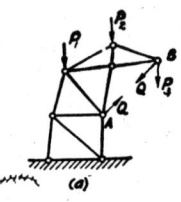

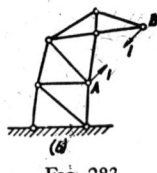

Fig. 283.

In investigating the deformation of trusses
it is sometimes necessary to know the change
in distance between two points of the system.
This can also be obtained by the Castigliano
method. Let us determine, for instance, what
decrease δ in the distance between the joints
A and b (Fig. 283a) is produced by the loads
P_1, P_2, P_3. At these joints two equal and
opposite imaginary forces Q are applied, as indicated in the figure. It follows from the
Castigliano theorem that the partial derivative
$(\partial U/\partial Q)_{Q=0}$ gives the shortening of the distance AB produced by the loads P_1, P_2, P_3. Using eq. (197),
this displacement is [23]

$$\delta = \left(\frac{\partial U}{\partial Q}\right)_{Q=0} = \sum_{i=1}^{i=m} \frac{S_i l_i}{A_i E} \frac{\partial S_i}{\partial Q_i} = \sum_{i=1}^{i=m} \frac{S_i l_i}{A_i E} \cdot S_i', \qquad (202)$$

in which S_i are the forces produced in the bars of the system
by the actual loads P_1, P_2, P_3; the quantities S_i' are to be
determined from the loading shown in Fig. 283b, in which all
actual loads are removed and two opposite unit forces are
applied at A and B; and m is the number of members.

---

[23] This problem was first solved by J. C. Maxwell, "On the Calculation
of the Equilibrium and Stiffness of Frames," *Phil. Mag.*, Vol. 27, p. 294,
1864; or *Scientific Papers*, Cambridge, Vol. 1, p. 598, 1890.

Problems

1. A system consisting of two prismatic bars of equal length and equal cross section (Fig. 284) carries a vertical load P. Determine the vertical displacement of the hinge A.

Solution. The tensile force in the bar AB and compressive force in the bar AC are equal to P. Hence the strain energy of the system is

$$U = 2\,\frac{P^2 l}{2AE}.$$

The vertical displacement of A is

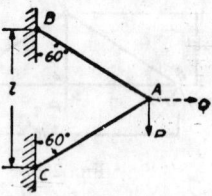

Fig. 284.

$$\delta = \frac{dU}{dP} = \frac{2Pl}{AE}.$$

2. Determine the horizontal displacement of the hinge A in the previous problem.

Solution. Apply a horizontal imaginary load Q as shown in Fig. 284 by the dotted line. The potential energy of the system is

$$U = \frac{(P + Q/\sqrt{3})^2 l}{2AE} + \frac{(P - Q/\sqrt{3})^2 l}{2AE}.$$

The derivative of this expression with respect to Q for $Q = 0$ gives the horizontal displacement

$$\delta_h = \left(\frac{\partial U}{\partial Q}\right)_{Q=0} = \left(\frac{2Ql}{3AE}\right)_{Q=0} = 0.$$

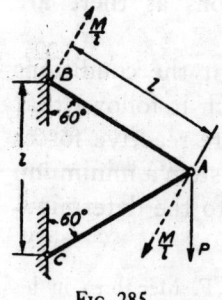

Fig. 285.

3. Determine the angular displacement of the bar AB produced by the load P in Fig. 285.

Solution. An imaginary couple M is applied to the system as shown in the figure by dotted lines. The displacement corresponding to this couple is the angular displacement φ of the bar AB due to the load P. The forces in the bars are $P + M/\sqrt{3}l$ in the bar AB and $-P - 2M/\sqrt{3}l$ in the bar AC. The strain energy is

$$U = \frac{l}{2AE}\left[\left(P + \frac{M}{\sqrt{3}l}\right)^2 + \left(-P - \frac{2M}{\sqrt{3}l}\right)^2\right].$$

from which

$$\varphi = \left(\frac{\partial U}{\partial M}\right)_{M=0} = \left(\frac{P\sqrt{3}}{AE} + \frac{5M}{3lAE}\right)_{M=0} = \frac{P\sqrt{3}}{AE}.$$

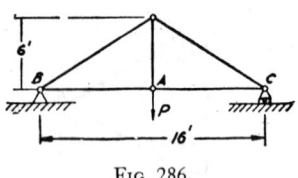

FIG. 286.

4. Determine the vertical displacement of the point A and horizontal displacement of the point C of the steel truss shown in Fig. 286 if $P = 2,000$ lb and the cross-sectional areas of the bars which act in compression are 5 sq in. and of the other bars 2 sq in.

74. Application of Castigliano Theorem in Solution of Statically Indeterminate Problems.

The Castigliano theorem is also very useful in the solution of statically indeterminate problems. Let us begin with problems in which the reactions at the supports are considered as the statically indeterminate quantities. Denoting by X, Y, Z, $\cdots$ the statically indeterminate reactive forces, the strain energy of the system can be represented as a function of these forces. For the immovable supports and for the supports whose motion is perpendicular to the direction of the reactions the partial derivatives of the strain energy with respect to the unknown reactive forces must be equal to zero by the Castigliano theorem. Hence

$$\frac{\partial U}{\partial X} = 0, \qquad \frac{\partial U}{\partial Y} = 0, \qquad \frac{\partial U}{\partial Z} = 0, \qquad \cdots. \qquad (203)$$

In this manner we obtain as many equations as there are statically indeterminate reactions.

It can be shown that eqs. (203) represent the conditions for a minimum of the function U, from which it follows that the magnitudes of the statically indeterminate reactive forces are such as to make the strain energy of the system a minimum. This is the *principle of least work* as applied to the determination of redundant reactions.[24]

[24] The principle of least work was stated first by F. Menabrea in his article, "Nouveau principe sur la distribution des tensions dans les systèmes élastiques," *Compt. rend.*, Vol. 46, p. 1056, 1858. See also *ibid.*, Vol. 98, p. 714, 1884. A complete proof of the principle was given by Castigliano, who made this principle the fundamental method of solution of statically

As an example of the application of the above principle let us consider a uniformly loaded beam built in at one end and supported at the other (Fig. 287). This is a problem with one statically indeterminate reaction. Taking the vertical reaction X at the right-hand support as the statically indeterminate quantity, this unknown force is found from the equation

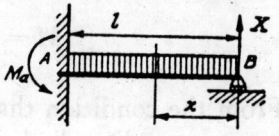

FIG. 287.

$$\frac{dU}{dX} = 0. \tag{a}$$

The strain energy of the beam, from eq. (190), is

$$U = \int_0^l \frac{M^2 dx}{2EI}, \tag{b}$$

in which

$$M = Xx - \frac{qx^2}{2}.$$

Substituting in (a), we obtain

$$\frac{dU}{dX} = \frac{1}{EI} \int_0^l M \frac{dM}{dX}\, dx = \frac{1}{EI} \int_0^l \left(Xx - \frac{qx^2}{2} \right) x\, dx$$

$$= \frac{1}{EI} \left(X \frac{l^3}{3} - q \frac{l^4}{8} \right) = 0,$$

from which

$$X = \tfrac{3}{8} ql.$$

Instead of the reactive force X the reactive couple M_a at the left end of the beam could have been taken as the statically indeterminate quantity. The strain energy will now be a function of M_a. Eq. (b) still holds, where now the bending

indeterminate systems. The application of strain energy methods in engineering was developed by O. Mohr, see his *Abhandlungen, loc. cit.*; by H. Müller-Breslau in his book, *Die neueren Methoden der Festigkeitslehre*; and by F. Engesser, "Über die Berechnung statisch unbestimmter Systeme," *Zentr. Bauverwalt.*, p. 606, 1907. A very complete bibliography of this subect is given in the article by M. Grüning, *Encyklopädie der Mathematischen Wissenschaft*, Vol. 4, p. 419.

moment at any cross section is

$$M = \left(\frac{ql}{2} - \frac{M_a}{l}\right) x - \frac{qx^2}{2}.$$

From the condition that the left end of the actual beam does not rotate when the beam is bent, the derivative of the strain energy with respect to M_a must be equal to zero. From this we obtain

$$\frac{dU}{dM_a} = \frac{1}{EI} \int_0^l M \frac{dM}{dM_a} dx$$

$$= -\frac{1}{EI} \int_0^l \left[\left(\frac{ql}{2} - \frac{M_a}{l}\right) x - \frac{qx^2}{2}\right] \frac{x}{l} dx$$

$$= -\frac{1}{EI} \left(\frac{ql^3}{24} - \frac{M_a l}{3}\right) = 0,$$

from which the absolute value of the

$$M_a = \tfrac{1}{8} q l^2.$$

Problems in which we consider the forces acting in redundant members of the system as the statically indeterminate quantities can also be solved by using the Castigliano theorem.

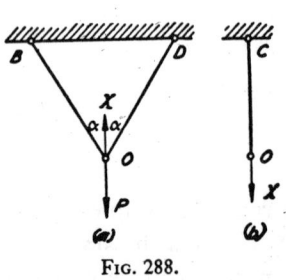

Take, as an example, the system represented in Fig. 18 which was discussed previously (see p. 20). Considering the force X in the vertical bar OC as the statically indeterminate quantity, the forces in the inclined bars OB and OD are $(P - X)/2 \cos \alpha$. Denoting by U_1 the strain energy of the inclined bars (Fig. 288a) and by U_2 the strain energy of the vertical bar (Fig. 288b), the total strain energy of the system is [25]

FIG. 288.

$$U = U_1 + U_2 = \left(\frac{P - X}{2 \cos \alpha}\right)^2 \frac{l}{AE \cos \alpha} + \frac{X^2 l}{2AE}. \qquad (c)$$

[25] It is assumed that all bars have tne same cross-sectional area A and the same modulus of elasticity E.

If δ is the actual displacement downwards of the joint O in Fig. 18, the derivative with respect to X of the energy U_1 of the system in Fig. 288a should be equal to $-\delta$, since the force X of the system has a direction opposite to that of the displacement δ. Also the derivative $\partial U_2/\partial X$ will be equal to δ, hence

$$\frac{\partial U}{\partial X} = \frac{\partial U_1}{\partial X} + \frac{\partial U_2}{\partial X} = -\delta + \delta = 0. \qquad (d)$$

It is seen that the true value of the force X in the redundant member is such as to make the total strain energy of the system a minimum. Substituting for U its expression (c) in eq. (d), we obtain

$$-\frac{(P - X)}{2 \cos^2 \alpha} \frac{l}{AE \cos \alpha} + \frac{Xl}{AE} = 0,$$

from which

$$X = \frac{P}{1 + 2 \cos^3 \alpha}.$$

Similar reasoning can be applied to any statically indeterminate system with one redundant member, and we can state that the force in that member is such as to make the strain energy of the system a minimum. To illustrate the procedure of calculating stresses in such systems let us consider the frame shown in Fig. 289a. The reactions here are statically deter-

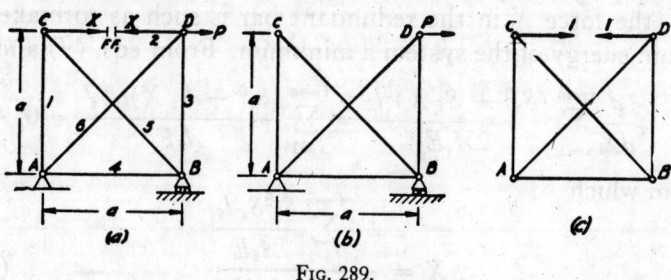

Fig. 289.

minate, but when we try to compute the forces in the bars, we find that there is one redundant member. Let us consider the bar CD as this redundant member. We cut this bar at any point and apply to each end F and F_1 a force X, equal to the force in the bar. Thus we arrive at a statically determinate system acted upon by the known force P, and in addition by the unknown forces X. The forces in the bars of this system will be found in two steps: (1) those produced by the external load P, assuming $X = 0$, Fig. 289b, and denoted by $S_i{}^0$, where i indicates the number of the bar; (2) those produced when the external force P is removed and unit forces replace the X forces (Fig. 289c). The latter forces are denoted by S_i'. Then the total force in any bar, when the force P and the forces X are all acting simultaneously, is

$$S_i = S_i{}^0 + S_i'X. \qquad (e)$$

The total strain energy of the system, from eq. (199), is

$$U = \sum_{i=1}^{i=m} \frac{S_i{}^2 l_i}{2A_i E} = \sum_{i=1}^{i=m} \frac{(S_i{}^0 + S_i'X)^2 l_i}{2A_i E}, \qquad (f)$$

in which the summation is extended over all the bars of the system, including the bar CD which is cut.[26] The Castigliano theorem is now applied and the derivative of U with respect to X gives the displacement of the ends F and F_1 towards each other. In the actual case the bar is continuous and this displacement is equal to zero. Hence

$$\frac{dU}{dX} = 0, \qquad (g)$$

i.e., the force X in the redundant bar is such as to make the strain energy of the system a minimum. From eqs. (f) and (g)

$$\frac{d}{dX} \sum_{i=1}^{i=m} \frac{(S_i{}^0 + S_i'X)^2 l_i}{2A_i E} = \sum_{i=1}^{i=m} \frac{(S_i{}^0 + S_i'X)l_i S_i'}{A_i E} = 0,$$

from which

$$X = - \frac{\displaystyle\sum_{i=1}^{i=m} \frac{S_i{}^0 S_i' l_i}{A_i E}}{\displaystyle\sum_{i=1}^{i=m} \frac{S_i'{}^2 l_i}{A_i E}}. \qquad (204)$$

This process may be extended to a system in which there are several redundant bars.

[26] For this bar $S_i{}^0 = 0$ and $S_i' = 1$.

The principle of least work can also be applied in those cases in which the statically unknown quantities are couples. Take, as an example, a uniformly loaded beam on three supports (Fig. 290). If the bending moment at the middle support is considered as the statically indeterminate quantity, we cut the beam at B and obtain two simply supported beams (Fig. 290*b*) carrying the unknown couples M_b in addition to the known uniform load q. There is no rotation of the end

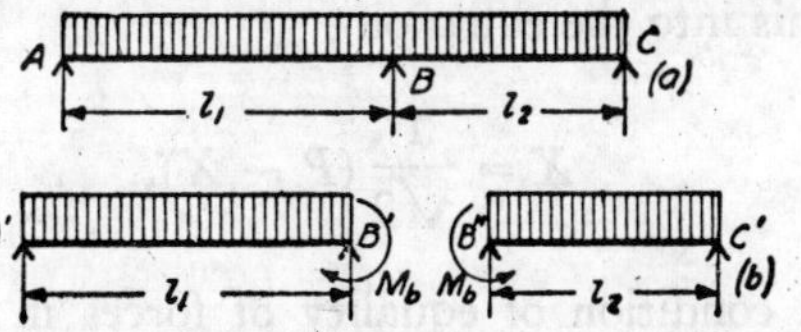

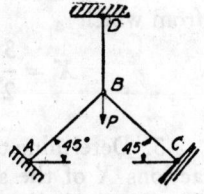

FIG. 290.

B' of the beam $A'B'$ with respect to the end B'' of the beam $B''C'$ because in the actual case (Fig. 290*a*) there is a continuous deflection curve. Hence

$$\frac{dU}{dM_b} = 0. \tag{205}$$

Again the magnitude of the statically indeterminate quantity is such as to make the strain energy of the system a minimum.

Problems

1. A vertical load P is supported by a vertical bar DB of length l and cross-sectional area A and by two equal inclined bars of length l and cross-sectional area A_1 (Fig. 291). Determine the forces in the bars and also the ratio A_1/A which will make the forces in all three bars numerically equal.

Solution. The system is statically indeterminate. Let X be the tensile force in the vertical bar. The compressive forces in the inclined bars then are $(P - X)/\sqrt{2}$ and the strain energy of the system is

FIG. 291.

$$U = \frac{X^2 l}{2AE} + \frac{(P - X)^2 l}{2A_1 E}.$$

The principle of least work gives the equation

$$\frac{dU}{dX} = \frac{Xl}{AE} - \frac{(P-X)l}{A_1E} = 0,$$

from which

$$X = \frac{P}{1 + \dfrac{A_1}{A}}.$$

Substituting this into the equation

$$X = \frac{1}{\sqrt{2}}(P - X),$$

expressing the condition of equality of forces in all three bars, we obtain

$$A_1 = \sqrt{2}\,A.$$

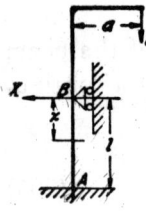

FIG. 292.

2. Determine the horizontal reaction X in the system shown in Fig. 292.

Solution. The unknown force X will enter into the expression for the potential energy of bending of the portion AB of the bar only. For this portion $M = Pa - Xx$, and the equation of least work gives [27]

$$\frac{dU}{dX} = \frac{d}{dX}\int_0^l \frac{M^2dx}{2EI} = \frac{1}{EI}\int_0^l M\frac{dM}{dX}\,dx = -\frac{1}{EI}\int_0^l (Pa - Xx)x\,dx$$

$$= \frac{1}{EI}\left(\frac{Xl^3}{3} - \frac{Pal^2}{2}\right) = 0,$$

from which

$$X = \frac{3}{2}P\frac{a}{l}.$$

3. Determine the horizontal reactions X of the system shown in Fig. 293. All dimensions are given in Table 6.

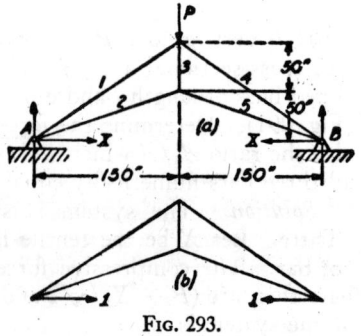

FIG. 293.

---

[27] The effect of the axial force on the bending of the bar AB is neglected in this case.

Solution. From the principle of least work we have

$$\frac{dU}{dX} = \frac{d}{dX} \Sigma \frac{S_i^2 l_i}{2A_i E} = \Sigma \frac{S_i l_i}{A_i E} \frac{dS_i}{dX} = 0. \qquad (h)$$

Let S_i^0 be the force in bar i produced by the known load P, assuming $X = 0$, and let S_i' be the force produced in the same bar by unit forces which replace the X forces (Fig. 293b). The values of S_i^0 and S_i' are determined from statics. They are given in columns 4 and 5 of Table 6. Then the total force in any bar is

$$S_i = S_i^0 + S_i'X. \qquad (i)$$

TABLE 6: DATA FOR THE TRUSS IN FIG. 293

	l_i in.	A_i in.^{2}	S_i^0	S_i'	$\dfrac{S_i^0 S_i' l_i}{A_i}$	$\dfrac{S_i'^2 l_i}{A_i}$
1	180.3	5	$-1.803P$	1.202	$-78.1P$	52.0
2	158.1	3	$1.581P$	-2.108	$-175.7P$	234
3	50.0	2	$1.000P$	-1.333	$-33.3P$	44.5
4	180.3	5	$-1.803P$	1.202	$-78.1P$	52.0
5	158.1	3	$1.581P$	-2.108	$-175.7P$	234

$$\Sigma = -540.9P \quad \Sigma = 616.5.$$

Substituting into eq. (h), we find

$$\sum_1^5 \frac{(S_i^0 + S_i'X)l_i}{A_i E} S_i' = 0,$$

from which

$$X = - \frac{\displaystyle\sum_{i=1}^{i=5} \frac{S_i^0 S_i' l_i}{A_i}}{\displaystyle\sum_{i=1}^{i=5} \frac{S_i'^2 l_i}{A_i}}.$$

The necessary figures for calculating X are given in columns 6 and 7 of the table. Substituting this data into eq. (j), we obtain

$$X = 0.877P.$$

4. Determine the force in the redundant horizontal bar of the system shown in Fig. 294, assuming that the length of this bar is

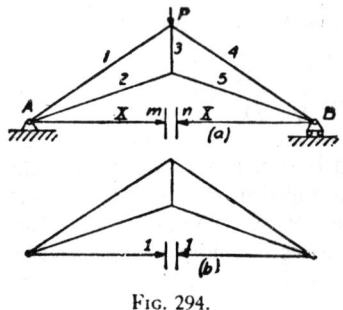

Fig. 294.

$l_0 = 300$ in. and the cross-sectional area is A_0. The other bars have the same dimensions as in Prob. 3.

Solution. The force in the horizontal bar (bar number 0) is calculated from eq. (204). This equation is similar to eq. (j) in Prob. 3, except that in the system of Fig. 294 there is the additional horizontal bar 0. The force produced in this bar by the force P alone $(X = 0)$ is zero, i.e., $S_0^0 = 0$. The force produced by two forces equal to unity (Fig. 294b) is $S_0' = 1$. The additional term in the numerator of eq. (j) is

$$\frac{S_0^0 S_0' l_0}{A_0} = 0.$$

The additional term in the denominator is

$$\frac{S_0'^2 l_0}{A_0} = \frac{1 \cdot l_0}{A_0} = \frac{300}{A_0}.$$

Then, by using the data of Prob. 3,

$$X = \frac{540.9P}{\dfrac{300}{A_0} + 616.5}.$$

Taking, for instance, $A_0 = 10$ sq in.,

$$X = \frac{540.9P}{30 + 616.5} = 0.836P,$$

which is only 4.7 per cent less than the value obtained in Prob. 3 for immovable supports.[28]

Taking the cross-sectional area $A_0 = 1$ sq in.,

$$X = \frac{540.9}{300 + 616.5} = 0.590P.$$

It can be seen that in statically indeterminate systems the forces in the bars depend not only on the applied loads, but also on the cross-sectional areas of the bars.

---

[28] Taking $A_0 = \infty$, we obtain the same condition as for immovable supports.

5. Determine the forces in the bars of the systems shown in Fig. 23 by using the principle of least work.

6. Determine the forces in the bars of the system shown in Fig. 295, assuming that all bars are of the same dimensions and material.

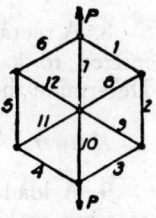

Solution. If one bar be removed, the forces in the remaining bars can be determined from statics. Hence the system has one redundant bar. Let 1 be this bar and X the force acting in it. Then all the

Fig. 295.

bars on the sides of the hexagon will have tensile forces X, bars 8, 9, 11 and 12 have compressive forces X and bars 7 and 10 have the force $P - X$. The strain energy of the system is

$$U = 10\frac{X^2 l}{2AE} + 2\frac{(P - X)^2 l}{2AE}.$$

From the equation $dU/dX = 0$ we obtain

$$X = \frac{P}{6}.$$

7. Determine the forces in the system shown in Fig. 289, assuming the cross-sectional areas of all bars equal and taking the force X in the diagonal AD as the statically indeterminate quantity.

Solution. Substituting the data given in Table 7 into eq. (204) we find

$$X = \frac{3 + 2\sqrt{2}}{4 + 2\sqrt{2}} P.$$

TABLE 7: DATA FOR THE SYSTEM IN FIG. 289

i	l_i	S_i^0	S_i'	$S_i^0 S_i' l_i$	$S_i'^2 l_i$
1	a	P	$-1/\sqrt{2}$	$-aP/\sqrt{2}$	$a/2$
2	a	P	$-1/\sqrt{2}$	$-aP/\sqrt{2}$	$a/2$
3	a	0	$-1/\sqrt{2}$	0	$a/2$
4	a	P	$-1/\sqrt{2}$	$-aP/\sqrt{2}$	$a/2$
5	$a\sqrt{2}$	$-P\sqrt{2}$	$+1$	$-2aP$	$a\sqrt{2}$
6	$a\sqrt{2}$	0	$+1$	0	$a\sqrt{2}$

$$\Sigma S_i^0 S_i' l_i = \frac{-(3 + 2\sqrt{2})aP}{\sqrt{2}} \; ; \; \Sigma S_i'^2 l_i = 2a(1 + \sqrt{2}).$$

8. A rectangular frame of uniform cross section (Fig. 296) is submitted to a uniformly distributed load of intensity q as shown. Determine the bending moments M at the corners.

Answer. $M = \dfrac{(a^3 + b^3)q}{12(a + b)}$.

9. A load P is supported by two beams of equal cross section, crossing each other as shown in Fig. 297. Determine the pressure X between the beams.

Answer. $X = \dfrac{Pl^3}{l^3 + l_1{}^3}$.

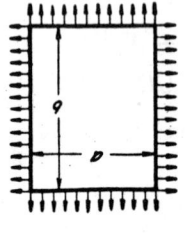

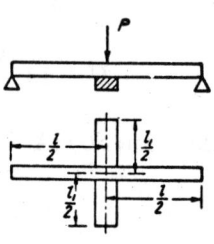

Fig. 296.   Fig. 297.

10. Find the statically indeterminate quantity H in the frame shown in Fig. 170, p. 193, by using the principle of least work.

Solution. The strain energy of bending of the frame is

$$U = 2\int_0^h \frac{H^2 x^2 dx}{2EI_1} + \int_0^l \frac{(M_0 - Hh)^2 dx}{2EI}, \qquad (k)$$

in which M_0 denotes the bending moment varying along the horizontal bar AB and calculated as for a beam simply supported at the ends. Substituting into the equation

$$\frac{dU}{dH} = 0, \qquad (l)$$

we find

$$\frac{2H}{EI_1}\frac{h^3}{3} + \frac{Hh^2 l}{EI} = \frac{h}{EI}\int_0^l M_0 dx. \qquad (m)$$

The integral on the right-hand side is the area of the triangular moment diagram for a beam carrying the load P. Hence

$$\int_0^l M_0 dx = \frac{1}{2} Pc(l - c).$$

Substituting into eq. (m), we obtain for H the same expression as in eq. (114), p. 193.

11. Find the statically indeterminate quantities in the frames shown in Figs. 169, 172 and 174 by using the principle of least work.

12. Find the bending moment M_b in the beam of Fig. 290, assuming that $l_1 = 2l_2$.

75. The Reciprocal Theorem.—Let us begin with the problem of a simply supported beam as shown in Fig. 298a and calculate the deflection at a point D when the load P is acting at C. This deflection is obtained by substituting $x = d$ into eq. (86), which gives

$$(y)_{x=d} = \frac{Pbd}{6l}(l^2 - b^2 - d^2).$$

It is seen that this expression does not change if we substitute d for b and b for d, which indicates that for the case shown in Fig. 298b the deflection at D_1 is the same as the deflection at D in Fig. 298a. From Fig. 298b we obtain Fig. 298c by simply rotating the beam through 180 degrees which brings point C_1 into coincidence with point D and point D_1 into coincidence

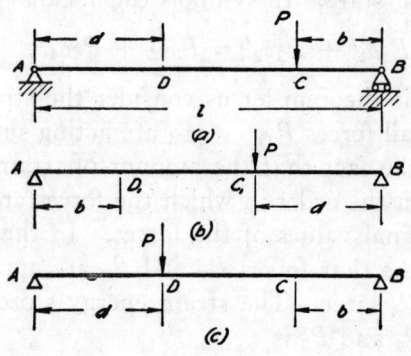

Fig. 298.

with point C. Hence the deflection at C in Fig. 298c is equal to the deflection at D in Fig. 298a. This means that if the load P is moved from point C to point D, the deflection measured at D in the first case of loading will now be obtained in the second case at point C. This is a particular case of the *reciprocal theorem*.

To establish the theorem in general form [29] we consider an elastic body, shown in Fig. 299, loaded in two different ways and supported in such a manner that any displacement as a rigid body is impossible. In the first state of stress the applied forces are P_1 and P_2, and in the second state P_3 and P_4. The displacements of the points of application in the directions of the forces are $\delta_1, \delta_2, \delta_3, \delta_4$ in the first state and $\delta_1', \delta_2', \delta_3', \delta_4'$

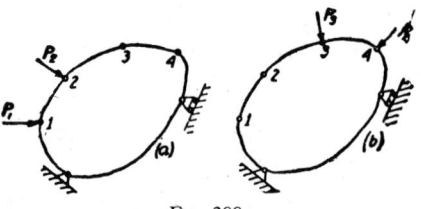

FIG. 299.

in the second state. The reciprocal theorem states that the work done by the forces of the first state on the corresponding displacements of the second state is equal to the work done by the forces of the second state on the corresponding displacements of the first state. In symbols this means

$$P_1\delta_1' + P_2\delta_2' = P_3\delta_3 + P_4\delta_4. \tag{206}$$

To prove this theorem let us consider the strain energy of the body when all forces $P_1, \cdots P_4$ are acting simultaneously, and let us use the fact that the amount of strain energy does not depend upon the order in which the forces are applied but only upon the final values of the forces. In the first manner of loading assume that forces P_1 and P_2 are applied first and forces P_3 and P_4 later. The strain energy stored during the application of P_1 and P_2 is

$$\frac{P_1\delta_1}{2} + \frac{P_2\delta_2}{2}. \tag{a}$$

[29] A particular case of this theorem was obtained by J. C. Maxwell, *loc. cit.*, p. 338. The theorem is due to E. Betti, *Nuovo cimento*, ser. 2, Vols. 7 and 8, 1872. In a more general form, the theorem was given by Lord Rayleigh, *Proc. London Math. Soc.*, Vol. 4, 1873; or *Scientific Papers*, Vol. 1, p. 179. Various applications of this theorem to the solution of engineering problems were made by O. Mohr and H. Müller-Breslau, *loc. cit.*, pp. 340 and 341.

Applying now P_3 and P_4, the work done by these forces is

$$\frac{P_3\delta_3'}{2} + \frac{P_4\delta_4'}{2}. \tag{b}$$

It must be noted, however, that during the application of P_3 and P_4 the points of application of the previously applied forces P_1 and P_2 will be displaced by δ_1' and δ_2'. Then P_1 and P_2 do the work

$$P_1\delta_1' + P_2\delta_2'.^{30} \tag{c}$$

Hence the total strain energy stored in the body, by summing (a), (b) and (c), is

$$U = \frac{P_1\delta_1}{2} + \frac{P_2\delta_2}{2} + \frac{P_3\delta_3'}{2} + \frac{P_4\delta_4'}{2} + P_1\delta_1' + P_2\delta_2'. \tag{d}$$

In the second manner of loading, let us apply the forces P_3 and P_4 first and P_1 and P_2 afterwards. Then, repeating the same reasoning as above, we obtain

$$U = \frac{P_3\delta_3'}{2} + \frac{P_4\delta_4'}{2} + \frac{P_1\delta_1}{2} + \frac{P_2\delta_2}{2} + P_3\delta_3 + P_4\delta_4. \tag{e}$$

Observing that expressions (d) and (e) must represent the same amount of strain energy, we obtain eq. (206) representing the reciprocal theorem. This theorem can be proved for any number of forces and also for couples, or for forces and couples. In the case of a couple, the corresponding angle of rotation is considered as the displacement.

For the particular case in which a single force P_1 acts in the first state of stress and a single force P_2 in the second state, eq. (206) becomes [31]

$$P_1\delta_1' = P_2\delta_2. \tag{207}$$

If $P_1 = P_2$, it follows that $\delta_1' = \delta_2$, i.e., the displacement of the point of application of the force P_2 in the direction of this

[30] These expressions are not divided by 2 because forces P_1 and P_2 remain constant during the time in which their points of application undergo the displacements δ_1' and δ_2'.

[31] This was proved first by J. C. Maxwell, and is frequently called *Maxwell's theorem*.

force, produced by the force P_1, is equal to the displacement of the point of application of the force P_1 in the direction of P_1, produced by the force P_2. A verification of this conclusion for the particular case of the beam shown in Fig. 298 has already been given.

As another example let us again consider the bending of a simply supported beam. In the first state of loading let it be bent by a load P at the middle and in the second state, by a couple M at the end. The load P produces the slope $\theta = Pl^2/16EI$ at the end. The couple M, applied at the end, produces the deflection $Ml^2/16EI$ at the middle. Eq. (207) becomes

$$P \frac{Ml^2}{16EI} = M \frac{Pl^2}{16EI}.$$

The reciprocal theorem is very useful in the problem of finding the most unfavorable position of moving loads on a statically indeterminate structure. An example is shown in

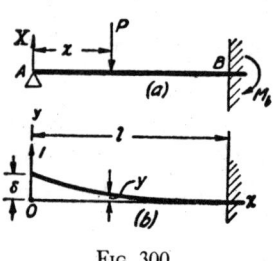

FIG. 300.

Fig. 300, which represents a beam built in at one end and simply supported at the other and carrying a concentrated load P. The problem is to find the variation in the magnitude of the reaction X at the left support as the distance x of the load from this support changes. Let us consider the *actual condition* of the beam (Fig. 300a) as the first state of loading. The second, or *fictitious state*, is shown in Fig. 300b. The external load and the redundant support are removed and a unit force upwards replaces the unknown reaction X. This second state of loading is statically determinate and the corresponding deflection curve is known (see eq. 97, p. 150). If the coordinate axes are taken as shown in Fig. 300b,

$$\bar{y} = \frac{1}{6EI} (l - x)^2 (2l + x). \tag{f}$$

Let δ denote the deflection at the end and y the deflection at distance x from the left support. Then, applying the reciprocal theorem, the work done by the forces of the first state on the corresponding displacements of the second state is

$$X\delta - Py.$$

In calculating now the work done by the forces of the second state, there is only the unit force on the end [32] and the corresponding displacement of the point A in the first state is equal to zero. Consequently this work is zero and the reciprocal theorem gives

$$X\delta - Py = 0,$$

from which

$$X = P\frac{y}{\delta}. \qquad (g)$$

It is seen that as the load P changes position on the beam of Fig. 300a, the reaction X is proportional to the corresponding values of y in Fig. 300b. Hence the deflection curve of the second state (eq. f) gives a complete picture of the manner in which X varies with x. Such a curve is called the *influence line* for the reaction X.[33]

If several loads act simultaneously on the beam of Fig. 300a, the use of eq. (g) together with the method of superposition gives

$$X = \frac{1}{\delta} \Sigma P_n y_n,$$

in which y_n is the ordinate of the influence line corresponding to the point of application of the load P_n and the summation is extended over all the loads.

Problems

1. Construct influence lines for the reactions at the supports of the beam on three supports (Fig. 301).

Solution. To obtain the influence line for the middle support, the actual state of loading shown in Fig. 301a is taken as the first

---

[32] The reactions at the built-in end are not considered in either case because the corresponding displacement is zero and the corresponding work vanishes.

[33] The use of models in determining influence lines was developed by G. E. Beggs, *J. Franklin Inst.*, 1927.

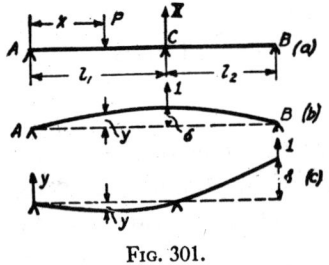

FIG. 301.

state. The second state is indicated in Fig. 301b, in which the load P is removed and the reaction X at the middle support is replaced by a unit force acting upward. This second state of loading is statically determinate and the deflection curve is known (eqs. 86 and 87, p. 145), hence the deflections δ and y can be calculated. Then the work done by the forces of the first state on the corresponding displacements of the second state is

$$X\delta - Py.$$

The work of the forces of the second state (force equal to unity) on the corresponding displacements of the first state (zero deflection at C) is zero. Hence

$$X\delta - Py = 0 \qquad X = P\frac{y}{\delta}.$$

Hence the deflection curve of the second state gives the shape of the influence line for the reaction X. In order to obtain the influence line for the reaction at B, the second state of loading should be taken as shown in Fig. 301c.

2. Determine the reaction at B by using the influence line of the preceding problem, if the load P is at the middle of the first span ($x = l_1/2$), Fig. 301a.

Answer. Reaction is downward and equal to

$$\frac{3P}{16} \frac{l_1^2}{l_2^2 + l_2 l_1}.$$

3. Find the influence line for the bending moment at the middle support C of the beam on three supports (Fig. 302). By using this influence line calculate the bending moment M_c when the load P is at the middle of the second span.

Solution. The first state of loading is the actual state (Fig. 302a) with an internal bending moment M_c acting at the cross section C. For the second state the load P is removed, the beam is cut at C and two equal and opposite unit couples replace M_c (Fig. 302b). This case is statically determinate. The angles θ_1 and θ_2 are given by eq. (104) and the deflection y by eq. (105). The sum of the angles $\theta_1 + \theta_2$ represents the displacement in the second state corresponding to the bending moment M_c acting in

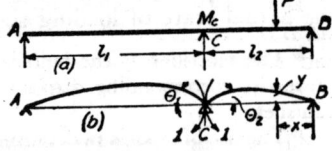

FIG. 302.

the first state. The work done by the forces of the first state on the corresponding displacement in the second state is [34]

$$M_c(\theta_1 + \theta_2) - Py.$$

The work done by the forces of the second state on the displacements of the first state is zero because there is no cut at the support C in the actual case and the displacement corresponding to the two unit couples of the second state is zero. Hence

$$M_c(\theta_1 + \theta_2) - Py = 0$$

and

$$M_c = P \frac{y}{\theta_1 + \theta_2}. \tag{h}$$

It will be seen that as the load P changes its position, the bending moment M_c changes in the same ratio as the deflection y. Hence the deflection curves of the second state represent the shape of the influence line for M_c. Noting that

$$\theta_1 + \theta_2 = \frac{l_1 + l_2}{3EI}$$

and that the deflection at the middle of the second span is

$$(y)_{x=l_2/2} = \frac{1 \cdot l_2^2}{16EI},$$

the bending moment at the support C when the load P is at the middle of the second span is, from eq. (h),

$$M_c = \frac{3}{16} \cdot \frac{Pl_2^2}{l_1 + l_2}.$$

The positive sign obtained for M_c indicates that the moment has the direction indicated in Fig. 302b. Following our general rule for the sign of moments (Fig. 63), we then consider M_c as a negative bending moment.

4. Find the influence line for the bending moment at the built-in end B of the beam AB, Fig. 300, and calculate this moment when the load is at the distance $x = l/3$ from the left support.

Answer. $M_b = -\frac{4}{27}Pl.$

---

[34] It is assumed that the bending moment M_c produces a deflection curve concave downward.

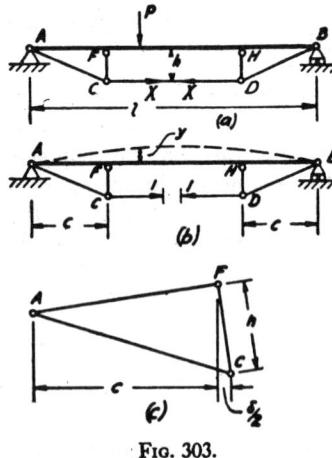

FIG. 303.

5. Construct the influence line for the horizontal reactions H of the frame shown in Fig. 170a as the load P moves along the bar AB.

Answer. The influence line has the same shape as the deflection curve of the bar AB for the loading condition shown in Fig. 169c.

6. Construct the influence line for the force X in the horizontal bar CD (Fig. 303a) as the load P moves along the beam $AB.$ Calculate X when the load is at the middle. The displacements due to elongation and contraction of the bars are to be neglected and only the displacement due to bending of the beam AB is to be taken into account.

Solution. The actual condition (Fig. 303a) is taken as the first state of loading. In the second state the load P is removed and the forces X are replaced by unit forces (Fig. 303b). Due to these forces, upward vertical pressures equal to $(1 \cdot h)/c$ are transmitted to the beam AB at the points F and H and the beam deflects as indicated by the dotted line. If y is the deflection of the beam at the point corresponding to the load P, and δ is the displacement of the points C and D towards one another in the second state, the reciprocal theorem gives

$$X\delta - Py = 0 \quad \text{and} \quad X = P\frac{y}{\delta}. \qquad (i)$$

Hence the deflection curve of the beam AB in the second state is the required influence line. The bending of the beam by the two symmetrically located loads is discussed in Prob. 1, p. 159. Substituting $(1 \cdot h)/c$ for P in the formulas obtained there, the deflections of the beam at F and at the middle are

$$(y)_{x=c} = \frac{ch}{6EI}(3l - 4c) \quad \text{and} \quad (y)_{x=l/2} = \frac{n}{24EI}(3l^2 - 4c^2),$$

respectively.

Considering now the rotation of the triangle AFC (Fig. 303c) as a rigid body, the horizontal displacement of the point C is equal to the vertical displacement of the point F multiplied by h/c. Hence

$$\delta = 2\frac{h}{c}(y)_{x=c} = \frac{h^2}{3EI}(3l - 4c).$$

Substituting this and the deflection at the middle into eq. (i), gives

$$X = \frac{P}{8h} \frac{3l^2 - 4c^2}{3l - 4c}.$$

7. Find the influence line for the force in the bar CD of the system shown in Fig. 304, neglecting displacements due to contractions and elongations and considering only the bending of the beam AB.

Answer. The line will be the same as that for the middle reaction of the beam on three supports (see Prob. 1, p. 355).

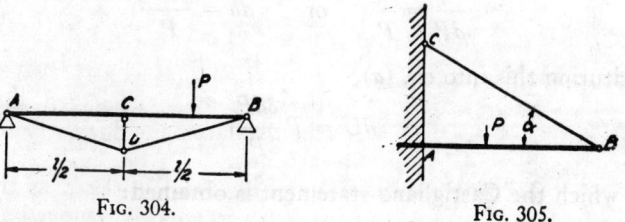

FIG. 304.   FIG. 305.

8. Construct the influence line for the bar BC which supports the beam AB, Fig. 305. Find the force in BC when P is at the middle.

Answer. Neglecting displacements due to elongation of the bar BC and contraction of the beam AB, the force in BC is $\frac{5}{16}(P/\sin \alpha)$.

76. Exceptional Cases.—In the derivation of both the Castigliano theorem and the reciprocal theorem it was assumed that the displacements due to deformation are proportional to the loads acting on the elastic system. There are cases in which the displacements are not proportional to the loads, although the material of the body may follow Hooke's law. This always occurs when the displacements due to deformations must be considered in discussing the action of external loads. In such cases, the strain energy is no longer a second degree function and the theorem of Castigliano does not hold.

In order to explain this limitation let us consider a simple case in which only one force P acts on the elastic system. Assume first that

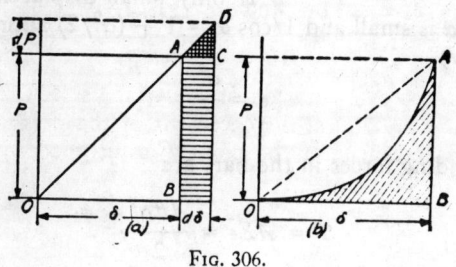

FIG. 306.

the displacement δ is proportional to the corresponding force P as represented by the straight line OA in Fig. 306a. Then the area of the triangle OAB represents the strain energy stored in the system during the application of the load P. For an infinitesimal increase $d\delta$ in the displacement the strain energy increases by an amount shown in the figure by the shaded area, and we obtain

$$dU = Pd\delta. \qquad (a)$$

With a linear relationship the infinitesimal triangle ADC is similar to the triangle OAB, therefore

$$\frac{d\delta}{dP} = \frac{\delta}{P} \quad \text{or} \quad d\delta = \frac{\delta dP}{P}. \qquad (b)$$

Substituting this into eq. (a),

$$dU = P \frac{\delta dP}{P},$$

from which the Castigliano statement is obtained:

$$\frac{dU}{dP} = \delta. \qquad (c)$$

An example to which the Castigliano theorem cannot be applied is shown in Fig. 307. Two equal horizontal bars AC and BC hinged at A, B and C are subjected to the action of the vertical force P at C. Let C_1 be the position of C after deformation and α the angle of inclination of either bar in its deformed condition. The unit elongation of the bars, from Fig. 307a, is

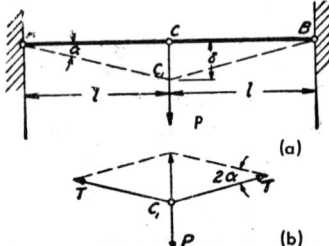

Fig. 307.

$$\epsilon = \left(\frac{l}{\cos\alpha} - l\right) \div l. \qquad (d)$$

If only small displacements are considered, then α is small and $1/\cos\alpha = 1 + (\alpha^2/2)$ approximately. Then, from (d),

$$\epsilon = \frac{\alpha^2}{2}.$$

The corresponding forces in the bars are

$$T = AE\epsilon = \frac{AE\alpha^2}{2}. \qquad (e)$$

From the condition of equilibrium of the point C_1 (Fig. 307b),

$$P = 2\alpha T, \qquad\qquad (f)$$

and for T, as given in eq. (e),

$$P = AE\alpha^3,$$

from which

$$\alpha = \sqrt[3]{\frac{P}{AE}} \qquad\qquad (g)$$

and

$$\delta = l\alpha = l\sqrt[3]{\frac{P}{AE}}. \qquad\qquad (208)$$

In this case the displacement is not proportional to the load P, although the material of the bars follows Hooke's law. The relation between δ and P is represented in Fig. 306b by the curve OA. The shaded area OAB in this figure represents the strain energy stored in the system. The amount of strain energy is

$$U = \int_0^\delta P d\delta. \qquad\qquad (h)$$

Substituting, from eq. (208),

$$P = AE\frac{\delta^3}{l^3}, \qquad\qquad (i)$$

we obtain

$$U = \frac{AE}{l^3}\int_0^\delta \delta^3 d\delta = \frac{AE\delta^4}{4l^3} = \frac{P\delta}{4} = \frac{Pl}{4}\sqrt[3]{\frac{P}{AE}}. \qquad (j)$$

This shows that the strain energy is no longer a function of the second degree in the force P. Also it is not one-half but only one-quarter of the product $P\delta$ (see Art. 71). The Castigliano theorem of course does not hold here:

$$\frac{dU}{dP} = \frac{d}{dP}\left(\frac{Pl}{4}\sqrt[3]{\frac{P}{AE}}\right) = \frac{1}{3}l\sqrt[3]{\frac{P}{AE}} = \frac{1}{3}\delta.$$

Analogous results are obtained in all cases in which the displacements are not proportional to the loads.[35]

[35] Such problems were discussed by F. Engesser, *Z. Architekt. u. Ing.-Ver.*, Vol. 35, p. 733, 1889; see also H. M. Westergaard, *Proc. Am. Soc. Civil Engrs.*, Feb. 1941.

CHAPTER XII

CURVED BARS

77. Pure Benaing of Curved Bars.—In the following discussion it is assumed that the line joining the centroids of the cross sections of the bar, called the *center line*, is a plane curve and that the cross sections have an axis of symmetry in this plane.[1] The bar is submitted to the action of forces lying in the plane of symmetry so that bending takes place in this plane.

Let us consider first the case of a bar of constant cross section subjected to *pure bending*, produced by couples M applied

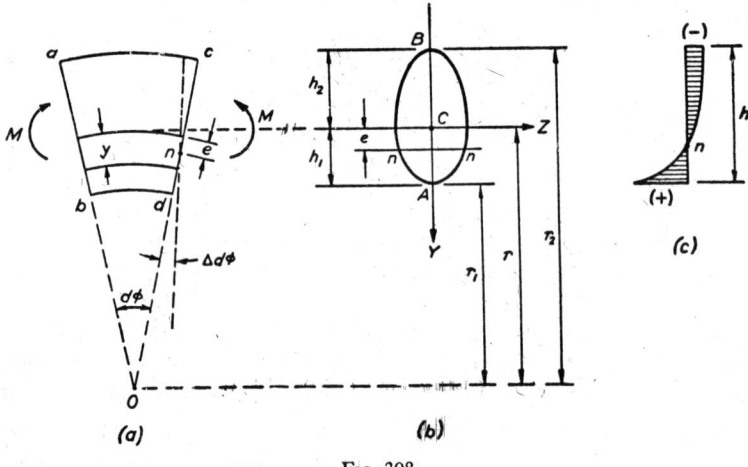

Fig. 308.

at the ends (Fig. 308). The stress distribution will be obtained by using the same assumption as in the case of straight bars, namely, that transverse cross sections which were originally plane and normal to the center line remain so after bend-

---

[1] The case of nonsymmetrical cross sections is discussed by F. K. G. Odqvist, *Publ. No. 107*, Inst. f. Festigkeitslehre Königl. Tech. Hochschule, Stockholm, 1953.

ing.[2] Let *ab* and *cd* be two adjacent cross sections of the bar (Fig. 308*a*) and let *dφ* be the small angle between them before bending. Due to bending the cross section *cd* rotates with respect to *ab* about the neutral axis *nn* by a small angle *Δdφ*. This angle and the corresponding moment *M* are taken positive if the initial curvature of the bar is diminished during bending. Due to this rotation the longitudinal fibers on the convex side of the bar are compressed and the fibers on the concave side are extended. Denoting by *y* the distances of the fibers from the centroidal axis perpendicular to the plane of bending, taken positive in the direction towards the center of curvature of the center line of the bar, and denoting by *e* the distance of the neutral axis *nn* from the centroid *C*, we find that the extension of any fiber during bending is $(y - e)\Delta d\varphi$ and the corresponding unit elongation of the fiber is

$$\epsilon = \frac{(y - e)\Delta d\varphi}{(r - y)d\varphi}. \tag{a}$$

In this expression, *r* denotes the radius of curvature of the center line of the bar (Fig. 308*b*) and the denominator in eq. (*a*) is the initial length of the fibers between the adjacent cross sections *ab* and *cd*.

Assuming that there is no lateral pressure between the longitudinal fibers,[3] the bending stress at a distance *y* from the centroidal axis and normal to the cross section is

$$\sigma = \frac{E(y - e)\Delta d\varphi}{(r - y)d\varphi}. \tag{b}$$

It is seen that the stress distribution is no longer linear as in

---

[2] This approximate theory was developed by E. Winkler, see *Civilingenieur*, Vol. 4, p. 232, 1858; also his book, *Die Lehre von der Elastizität und Festigkeit*, Prague, Chap. 15, 1867. A similar theory was also developed by H. Résal, *Ann. mines*, p. 617, 1862. Further development of the theory was made by F. Grashof in his book, *Elastizität und Festigkeit*, p. 251, 1878; and by Todhunter and Pearson, *History of the Theory of Elasticity*, Vol. 2, p. 422, 1893. For publications dealing with the rigorous solution of curved bar problems see *Theory of Elasticity*, by the writer and J. N. Goodier, p. 63, 1951.

[3] The rigorous theory shows that there is a certain radial pressure but that it has no substantial effect on the normal bending stress *σ*.

the case of straight bars, but that it follows a hyperbolic law as shown in Fig. 308c. From the fact that the sum of the normal forces distributed over the cross section is zero in the case of pure bending, it can be concluded that the neutral axis is displaced from the centroid of the cross section towards the center of curvature of the center line of the bar. In the case of a rectangular cross section, the shaded area (Fig. 308c) in tension must equal that in compression and we see at once that the greatest stress acts on the concave side. To make the stresses in the most remote fibers in tension and in compression numerically equal, it is necessary to use cross-sectional shapes which have the centroid nearer the concave side of the bar.

Eq. (b) contains two unknowns, the distance e of the neutral axis nn from the centroid C (Fig. 308b) and the angle of rotation $\Delta d\varphi$. To determine these quantities we use the two equations of statics which state that the sum of the normal forces distributed over a cross section is equal to zero and the moment of these forces is equal to the external moment M. These equations are

$$\int_A \sigma dA = \frac{E\Delta d\varphi}{d\varphi} \int_A \frac{(y - e)dA}{r - y} = 0, \qquad (c)$$

$$\int_A \sigma y dA = \frac{E\Delta d\varphi}{d\varphi} \int_A \frac{(y - e)y dA}{r - y} = M. \qquad (d)$$

From the first of these equations it follows that

$$\int_A \frac{y dA}{r - y} - e \int_A \frac{dA}{r - y} = 0. \qquad (e)$$

The first integral on the left side of this equation has the dimensions of area and may be expressed as

$$\int_A \frac{y dA}{r - y} = mA, \qquad (f)$$

where m is a pure number to be determined for each particular shape of the cross section by performing the indicated integration. The quantity mA is called the *modified area* of the cross section. The second integral may be transformed as follows:

$$\int_A \frac{dA}{r-y} = \int_A \frac{(y+r-y)dA}{r(r-y)} = \int_A \frac{ydA}{r(r-y)} + \int_A \frac{dA}{r}$$

$$= (m+1)\frac{A}{r},$$

and eq. (*e*) becomes

$$mA - (m+1)\frac{Ae}{r} = 0,$$

from which

$$e = r\frac{m}{m+1} \quad \text{or} \quad m = \frac{e}{r-e}. \tag{209}$$

Now taking eq. (*d*) and using the transformation

$$\int_A \frac{y^2 dA}{r-y} = -\int_A \left(y - \frac{ry}{r-y}\right) dA$$

$$= -\int_A ydA + r\int_A \frac{ydA}{r-y} = mrA, \tag{g}$$

we obtain

$$\frac{E\Delta d\varphi}{d\varphi}(mrA - meA) = M$$

and

$$\frac{E\Delta d\varphi}{d\varphi} = \frac{M}{m(r-e)A} = \frac{M}{Ae}. \tag{210}$$

Substituting into expression (*b*), we obtain the following formula for bending stresses:

$$\sigma = \frac{M(y-e)}{m(r-e)A(r-y)} = \frac{M(y-e)}{Ae(r-y)}. \tag{211}$$

To obtain the stresses in the most remote fibers we substitute for the points A and B (Fig. 308*b*) the values $y = h_1$ and $y = -h_2$, which gives

$$\sigma_A = \frac{M(h_1 - e)}{Aer_1}, \qquad \sigma_B = \frac{-M(h_2 + e)}{Aer_2}, \tag{212}$$

where r_1 and r_2 denote the inner and the outer radii of the curved bar. By determining from eqs. (f) and (209) the quantities m and e for any given shape of the cross section, we can readily calculate bending stresses by using eq. (211).

The change $\Delta d\varphi$ of the angle $d\varphi$ between the two consecutive cross sections is obtained from eq. (210) which gives

$$\Delta d\varphi = \frac{M d\varphi}{e A E} = \frac{M ds}{e r A E},$$

and the corresponding change in curvature of the center line of the bar is

$$\frac{\Delta d\varphi}{ds} = \frac{M}{e r A E} = \frac{M(m+1)}{m r^2 A E}. \tag{213}$$

If the radial dimension h of the curved bar is small in comparison with the radius of curvature r of the center line, we can neglect y in comparison with r in eqs. (f) and (g), and we conclude that as the radius of curvature becomes larger and larger, the number m approaches zero and the quantity $m r^2 A$ approaches the value of the centroidal moment of inertia I_z of the cross section. Then expression (213) for the change in curvature approaches the value

$$\frac{\Delta d\varphi}{ds} = \frac{M}{E I_z}, \tag{214}$$

which is the same as obtained previously for the curvature of initially straight bars (see p. 138).

78 Bending of Curved Bars by Forces Acting in the Plane of Symmetry.—Let us consider now a more general case of bending of a curved bar, shown in Fig. 309a. It is assumed that the forces $P_1, \cdots P_4$ represent a system of forces in equilibrium and are acting in the plane of the center line, which is the plane of symmetry of the bar. The deflection of the bar will evidently occur in the same plane. To find the stresses at any cross section mn of the bar, Fig. 309a, we assume that the portion of the bar to the right of cross section mn is removed and its action on the left portion of the bar is replaced by a force applied at the centroid C of the cross section and by the couple M. Resolving the force into two components, N and

V, one normal to the cross section and the other directed radially, we finally obtain the *bending moment M*, the *longitudinal force N* and the *shearing force V*. The positive directions of these forces are shown in Fig. 309*b*.

The stresses and deformations produced by a couple were discussed in the preceding article in considering the pure bending of a curved bar. The stresses corresponding to the longitudinal force will be uniformly distributed over the cross section and their magnitude will be N/A. These stresses will

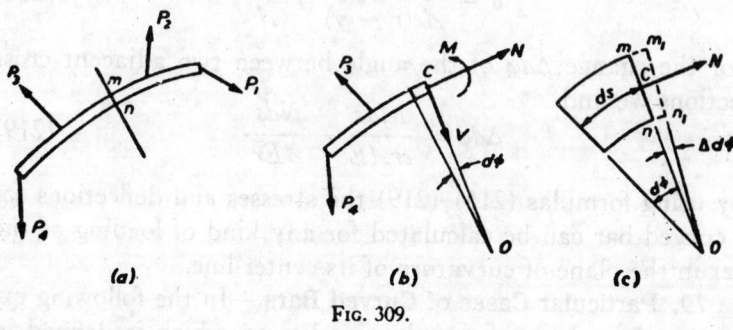

| (a) | (b) | (c) |

Fig. 309.

produce uniformly distributed unit elongations of the fibers, but the total elongations, proportional to the initial length of the fibers between any two adjacent cross sections, will be proportional to the distance from the center of curvature O of the center line, Fig. 309*c*. Thus, due to the action of the longitudinal force, the initial angle $d\varphi$ will be increased by the amount

$$\Delta d\varphi = \frac{Nds}{AEr}. \tag{215}$$

At the same time the initial length ds of the element of the center line will increase by the amount

$$\Delta ds = \frac{Nds}{AE}. \tag{216}$$

The transverse force V produces shearing stresses and some warping of the cross section. It is usually assumed that the distribution of shearing stresses over the cross section is the same as for a straight bar.[4] In such a case the relative radial displacement of the two adjacent cross sections will be the same as for straight bars and will be equal to (see p. 171)

$$\frac{\alpha V}{AG} ds. \tag{217}$$

Combining the bending stresses produced by the couple M with those produced by the force N, we obtain

$$\sigma = \frac{M(y - e)}{Ae(r - y)} + \frac{N}{A}. \tag{218}$$

For the change $\Delta d\varphi$ of the angle between two adjacent cross sections we find

$$\Delta d\varphi = \frac{Mds}{erAE} - \frac{Nds}{AEr}. \tag{219}$$

By using formulas (215)–(219) the stresses and deflections for a curved bar can be calculated for any kind of loading of the bar in the plane of curvature of its center line.

79. Particular Cases of Curved Bars.—In the following examples the calculation of the number m, which is defined in eq. (f) of Art. 77, and of the distance e of the neutral axis from the centroid of the cross section is given for several particular cases. Knowing these quantities the stresses in curved bars can be calculated by using the formulas of the two preceding articles.

Rectangular Cross Section.—In this case the width b of the cross section, Fig. 310, is constant and we obtain

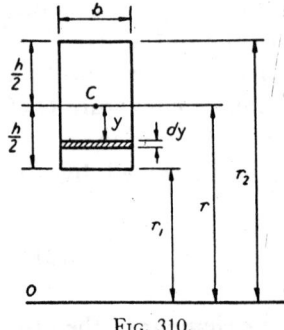

Fig. 310.

$$mA = \int_A \frac{ydA}{r - y} = b \int_{-h/2}^{h/2} \frac{(y - r + r)dy}{r - y}$$

$$= br \int_{-h/2}^{h/2} \frac{dy}{r - y} - bh = br \log_e \frac{r_2}{r_1} - bh,$$

---

[4] This assumption is in satisfactory agreement with a rigorous solution for a narrow rectangular cross section; see author's *Theory of Elasticity*, p. 75, 1951.

so that

$$m = \frac{r}{h} \log_e \frac{r_2}{r_1} - 1 \qquad (a)$$

and

$$e = \frac{mr}{m+1} = \frac{r[(r/h)\log_e (r_2/r_1) - 1)]}{(r/h)\log_e (r_2/r_1)} = r - \frac{h}{\log_e (r_2/r_1)}. \qquad (b)$$

For small values of h/r the distance e is small in comparison with h and to calculate it from eq. (b) with sufficient accuracy it is necessary to take $\log_e (r_2/r_1)$ with a high degree of accuracy. To eliminate this inconvenience we can use the known series

$$\log_e (r_2/r_1) = \log_e \frac{r + h/2}{r - h/2} = \frac{h}{r}\left[1 + \frac{1}{3}\left(\frac{h}{2r}\right)^2 + \frac{1}{5}\left(\frac{h}{2r}\right)^4 + \cdots \right].$$

Then, from eq. (a),

$$m = \frac{1}{3}\left(\frac{h}{2r}\right)^2 + \frac{1}{5}\left(\frac{h}{2r}\right)^4 + \cdots. \qquad (c)$$

We thus obtain a rapidly converging series, from which the quantities m and e can be readily calculated with any desired accuracy. Taking only the first term of the series, we obtain

$$m \approx h^2/12r^2 \qquad \text{and} \qquad e \approx h^2/12r.$$

With two terms of the series we find

$$m \approx \frac{h^2}{12r^2}\left(1 + \frac{3}{5}\frac{h^2}{4r^2} \right) \qquad \text{and} \qquad e \approx \frac{h^2}{12r}\left(1 + \frac{4}{15}\frac{h^2}{4r^2} \right).$$

It is seen that for small values of h/r the distance e is very small. Thus a linear stress distribution instead of a hyperbolic one may be assumed with sufficient accuracy. To compare the results obtained for the two types of stress distribution, for the case of a rectangular cross section, Table 8 has been calculated. In this table, the ratios

$$\alpha = \frac{\sigma_{\max}}{M/Ar} \qquad \text{and} \qquad \beta = \frac{\sigma_{\min}}{M/Ar}$$

are given for various values of the ratio r/h. It is seen from this table that for $r/h > 10$ a linear stress distribution can be

TABLE 8: COMPARISON OF HYPERBOLIC AND LINEAR STRESS DISTRIBUTIONS

r/h	Hyperbolic Stress Distribution		Linear Stress Distribution		Error in σ_{max} %
	α	β	α	β	
1	9.2	−4 4	6	−6	35.0
2	14.4	−10.3	12	−12	17.0
3	20.2	−16.1	18	−18	10.9
4	26.2	−22.2	24	−24	9.2
10	62.0	−58	60	−60	3.2

assumed and the *straight bar formula* for maximum stress may be used with sufficient accuracy.

Trapezoidal Cross Section.—Using again the equation

$$mA \doteq \int_A \frac{ydA}{r-y} \tag{d}$$

and introducing the notation

$$v = r - y, \tag{e}$$

where v is the distance of the shaded element (Fig. 311) from the axis O–O through the center of curvature of the center line of the bar, we obtain

$$mA = \int_{r_1}^{r_2} \frac{(r-v)dA}{v} = r \int_{r_1}^{r_2} \frac{dA}{v} - A. \tag{f}$$

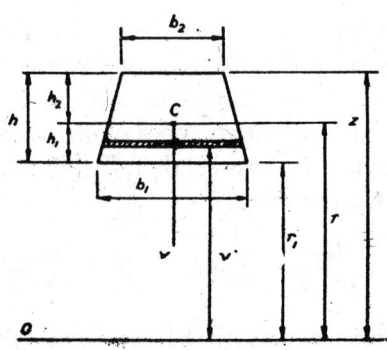

FIG. 311.

The variable width of the cross section (Fig. 311) is

$$b = b_2 + (b_1 - b_2)(r_2 - v)/h$$

and

$$dA = dv[b_2 + (b_1 - b_2)(r_2 - v)/h].$$

Substituting into eq. (f), we obtain

$$m = \frac{r}{A} \int_{r_1}^{r_2} [b_2 + (b_1 - b_2)(r_2 - v)/h] \frac{dv}{v} - 1$$

$$= \frac{r}{A} \left\{ [b_2 + r_2(b_1 - b_2)/h] \log_e \frac{r_2}{r_1} - (b_1 - b_2) \right\} - 1. \quad (g)$$

When $b_1 = b_2 = b$, this formula coincides with formula (a), which was obtained for a rectangular cross section. If we take $b_2 = 0$ we obtain from formula (g) the value of m for a *triangular cross section*.

⊥ *Cross Section.*—Proceeding as in the previous case, we obtain for the cross section in Fig. 312

$$m = \frac{r}{A} \left(b_1 \log_e \frac{r_3}{r_1} + b_2 \log_e \frac{r_2}{r_3} \right) - 1. \quad (h)$$

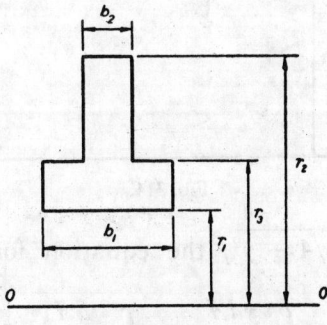

FIG. 312.

I *Cross Section.*—The same procedure as in the preceding case gives (Fig. 313):

$$m = \frac{r}{A} \left(b_1 \log_e \frac{r_3}{r_1} + b_2 \log_e \frac{r_4}{r_3} + b_3 \log_e \frac{r_2}{r_4} \right) - 1. \quad (i)$$

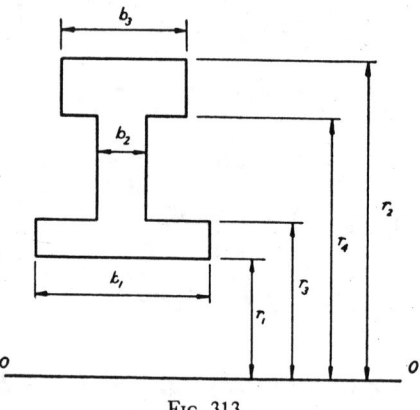

FIG. 313.

Circular Cross Section.—Observing that in this case (Fig. 314) the width of the cross section at the distance y from the

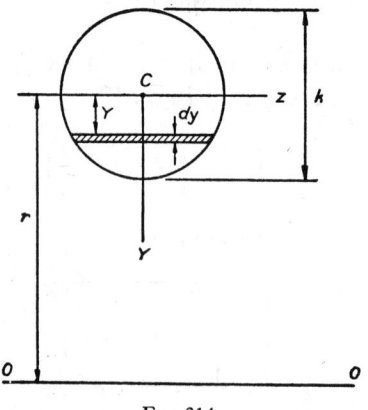

FIG. 314.

centroid is $2\sqrt{h^2/4 - y^2}$, the equation for determining m becomes

$$mA = \int_A \frac{y\,dA}{r-y} = r\int_A \frac{dA}{r-y} - A$$

$$= 2r\int_{-h/2}^{h/2} \frac{\sqrt{h^2/4 - y^2}}{r - y}\,dy - A$$

$$= 2\pi r(r - \sqrt{r^2 - h^2/4}) - A. \qquad (j)$$

Using the series

$$\sqrt{1 - h^2/4r^2}$$

$$= 1 - \frac{1}{2}\frac{h^2}{4r^2} - \frac{1}{8}\left(\frac{h^2}{4r^2}\right)^2 - \frac{1}{16}\left(\frac{h^2}{4r^2}\right)^3 - \frac{5}{128}\left(\frac{h^2}{4r^2}\right)^4 - \cdots,$$

we then obtain

$$m = \frac{1}{4}\left(\frac{h}{2r}\right)^2 + \frac{1}{8}\left(\frac{h}{2r}\right)^4 + \frac{5}{64}\left(\frac{h}{2r}\right)^6 + \cdots. \qquad (220)$$

This is a rapidly converging series from which m may easily be calculated to any desired accuracy.

It can be seen that in calculating m from eq. (d) the magnitude of m does not change if all elements dA are multiplied by some constant, since in this way the integral of eq. (d) and the area A of the same equation will be increased in the same proportion. From this it follows that the value of m obtained for a circular cross section from eq. (j) can be used also for an ellipse with the axes h and h_1, since in this case each elemental area obtained for a circle is to be multiplied by the constant ratio h_1/h.

The calculation of the integral in eq. (d) can sometimes be simplified by dividing the cross section into several parts, integrating for each part and adding the results of these integrations. Taking, for example, a circular ring cross section with outer diameter h and inner diameter h_1, and using eq. (j) for the outer and inner circles, we find for the ring cross section:

$$m = \frac{1}{h^2 - h_1^2}\left\{h^2\left[\frac{1}{4}\left(\frac{h}{2r}\right)^2 + \frac{1}{8}\left(\frac{h}{2r}\right)^4 + \cdots\right]\right.$$

$$\left. - h_1^2\left[\frac{1}{4}\left(\frac{h_1}{2r}\right)^2 + \frac{1}{8}\left(\frac{h_1}{2r}\right)^4 + \cdots\right]\right\}. \qquad (k)$$

In a similar manner we can develop formulas for the cross sections shown in Figs. 312 and 313. When m is calculated, we find e from eq. (209) and the maximum stress from eq. (212).

Eq. (d) is the basis of a graphical determination of the quantity m for cases in which the shape of the cross section cannot be simply expressed analytically. It is seen that in calculating the modified area from eq. (d) every elemental area must be diminished in the ratio $y/(r - y)$. This can be done by retaining the width of the elemental strips but diminishing their lengths in the above ratio (Fig. 315). In this

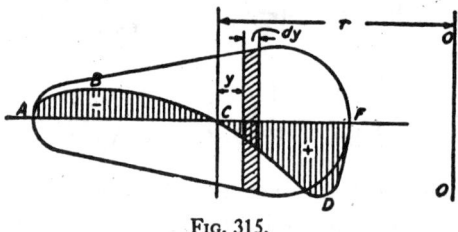

FIG. 315.

manner the shaded area in the figure is obtained. The difference between the areas CDF and ABC gives the modified area mA. Knowing this, the quantities m and e can readily be calculated.

Since the value mA is obtained as the difference of the two areas, the accuracy of the result is low and the method gives only a rough approximation. A much better accuracy can be obtained by dividing the cross section into strips of equal width and then using Simpson's rule for calculating the integral of eq. (d).[5]

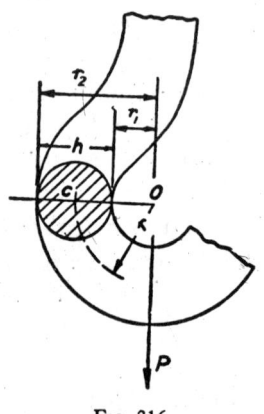

FIG. 316.

The theory of curved bars developed above is applied in designing *crane hooks*.[6] In Fig. 316 is represented the working portion of a hook of constant circular cross section. It is assumed that the vertical force P passes through the center of curvature O of the axis of the hook. The maximum bending stress occurs in

[5] Examples of such calculations are given in a paper by A. M. Wahl, *J. Appl. Mech.*, Vol. 13, p. 239, 1946.

[6] A theoretical and experimental investigation of crane hooks was made by the National Physical Laboratory in England; see paper by H. J. Gough, H. L. Cox and D. G. Sopwith, *Proc. Inst. Mech. Engrs. (London)*, Dec. 1934. For a comparison of theoretical stresses in hooks of a rectangular cross section with experimental results see K. Böttcher, *Forschungsarb.*, *No. 337*, 1931.

the cross section perpendicular to the load P. Then proceeding as explained in Art. 78, we find that on the horizontal cross section of the hook the tensile force P, applied at the center C of the cross section, and the bending moment $M = Pr$ are acting. Combining direct and bending stresses and using eq. (218) for the latter, we obtain

$$\sigma = \frac{M(y - e)}{Ae(r - y)} + \frac{P}{A} = \frac{Py}{Am(r - y)},$$

Applying this formula to the most remote points, for which. $y = \pm \dfrac{h}{2}$, we find that

$$\sigma_{max} = \frac{P}{A} \frac{h}{2mr_1}, \qquad \sigma_{min} = -\frac{P}{A} \frac{h}{2mr_2}. \qquad (221)$$

It is seen that the numerically largest stress is the tensile stress at the inner surface, which is obtained by multiplying the stress P/A by the *stress factor:*

$$k = \frac{h}{2mr_1}, \qquad (222)$$

the magnitude of which depends on the ratio $h/2r$. Using expression (220) for m, we find that k varies from 13.5 to 15.4 as the ratio $h/2r$ changes from 0.6 to 0.4.[7] The calculation of stresses in crane hooks of irregular section may be carried out by using Simpson's rule to determine m.[8]

Problems

1. Determine the ratio of the numerical values of σ_{max} and σ_{min} for a curved bar of rectangular cross section in pure bending if $r = 5$ in. and $h = 4$ in.

Answer. $\left| \dfrac{\sigma_{max}}{\sigma_{min}} \right| = 1.75$.

2. Solve the previous problem, assuming a circular cross section.

Answer. $e = 0.208$ in.; $\left| \dfrac{\sigma_{max}}{\sigma_{min}} \right| = \dfrac{1.792}{2.208} \cdot \dfrac{7}{3} = 1.89$.

---

[7] At $h/2r = 0.6$ the factor k has its minimum value.

[8] This method is described by A. M. Wahl, *loc. cit.*, p. 374.

3. Determine the dimensions b_1 and b_3 of an I cross section (Fig. 313) to make σ_{max} and σ_{min} numerically equal in pure bending. The dimensions are $r_1 = 3$ in., $r_3 = 4$ in., $r_4 = 6$ in., $r_2 = 7$ in., $b_2 = 1$ in., $b_1 + b_3 = 5$ in.

Solution. From eqs. (212),

$$\frac{h_1 - e}{r_1} = \frac{h_2 + e}{r_2} \quad \text{or} \quad \frac{r - r_1 - e}{r_1} = \frac{r_2 - r + e}{r_2},$$

from which

$$r - e = \frac{r}{m + 1} = \frac{2r_1 r_2}{r_1 + r_2} = 4.20 \text{ in.}$$

Substituting for m the value (i) and noting that $A = 7$ in.^{2}, we have

$$4.20 = \frac{7}{b_1 \log_e \frac{4}{3} + 1 \cdot \log_e \frac{6}{4} + (5 - b_1) \log_e \frac{7}{6}},$$

from which

$$b_1 \log_e \frac{4}{3} + (5 - b_1) \log_e \frac{7}{6} = \frac{7}{4.20} - 1 \cdot \log_e \frac{6}{4},$$

$$0.288 b_1 + 0.154(5 - b_1) = 1.667 - 0.406 = 1.261 \text{ in.,}$$

$$b_1 = 3.67 \text{ in.,} \qquad b_3 = 5 - 3.67 = 1.33 \text{ in.}$$

4. Determine the dimension b_1 of the $\perp$ section (Fig. 312) to make σ_{max} and σ_{min} numerically equal in pure bending. The dimensions are $r_1 = 3$ in., $r_2 = 7$ in., $r_3 = 4$ in., $b_2 = 1$ in.

Answer. $b_1 = 3.09$ in.

5. Determine σ_{max} and σ_{min} for the trapezoidal cross section mn of the hook represented in Fig. 317 if $P = 4,500$ lb, $b_1 = 1\frac{5}{8}$ in., $b_2 = \frac{3}{8}$ in., $r_1 = 1\frac{1}{4}$ in., $r_2 = 5$ in.

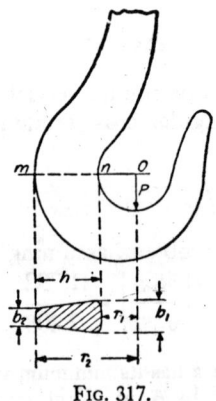

FIG. 317.

Solution: From eq. (g),

$$r - e = \frac{r}{m+1} = \frac{3.750}{\dfrac{1\frac{5}{8} \times 5 - \frac{3}{8} \times 1\frac{1}{4}}{3\frac{3}{4}} \log_e 5/1\frac{1}{4} - (1\frac{5}{8} - \frac{3}{8})}$$

$$= \frac{3.750}{1.580} = 2.373 \text{ in}$$

The radius of the center line

$$r = r_1 + \frac{b_1 + 2b_2}{b_1 + b_2} \cdot \frac{h}{3} = 2.734 \text{ in.}$$

Therefore, $e = 0.361$ in., $h_1 - e = r - e - r_1 = 2.373 - 1.250 = 1.123$ in., $h_2 + e = r_2 - r + e = 5 - 2.373 = 2.627$ in., $Ae = 3.75 \times 0.361 = 1.35$, $M = Pr = 12,300$ in. lb. The bending stresses, from eqs. (212), are

$$\sigma_{\max} = \frac{12,300 \times 1.123}{1.35 \times 1.25} = 8,200 \text{ lb per sq in.}$$

$$\sigma_{\min} = -\frac{12,300 \times 2.627}{1.35 \times 5} = -4,800 \text{ lb per sq in.}$$

On these bending stresses, a uniformly distributed tensile stress $P/A = 4,500/3.75 = 1,200$ lb per sq in. must be superposed. The total stresses are

$$\sigma_{\max} = 8,200 + 1,200 = 9,400 \text{ lb per sq in.,}$$

$$\sigma_{\min} = -4,800 + 1,200 = -3,600 \text{ lb per sq in.}$$

6. Find the maximum stress in a hook of circular cross section if the diameter of the cross section is $h = 1$ in., radius of the central axis $r = 1$ in. and $P = 1,000$ lb.

Answer. $\sigma_{\max} = 17,700$ lb per sq in.

7. Find $\sigma_{\max}$ and $\sigma_{\min}$ for the curved bar of circular cross section, loaded as shown in Fig. 318, if $h = 4$ in., $r = 4$ in., $a = 4$ in., and $P = 5,000$ lb.

Answer. $\sigma_{\max} = 10,650$ lb per sq in., $\sigma_{\min} = -4,080$ lb per sq in.

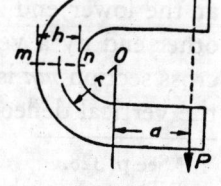

Fig. 318.

8. Solve the preceding problem, assuming that the cross section mn has the form shown in Fig. 312 with the following dimensions:

$r_1 = 2$ in., $r_3 = 3$ in., $r_2 = 9$ in., $b_1 = 4$ in., $b_2 = 1$ in., $a = 4$ in. and $P = 4{,}000$ lb.

Answer. $\sigma_{max} = 3{,}510$ lb per sq in., $\sigma_{min} = -1{,}800$ lb per sq in.

9. Solve Prob. 7, assuming that the cross section mn is trapezoidal, as in Fig. 311, with the dimensions $r_1 = 2$ in., $r_2 = 4\frac{1}{4}$ in., $b_1 = 2$ in., $b_2 = 1$ in., $a = 0$ and $P = 1.25$ tons.

Answer. $\sigma_{max} = 3.97$ tons per sq in., $\sigma_{min} = -2.33$ tons per sq in.

80. Deflection of Curved Bars.—The deflections of curved bars will be calculated by the use of Castigliano's theorem.[9] We start with the simplest case in which the cross-sectional dimensions of the bar are small in comparison with the radius of curvature of its center line.[10] Then the change in the angle between two adjacent cross sections is given by eq. (214),

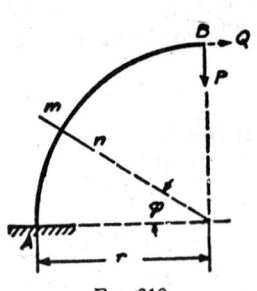

FIG. 319.

analogous to eq. (a), p. 138, for straight bars, and the strain energy of bending is given by the equation

$$U = \int_0^s \frac{M^2 ds}{2EI_z}, \qquad (223)$$

in which the integration is extended along the total length s of the bar. Eq. (223) is analogous to eq. (190) for straight beams,[11] and the deflection of the point of application of any load P in the direction of the load is

$$\delta = \frac{\partial U}{\partial P}.$$

As an example, take a curved bar of uniform cross section whose center line is a quarter of a circle (Fig. 319), built in at the lower end A with a vertical tangent and loaded at the other end by a vertical load P. The bending moment at any cross section mn is $M = -Pr \cos \varphi$. Substituting in eq. (223), the vertical deflection of the end B is

---

[9] See p. 328.

[10] The case in which the cross-sectional dimensions are not small is discussed in Prob. 6, p. 387.

[11] The strain energy due to longitudinal and shearing forces can be neglected in the case of thin curved bars. See p. 383.

$$\delta = \frac{d}{dP} \int_0^{\pi/2} \frac{M^2 r d\varphi}{2EI_z} = \frac{1}{EI_z} \int_0^{\pi/2} M \frac{dM}{dP} r d\varphi$$

$$= \frac{1}{EI_z} \int_0^{\pi/2} Pr^3 \cos^2 \varphi d\varphi = \frac{\pi}{4} \frac{Pr^3}{EI_z}.$$

If the horizontal displacement of the end B is required, a horizontal fictitious load Q must be added as shown in the figure by the dotted line. Then

$$M = -[Pr \cos \varphi + Qr (1 - \sin \varphi)]$$

and

$$\frac{\partial M}{\partial Q} = -r(1 - \sin \varphi).$$

The horizontal deflection is

$$\delta_1 = \left(\frac{\partial U}{\partial Q}\right)_{Q=0} = \frac{\partial}{\partial Q} \int_0^{\pi/2} \frac{M^2 r d\varphi}{2EI_z} = \frac{1}{EI_z} \int_0^{\pi/2} M \frac{\partial M}{\partial Q} r d\varphi.$$

$Q = 0$ must be substituted in the expression for M, giving

$$\delta_1 = \frac{1}{EI_z} \int_0^{\pi/2} Pr^3 \cos \varphi(1 - \sin \varphi)d\varphi = \frac{Pr^3}{2EI_z}.$$

Thin Ring.—As a second example consider the case of a thin circular ring submitted to the action of two equal and opposite forces P acting along the vertical diameter (Fig. 320). Due to symmetry only one quadrant of the ring (Fig. 320b) need be considered, and we can also conclude that there are no shearing stresses over the cross section mn and that the tensile force on this cross section is equal to $P/2$. The magnitude of the bending moment M_0 acting on this cross section is statically indeterminate and may be found by

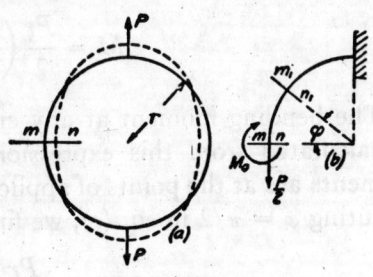

FIG. 320.

the Castigliano theorem. It is seen from the condition of symmetry that the cross section mn does not rotate during the bending of the ring. Hence the displacement corresponding to M_0 in Fig. 320b is zero and

$$\frac{dU}{dM_0} = 0, \qquad (a)$$

in which U is the strain energy of the quadrant of the ring which we are considering. For any cross section $m_1 n_1$ at an angle φ with the horizontal the bending moment is [12]

$$M = M_0 - \frac{P}{2} r(1 - \cos \varphi) \qquad (b)$$

and

$$\frac{dM}{dM_0} = 1.$$

Substituting this into expression (223) for the potential energy and using eq. (a), we find

$$0 = \frac{d}{dM_0} \int_0^{\pi/2} \frac{M^2 r d\varphi}{2EI_z} = \frac{1}{EI_z} \int_0^{\pi/2} M \frac{dM}{dM_0} r d\varphi$$

$$= \frac{1}{EI_z} \int_0^{\pi/2} \left[M_0 - \frac{P}{2} r(1 - \cos \varphi) \right] r d\varphi,$$

from which

$$M_0 = \frac{Pr}{2} \left(1 - \frac{2}{\pi} \right) = 0.182 Pr. \qquad (224)$$

Substituting into eq. (b), we obtain

$$M = \frac{Pr}{2} \left(\cos \varphi - \frac{2}{\pi} \right). \qquad (c)$$

The bending moment at any cross section of the ring may be calculated from this expression. The greatest bending moments are at the points of application of the forces P. Substituting $\varphi = \pi/2$ in eq. (c), we find

$$M = -\frac{Pr}{\pi} = -0.318 Pr. \qquad (225)$$

---

[12] Moments which tend to decrease the initial curvature of the bar are taken as positive.

Th⟨ minus sign indicates that the bending moments at the
po⟨ts of application of the forces P tend to increase the
c⟨rvature, while the moment M_0 at the cross section mn tends
to decrease the curvature of the ring, and the shape of the ring
after bending is that indicated in the figure by the dotted line.

The increase in the vertical diameter of the ring may also
be calculated by the Castigliano theorem. The total strain
energy stored in the ring is

$$U = 4 \int_0^{\pi/2} \frac{M^2 r d\varphi}{2EI_z},$$

in which M is given by eq. (c). Then the increase in the vertical diameter is

$$\delta = \frac{dU}{dP} = \frac{4}{EI_z} \int_0^{\pi/2} M \frac{dM}{dP} r d\varphi = \frac{Pr^3}{EI_z} \int_0^{\pi/2} \left(\cos \varphi - \frac{2}{\pi} \right)^2 d\varphi$$

$$= \left(\frac{\pi}{4} - \frac{2}{\pi} \right) \frac{Pr^3}{EI_z} = 0.149 \frac{Pr^3}{EI_z}. \qquad (226)$$

For calculating the decrease of the horizontal diameter of the
ring in Fig. 320, two oppositely directed fictitious forces Q are
applied at the ends of the horizontal diameter. Then by calculating $(\partial U/\partial Q)_{Q=0}$, we find that the decrease in the horizontal diameter is [13]

$$\delta_1 = \left(\frac{2}{\pi} - \frac{1}{2} \right) \frac{Pr^3}{EI_z} = 0.137 \frac{Pr^3}{EI_z}. \qquad (227)$$

Thick Ring.—When the cross-sectional dimensions of a curved
bar are not small in comparison with the
radius of the center line, not only the strain
energy due to bending moment but also that
due to longitudinal and shearing forces must
be taken into account. The change in the
angle between two adjacent cross sections
(Fig. 321) in this case, from eq. (213), is

$$\Delta d\varphi = \frac{Md\varphi}{AEe} = \frac{Mds}{AEer},$$

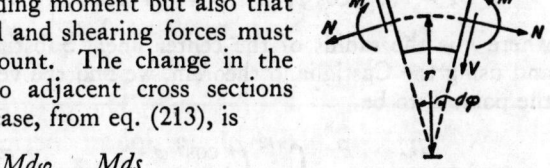

Fig. 321.

[13] A very complete study of circular rings under various kinds of loading
was made by C. B. Biezeno and his collaborator, J. J. Koch. The principal
results of their publications are presented in the book by Biezeno and R.
Grammel, *Technische Dynamik*, 2d Ed., Vol. 1, pp. 362–95, 1953.

and the energy due to bending for the element between the two adjacent cross sections is

$$dU_1 = \frac{1}{2} M\Delta d\varphi = \frac{M^2 ds}{2AEer}. \tag{d}$$

The longitudinal force N produces an elongation of the element between the two adjacent cross sections in the direction of the center line of the bar equal to Nds/AE and increases the angle $d\varphi$ by Nds/AEr (eq. 219). The work done by the forces N during their application is $N^2ds/2AE$. During the application of the forces N the couples M do the negative work $-MNds/AEr$. Hence the total energy stored in an element of the bar during the application of the forces N is

$$dU_2 = \frac{N^2 ds}{2AE} - \frac{MNds}{AEr}. \tag{e}$$

The shearing force V produces sliding of one cross section with respect to another, of the amount $\alpha Vds/AG$, where α is a numerical factor depending upon the shape of the cross section (see eq. 217). The corresponding amount of strain energy is

$$dU_3 = \frac{\alpha V^2 ds}{2AG}. \tag{f}$$

Adding (d), (e) and (f) and integrating along the length of the bar, the total strain energy of a curved bar becomes

$$U = \int_0^s \left(\frac{M^2}{2AEer} + \frac{N^2}{2AE} - \frac{MN}{AEr} + \frac{\alpha V^2}{2AG} \right) ds. \tag{228}$$

Let us use this equation to solve the problem represented in Fig. 319. Taking as positive the directions shown in Fig. 321, we have

$$M = -Pr\cos\varphi, \qquad N = -P\cos\varphi, \qquad V = P\sin\varphi,$$

where r is the radius of the center line. Substituting in eq. (228) and using the Castigliano theorem, we find the vertical deflection of the point B to be

$$\delta = \frac{dU}{dP} = \frac{Pr}{AE} \int_0^{\pi/2} \left(\frac{r\cos^2\varphi}{e} - \cos^2\varphi + \frac{\alpha E}{G}\sin^2\varphi \right) d\varphi$$

$$= \frac{\pi Pr}{4AE} \left(\frac{r}{e} + \frac{\alpha E}{G} - 1 \right).$$

If the cross section of the bar is a rectangle of the width b and depth

h, using for e the approximate value $h^2/12r$ (see p. 369) and taking $\alpha = 1.2$ and $E/G = 2.6$,

$$\delta = \frac{\pi P r}{4AE}\left(\frac{12r^2}{h^2} + 2.12\right).$$

When h is small in comparison with r, the second term in the parentheses, representing the influence on the deflection of N and V, can be neglected and we arrive at the equation obtained before (see p. 379).

The above theory of thick curved bars is often applied in calculating stresses in such machine elements as links and eye-shaped ends of bars (Fig. 322). In such cases a difficulty arises in determining the load distribution over the surface of the curved bar. This distribution depends on the amount of clearance between the bolt and the curved bar. A satisfactory solution of the problem may be expected only by combining analytical and experimental methods of investigation.[14]

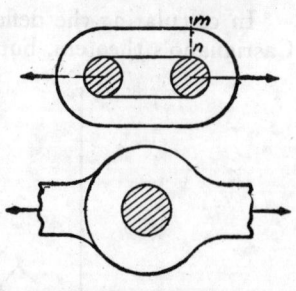

Fig. 322.

The particular case of an eye-shaped end of rectangular cross section, Fig. 322b, was rigorously investigated.[15] In this discussion it was assumed that there are no clearances and that the bolt is absolutely rigid. The maximum tensile stress occurs at the inner surface in the cross sections perpendicular to the axis of the bar, and its magnitude can be represented by the formula:

$$\sigma_{max} = \alpha \cdot \frac{8P}{\pi^2 r_2 t}, \qquad (g)$$

in which P is the total tensile force transmitted by the bar, α is a numerical factor depending on the ratio r_2/r_1 between

[14] For a theoretical investigation of the problem, see H. Reissner, *Jahrb. wiss. Ges. f. Luftfahrt* 1928; also J. Beke, *Eisenbau*, p. 233, 1921; F. Bleich, *Theorie und Berechnung der eisern Brücken*, W 256, 1924; Blumenfeld, *Z. Ver. deut. Ing.*, 1907; and Baumann, *ibid.*, p. 397, 1908. Experiments have been made by Dr. Mathar, *Forschungsarb.*, No. 306, 1928; see also D. Rühl, dissertation, Danzig, 1920; Preuss, *Z. Ver. deut. Ing.*, Vol. 55, p. 2173, 1911; M. Voropaeff, *Bull. Polytech. Inst.* (Kiev), 1910; E. G. Coker, "Photoelasticity," *J. Franklin Inst.*, 1925.

[15] H. Reissner and F. Strauch, *Ing. Arch.*, Vol. 4, p. 481, 1933.

the outer and the inner radii of the eye and t is the thickness of the eye perpendicular to the plane of the figure. For r_2/r_1 equal to 2 and 4 the values of α are, respectively, 4.30 and 4.39. The values obtained from formula (g) are in satisfactory agreement with experiments.[16]

The theory of thin rings has found application in the analysis of fuselage rings in airplane structures [17] and stiffening rings in submarines.[18]

In calculating the deflections of curved bars we used up to now Castigliano's theorem, but the problem can also be solved, as in the

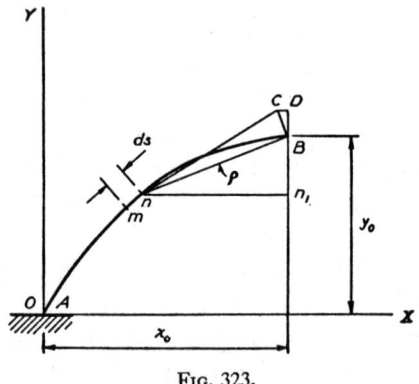

FIG. 323.

case of straight bars, by introducing fictitious loads. The calculations are especially simple in the case of thin bars when the effect of longitudinal and shearing forces on the deflections can be neglected. Let us consider a bar AB, Fig. 323, built in at the end A and loaded in its plane of symmetry xy. To calculate the deflection of the end B we consider the infinitesimal displacement $\overline{BC}$ of that end due to

[16] See G. Bierett, *Mitt. deut. Materialprüfungsanstalt.*, *Spec. No. 15*, 1931. A photoelastic investigation of the eye-shaped end was made by K. Takemura and Y. Hosokawa, *Rept. Aeronaut. Inst.* (*Tokyo*), Vol. 18, p. 128, 1926. See also M. M. Frocht and H. N. Hill, *J. Appl. Mech.*, Vol. 7, p. 5, 1940. In the latter paper the effect of clearance between the bolt and the hole is investigated.

[17] See W. Stieda, *Luftfahrt-Forsch.*, Vol. 18, p. 214, 1941. For English translation see *Nat. Advisory Comm. Aeronaut.*, *Tech. Mem. No. 1004*, 1942. See also D. A. Du Plantier, *J. Aeronaut. Sci.*, Vol. 11, p. 136, 1944; R. Benjamin, *ibid.*, Vol. 19, p. 585, 1952.

[18] A very complete study of such rings can be found in the book by P. F. Papkovitch, *Structural Mechanics of Ships*, Moscow, Vol. 2, pp. 1–816, 1947.

bending of an element mn of the bar. Using eq. (214) for the change of the angle between the two adjacent cross sections m and n, we find

$$\overline{BC} = \frac{Mds}{EI_z}\,\rho. \qquad (h)$$

Observing that the infinitesimal triangle BCD is similar to the triangle nBn_1, we find the two components of this displacement:

$$\overline{CD} = \frac{Mds}{EI_z}\,(y_0 - y), \qquad \overline{DB} = \frac{Mds}{EI_z}\,(x_0 - x). \qquad (i)$$

To obtain the two components of the total deflection of the end B, we have only to sum up the elemental displacements (i) for all elements of the bar. Denoting these components by u and v and taking them positive when they are in the positive directions of the x and y axes, we obtain

$$u = -\int_0^s \frac{Mds}{EI_z}\,(y_0 - y), \qquad v = \int_0^s \frac{Mds}{EI_z}\,(x_0 - x). \qquad (j)$$

Considering, for example, the bar represented in Fig. 319, we have

$$M = -Pr\cos\varphi, \qquad y_0 - y = r(1 - \sin\varphi), \qquad x_0 - x = r\cos\varphi.$$

Substituting into eqs. (j) and performing the integrations, we obtain results coinciding with those obtained on p. 379.

It is seen from eqs. (j) that if we apply to each element of the bar a fictitious horizontal load of magnitude Mds/EI_z, the moment of this load distribution about the end B gives the value of the deflection u. If, instead of horizontal, we take the fictitious loads in the vertical direction and of the same intensity, the moment of those loads gives the deflection v.

Problems

1. Determine the vertical deflection of the end B of the thin curved bar of uniform cross section and semicircular center line (Fig. 324).

Solution. The strain energy of bending is

$$U = \int_0^\pi \frac{M^2 r d\varphi}{2EI_z} = \int_0^\pi \frac{P^2 r^2 (1 - \cos\varphi)^2 r d\varphi}{2EI_z}.$$

The deflection at the end is

$$\delta = \frac{dU}{dP} = \frac{Pr^3}{EI_z}\int_0^\pi (1 - \cos\varphi)^2 d\varphi = \frac{3\pi}{2}\frac{Pr^3}{EI_z}.$$

2. Determine the horizontal displacement of the end B in the previous problem.

Answer. $\delta = \dfrac{2Pr^3}{EI_z}$.

3. Determine the increase in the distance between the ends A and B of a thin bar of uniform cross section consisting of a semicircular portion CD and two straight portions AC and BD (Fig. 325).

Answer. $\delta = \dfrac{2P}{EI_z}\left[\dfrac{l^3}{3} + r\left(\dfrac{\pi}{2}l^2 + \dfrac{\pi}{4}r^2 + 2lr\right)\right]$.

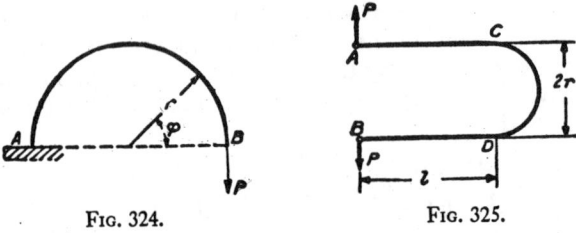

FIG. 324.   FIG. 325.

4. A link consisting of two semicircles and of two straight portions is submitted to the action of two equal and opposite forces acting along the vertical axis of symmetry (Fig. 326). Determine the max-

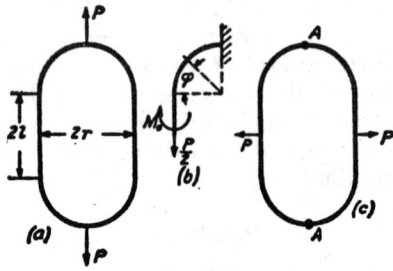

FIG. 326.

imum bending moment, assuming that the cross-sectional dimensions of the link are small in comparison with the radius r.

Solution. Considering only one-quarter of the link (Fig. 326b), we find the statically indeterminate moment M_0 from the condition that the cross section on which this moment acts does not rotate. Then,

$$\frac{dU}{dM_0} = 0.$$

Noting that for the straight portion $M = M_0$ and that for a curved portion $M = M_0 - (P/2)r(1 - \cos \varphi)$ and taking into consideration the strain energy of bending only, we find

$$\frac{dU}{dM_0} = \frac{d}{dM_0} \left\{ \frac{1}{2EI_z} \int_0^l M_0^2 dx \right.$$

$$\left. + \frac{1}{2EI_z} \int_0^{\pi/2} \left[M_0 - \frac{Pr}{2}(1 - \cos \varphi) \right]^2 rd\varphi \right\} = 0,$$

from which

$$M_0 = \frac{Pr^2}{2} \frac{\pi - 2}{2l + \pi r}.$$

For $l = 0$, this coincides with eq. (224) obtained before for a circular ring. The largest moment is at the points of application of forces P and is equal to

$$M_1 = M_0 - \frac{Pr}{2}.$$

5. Solve the previous problem, assuming that forces P are applied as shown in Fig. 326c.

Answer. The bending moment at points A is

$$M_1 = \frac{P}{2} \frac{r^2(\pi - 2) + 2rl + l^2}{\pi r + 2l}.$$

For $l = 0$, the equation coincides with that for a circular ring. For $r = 0$, $M_1 = Pl/4$ as for a bar with built-in ends.

6. Determine the bending moment M_0 and the increase in the vertical diameter of the circular ring shown in Fig. 320, assuming that the cross section of the ring is a rectangle of width b and depth h, dimensions which are not small in comparison with the radius r of the center line.

Solution. If we use eq. (228) for the potential energy and eq. (b) for the bending moment, the equation for determining M_0 is

$$\frac{dU}{dM_0} = \int_0^{\pi/2} \left(\frac{M}{AEe} - \frac{N}{AE} \right) d\varphi = 0,$$

from which

$$M_0 = \frac{Pr}{2} \left(1 - \frac{2}{\pi} + \frac{2e}{\pi r} \right).$$

Comparing this with eq. (224), we see that the third term in the parentheses represents the effect of the longitudinal force and of the non-linear stress distribution. The magnitudes of the errors in using the approximate eq. (224) instead of the above accurate equation are given in Table 9 below:

TABLE 9

$r/h =$	1	1.5	2	3
$e/r =$	0.090	0.038	0.021	0.009
Error in % =	15.8	6.7	3.7	1.6

It can be seen that in the majority of cases the approximate eq. (224) can be used for calculating M_0 and that the error is substantial only when h approaches r or becomes larger than r.

The increase in the vertical diameter of the ring is obtained from the equation

$$\delta = \frac{dU}{dP}.$$

Using eq. (228) for U and substituting in this equation

$$M = M_0 - \frac{Pr}{2}(1 - \cos\varphi), \qquad N = \frac{P}{2}\cos\varphi, \qquad V = -\frac{P}{2}\sin\varphi,$$

we find

$$\delta = \frac{Pr^2}{AEe}\left\{\frac{\pi}{4} - \frac{2}{\pi}\left(1 - \frac{e^2}{r^2}\right) + \frac{2e}{r}\left[\frac{2}{\pi}\left(1 - \frac{e}{r}\right) - \frac{\pi}{8}\right] + \frac{\pi\alpha}{4}\frac{E}{G}\frac{e}{r}\right\}.$$

Comparison with eq. (226) shows that the effect of the longitudinal and shearing forces on the magnitude of δ is usually very small.[19]

7. Determine the bending moments in a thin ring with two axes of symmetry submitted to the action of a uniform internal pressure p.

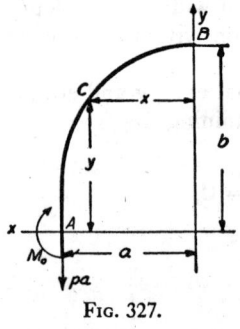

FIG. 327.

Solution. Consider one quadrant of the ring (Fig. 327), with semi-axes a and b. If M_0 represents the statically indeterminate moment at A, the bending moment at any cross section C with coordinates x and y is

$$M = M_0 - pa(a - x) + \frac{p(a - x)^2}{2} + \frac{py^2}{2}$$

$$= M_0 - \frac{pa^2}{2} + \frac{px^2}{2} + \frac{py^2}{2}. \qquad (k)$$

---

[19] A more accurate solution of the problem shown in Fig. 320 is given by the author; see *Bull. Polytech. Inst. (Kiev)*, 1910; see also *Phil. Mag.*, Vol. 44, p. 1014, 1922; and *Theory of Elasticity*, p. 121, 1951. This solution shows

Substituting into the equation $dU/dM_0 = 0$, we find

$$\left(M_0 - \frac{pa^2}{2}\right)s + \frac{p}{2}(I_x + I_y) = 0,$$

in which s denotes the length of the quadrant of the ring and

$$I_x = \int_0^s y^2 ds \quad \text{and} \quad I_y = \int_0^s x^2 ds.$$

Then

$$M_0 = \frac{pa^2}{2} - \frac{p}{2s}(I_x + I_y). \tag{l}$$

If the ring has the shape of the link shown in Fig. 326, with $a = r$ and $b = l + r$, we obtain

$$s = b - a + \frac{\pi a}{2},$$

$$I_x = \frac{1}{3}(b - a)^3 + \frac{\pi a}{2}(b - a)^2 + \frac{\pi}{4}a^3 + 2a^2(b - a),$$

$$I_y = (b - a)a^2 + \frac{\pi a^3}{4}.$$

Substituting into eq. (l), we obtain

$$M_0 = \frac{pa^2}{2} -$$

$$\frac{p}{2b + (\pi - 2)a}\left[\frac{1}{3}(b - a)^3 + \frac{\pi}{2}a^3 + 3a^2(b - a) + \frac{\pi}{2}a(b - a)^2\right].$$

The bending moment at any other cross section may now be obtained from eq. (k).

For an elliptical ring the calculations are more complicated.[20] Using the notations $I_x + I_y = \alpha a^2 b$, $M_0 = -\beta pa^2$, and moment at B (Fig. 327) is $M_1 = \gamma pa^2$, then the values of the numerical factors

---

that the above theory, based on the assumption that cross sections remain plane during bending, gives very satisfactory results.

[20] See J. A. C. H. Bresse, *Cours de mécanique appliquée*, Paris, 3d Ed., p. 493, 1880.   See also H. Résal, *J. math.* (Liouville), Vol. 3, 1877; M. Marbec, *Bull. assoc. tech. maritime*, Vol. 19, 1908; M. Goupil, *Ann. ponts et chaussées*, Vol. 2, p. 386, 1912; Mayer Mita, *Z. Ver. deut. Ing.*, Vol. 58, p. 649, 1914; W. F. Burke, *Nat. Advisory Comm. Aeronaut.*, *Tech. Notes*, *444*, 1933.

α, β and γ, for different values of the ratio a/b, are as given in Table 10 below:

TABLE 10: CONSTANTS FOR CALCULATING ELLIPTICAL RINGS

$a/b =$	1	0.9	0.8	0.7	0.6	0.5	0.4	0.3
α......	1.571	1.663	1.795	1.982	2.273	2.736	3.559	5.327
β......	0	0.057	0.133	0.237	0.391	0.629	1.049	1.927
γ......	0	0.060	0.148	0.283	0.498	0.870	1.576	3.128

8. A flat *spiral spring* (Fig. 328) is attached at the center to a spindle C. A couple M_0 is applied to this spindle to wind up the spring. It is balanced by a horizontal force P at the outer end of the spring A and by the reaction at the axis of the spindle. Establish the relation between M_0 and the angle of rotation of the spindle if all the dimensions of the spring are given.

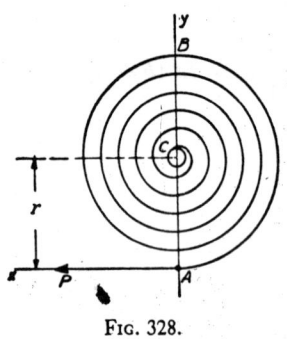

It is assumed that the angle of twist is not large enough to cause adjacent coils to touch each other.

Solution. Taking the origin of coordinates at A, the bending moment at any point of the spring at distance y from the force P is $M = Py$. The change in the angle between two adjacent cross sections at the point taken, from eq. (214), is

FIG. 328.

$$\Delta d\varphi = \frac{Mds}{EI_z} = \frac{Pyds}{EI_z}.$$

The total angle of rotation of one end of the spring with respect to the other during winding is

$$\varphi = \int_0^s \frac{Pyds}{EI_z} = \frac{P}{EI_z} \int_0^s yds. \qquad (m)$$

The integral on the right side of this equation represents the moment of the center line of the spring with respect to the x axis. This moment is obtained by multiplying the total length s of the spiral by the distance of its center of gravity from the x axis. In the usual case, it is sufficiently accurate to take this distance equal to r, the distance from the center of the spindle to the force P. Then, from eq. (m),

$$\varphi = \frac{Prs}{EI_z} = \frac{M_0 s}{EI_z}. \qquad (n)$$

If the end A is pin-connected, the turning moment M_0 applied at C produces a reactive force P at the fixed end A of the spring. As long as the thickness of the spring is very small and the number of windings of the spiral is large and the coils do not touch, the above assumption that the force P remains horizontal can be considered as sufficiently accurate. Hence eq. (n) holds.[21]

9. Assuming that the spring represented in Fig. 328 is in an unstressed condition and pin-connected at A, determine the maximum stress produced and the amount of energy stored in the spring by three complete turns of the spindle. Take the spring to be of steel, $\frac{1}{2}$ in. wide, $\frac{1}{40}$ in. thick and 120 in. long.

Solution. Substituting the above figures into eq. (n),

$$6\pi = M_0 \frac{120 \times 40^3 \times 12}{30 \times 10^6 \times \frac{1}{2}},$$

from which $M_0 = 3.07$ in. lb.

The amount of energy stored is

$$U = \int_0^s \frac{M^2 ds}{2EI} = \frac{P^2}{2EI} \int_0^s y^2 ds = \frac{P^2}{2EI}\left(sr^2 + \frac{sr^2}{4}\right)$$

$$= \frac{5}{8}\frac{M_0^2 s}{EI} = 36.1 \text{ in. lb.}$$

The maximum bending stress is at point B, where the bending moment can be taken equal to $2Pr = 2M_0$, so that

$$\sigma_{\max} = 3.07 \times 2 \times 40^2 \times 6 \times 2 = 118,000 \text{ lb per sq in.}$$

10. A *piston ring* of a circular outer boundary has a rectangular cross section of constant width b and of a variable depth h (Fig. 329). Determine the law of variation of the depth h in order to obtain a ring which, when assembled with the piston in the cylinder, produces a uniformly distributed pressure on the cylinder wall.

Solution. Let r denote the inner radius of the cylinder and $r + \delta$ the outer radius of the ring in the unstrained state. An approximate solution of the problem is obtained by using the outer radius of the ring instead of the

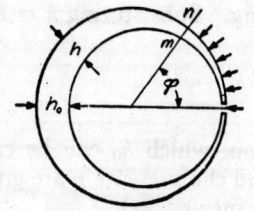

FIG. 329.

---

[21] A more complete discussion of the problem is given in the book by A. Castigliano, *loc. cit.*, p. 212. See also E. C. Wadlow, *Engineer*, Vol. 150, p. 474, 1930; and J. A. Van den Broek, *Trans. A.S.M.E.*, Vol. 53, p. 247, 1931.

variable radius of curvature of its center line. Then, by using eq. (214), the change in curvature due to bending is

$$\frac{\Delta d\varphi}{ds} = \frac{1}{r} - \frac{1}{r+\delta} = -\frac{M}{EI}. \tag{o}$$

The bending moment M produced at any cross section mn of the ring by the pressure p uniformly distributed over the outer surface of the ring is

$$M = -2pbr^2 \sin^2 \frac{\varphi}{2}. \tag{p}$$

If we substitute this into eq. (o) and take $bh^3/12$ for I and, for small δ, use δ/r^2 instead of $(1/r) - 1/(r+\delta)$, then the following equation for calculating h is obtained:

$$\frac{\delta}{r^2} = \frac{p}{E} \frac{24r^2}{h^3} \sin^2 \frac{\varphi}{2}, \tag{q}$$

from which

$$h^3 = \frac{p}{E} \frac{24r^4}{\delta} \sin^2 \frac{\varphi}{2}. \tag{r}$$

Letting $\varphi = \pi$, the maximum value of h^3, denoted by h_0^3, is

$$h_0^3 = \frac{p}{E} \frac{24r^4}{\delta}. \tag{s}$$

The maximum bending stress at any cross section mn is

$$\sigma = \frac{M}{Z} = \frac{12pr^2 \sin^2 (\varphi/2)}{h^2}. \tag{t}$$

From (t) and (r) it may be seen that the maximum bending stress occurs at $\varphi = \pi$, i.e., at the cross section opposite to the slot of the ring. Substituting $h = h_0$ and $\varphi = \pi$ in eq. (t),

$$\sigma_{\max} = \frac{12pr^2}{h_0^2}, \tag{u}$$

from which h_0 can be calculated if the working stress for the ring and the pressure p are given. The value of δ is found by substituting h_0 into eq. (s).

It may be noted that if two equal and opposite tensile forces P be applied tangentially to the ends of the ring at the slot, they produce at any cross section mn the bending moment

$$-Pr(1 - \cos \varphi) = -2Pr \sin^2 \varphi/2,$$

i.e., the bending moment varies with φ exactly in the same manner as that given by eq. (p). Therefore, if the ends of the open ring be pulled together and in this condition it be machined to the outer radius r, such a ring will, when assembled, produce a uniform pressure against the wall of the cylinder.[22]

Determine, for example, δ and h_0 for a cast-iron piston ring if $r = 10$ in., $\sigma_W = 4,200$ lb per sq in., $p = 1.4$ lb per sq in. and $E = 12 \times 10^6$ lb per sq in. Substituting in eq. (u), we find $h_0 = 0.632$ in. From eq. (s), $\delta = 0.111$ in.

11. Derive formula (227), given on p. 381.

12. A frame consisting of two vertical bars and a semicircular portion, Fig. 330, is acted upon by the force P directed along the axis of symmetry of the frame. Find the horizontal reactions H and the bending moment M at the point of application of the force P, assuming that the frame has a constant cross section and that it is hinged at the supports

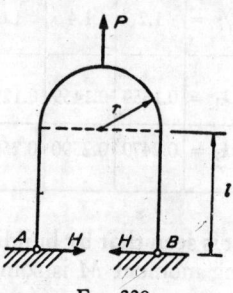

Fig. 330.

A and B. Neglect the influence of longitudinal and shear force on the deformation.

Solution. The magnitude of the forces H is obtained from the equation

$$\frac{dU}{dH} = 0.$$

The results can be put in the form

$$H = k_1 P, \qquad M = k_2 P r, \tag{v}$$

where k_1 and k_2 are numerical factors depending on the magnitude of the ratio l/r. Several values of these factors are given in Table 11 below.

TABLE 11

$l/r =$	1.2	1.4	1.6	1.8	2.0	2.2	2.4	2.6	2.8	3.0
$k_1 =$	0.0984	0.0857	0.0754	0.0669	0.0598	0.0537	0.0486	0.0442	0.0403	0.0370
$k_2 =$	0.284	0.294	0.304	0.313	0.321	0.328	0.335	0.341	0.347	0.352

[22] This theory was developed by H. Résal, *Ann. mines*, Vol. 5, p. 38, 1874; *Compt. rend.*, Vol. 73, p. 542, 1871. See also E. Reinhardt, *Z. Ver. deut. Ing.*, Vol. 45, p. 232, 1901; H. Friedmann, *Z. österr. Ing. Architekt.-Ver.*,

13. Solve the preceding problem, assuming that at the supports A and B the ends of the frame are built in.

Answer. The values H and M are given by formulas (v). The numerical values of the factors k_1 and k_2 are given in Table 12.

TABLE 12

$l/r =$	1.2	1.4	1.6	1.8	2.0	2.2	2.4	2.6	2.8	3.0
$k_1 =$	0.1659	0.1459	0.1295	0.1157	0.1040	0.0941	0.0855	0.0781	0.0716	0.0659
$k_2 =$	0.2479	0.2590	0.2691	0.2784	0.2870	0.2948	0.3021	0.3067	0.3152	0.3211

It is seen that by building in the ends of the frame the maximum bending moment M is somewhat smaller than in the preceding problem.

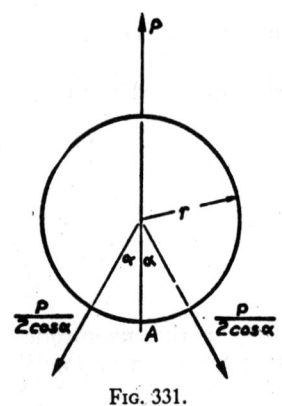

FIG. 331.

The results obtained can be utilized for analyzing stresses in the chain link of Fig. 326, provided a stud is inserted to prevent any change in the horizontal dimension of the link.

14. Find bending moment M_0 and tensile force H in the cross section A of the symmetrically loaded circular ring shown in Fig. 331.

Answer.

$$H = P \cdot \frac{\pi - \alpha}{2\pi} \tan \alpha,$$

$$M_0 = -\frac{Pr}{2\pi} [1 + \sec \alpha - (\pi - \alpha) \tan \alpha].$$

81. Arch Hinged at the Ends.—Figure 332 shows an arch with hinged ends at the same level carrying a vertical load. The vertical components of the reactions at A and B may be determined from equations of equilibrium in the same manner as for a simply supported beam, and the horizontal components must be equal and opposite in direction. The magnitude

---

Vol. 60, p. 632, 1908, and Z. *Ver. deut. Ing.*, Vol. 68, p. 254, 1924. Regarding rings in space, see N. J. Hoff, "Edge Reinforcements of Cutouts in Monocoques," *J. Appl. Mech.*, Vol. 10, p. 161, 1943.

H of these components is called the *thrust of the arch*. It cannot be obtained statically, but may be determined by use of the theorem of Castigliano. In the case of a flat arch, the two last terms in the general expression (228) for the strain energy

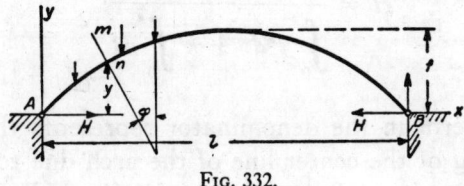

FIG. 332.

can be neglected and for usual proportions of arches the product Aer can be replaced by the moment of inertia I_z of the cross section. The equation for calculating H is then

$$\frac{dU}{dH} = \frac{d}{dH} \int_0^s \left(\frac{M^2}{2EI_z} + \frac{N^2}{2AE} \right) ds = 0. \qquad (a)$$

The bending moment at any cross section mn of the arch is

$$M = M_0 - Hy, \qquad (b)$$

in which M_0 is the bending moment calculated for the corresponding section of a simply supported beam having the same load and the same span as the arch. The second term under the integral sign of eq. (a) represents the strain energy due to compression in the tangential direction and is of secondary importance. A satisfactory approximation for flat arches is obtained by assuming this compression equal to the thrust H. Substituting expression (b) and $N = H$ in eq. (a), we obtain

$$-\int_0^s \frac{(M_0 - Hy)y\,ds}{EI_z} + \int_0^s \frac{H\,ds}{AE} = 0,$$

from which

$$H = \frac{\displaystyle\int_0^s \frac{M_0 y\,ds}{EI_z}}{\displaystyle\int_0^s \frac{y^2\,ds}{EI_z} + \int_0^s \frac{ds}{AE}}. \qquad (229)$$

For an arch of constant cross section, using the notation $k^2 = I_z/A$, eq. (229) becomes

$$H = \frac{\int_0^s M_0 y \, ds}{\int_0^s y^2 \, ds + k^2 \int_0^s ds}. \tag{230}$$

The second term in the denominator represents the effect of the shortening of the center line of the arch due to the longitudinal compression. In many cases it is small and can be neglected. Then

$$H = \frac{\int_0^s M_0 y \, ds}{\int_0^s y^2 \, ds}. \tag{231}$$

Take, for example, the case of a parabolic arch carrying a continuous load uniformly distributed along the length of the span with a center line given by the equation:

$$y = \frac{4 f x (l - x)}{l^2}. \tag{c}$$

Then

$$M_0 = \frac{q}{2} x (l - x). \tag{d}$$

Substituting (c) and (d) into eq. (231), we obtain

$$H = \frac{q l^2}{8 f}. \tag{e}$$

The actual thrust, H, will be less than that obtained from eq. (e). To give some idea of the possible error ΔH, the ratios $(\Delta H)/H$ for various proportions of arches are given in Table 13 below.[23] In calculating this table the complete expression

---

[23] See author's paper, "Calcul des arcs élastiques," Paris, Béranger Ed., 1922.

TABLE 13

=	$\frac{1}{12}$			$\frac{1}{8}$			$\frac{1}{4}$		
$\dfrac{h}{l} =$	$\frac{1}{10}$	$\frac{1}{20}$	$\frac{1}{30}$	$\frac{1}{10}$	$\frac{1}{20}$	$\frac{1}{30}$	$\frac{1}{10}$	$\frac{1}{20}$	$\frac{1}{30}$
$\dfrac{\Delta H}{H} =$	0.1771	0.0513	0.0235	0.0837	0.0224	0.0101	0.0175	0.00444	0.00198

(228) for strain energy was used and it was assumed that for any cross section of the arch

$$A = \frac{A_0}{\cos \varphi} \quad \text{and} \quad EI_z = \frac{EI_0}{\cos \varphi},$$

where A_0 and EI_0 are, respectively, the cross-sectional area and the flexural rigidity of the arch at the top, φ is the angle between the cross section and the y axis and h is the depth of the cross section at the top. Eq. (e) was used in calculating the value of H in the ratio $\Delta H/H$. The table shows that the error in using eq. (e) has a substantial magnitude only for flat arches of considerable thickness.

Since the supports of the arch are a fixed distance apart, a change in temperature may produce appreciable stresses in the structure. To calculate the thrust due to an increase in temperature of t degrees, we assume that one of the supports is movable. Then, thermal expansion would increase the span of the arch by $l\alpha t$, where α is the coefficient of thermal expansion of the material of the arch. The thrust is then found from the condition that it prevents such an expansion by producing a decrease in the span equal to $\alpha l t$. Using the Castigliano theorem, we obtain

$$\frac{dU}{dH} = \frac{d}{dH} \int_0^s \left(\frac{M^2}{2EI_z} + \frac{N^2}{2AE} \right) ds = \alpha l t. \qquad (f)$$

Taking only the thermal effect and putting $M_0 = 0$ and $N = H$, we obtain from (f)

$$H = \frac{\alpha l t}{\int_0^s \frac{y^2 ds}{EI_z} + \int_0^s \frac{ds}{AE}}. \qquad (232)$$

A more detailed study of stresses in arches may be found in books on the theory of structures.

82. Stresses in a Flywheel.[24] —Due to the effect of the spokes, the rim of a rotating flywheel undergoes not only extension but also bending. We take as the free body a portion of the rim (Fig. 333b) between two cross sections which bisect the angles between the spokes. Let

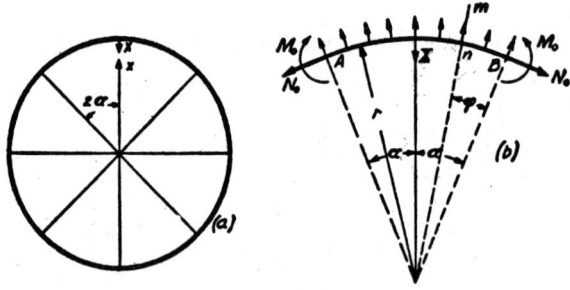

Fig. 333.

r = the radius of the center line of the rim,
A = the cross-sectional area of the rim,
A_1 = the cross-sectional area of a spoke,
I = moment of inertia of the cross section of the rim,
2α = the angle between two consecutive spokes,
q = the weight of the rim per unit length of the center line,
q_1 = the weight of a spoke per unit length,
ω = the angular velocity of the wheel.

From the condition of symmetry, there can be no shearing stresses over the cross sections A and B and the forces acting on these cross sections are reducible to the longitudinal force N_0 and the bending moment M_0. If X denotes the force exerted by the spoke on the rim, the equation of equilibrium of the portion AB of the rim is

---

[24] A very complete study of the bending of circular rings with spokes is given by C. B. Biezeno and K. Grammel, *loc. cit.*, p. 381.

$$2N_0 \sin \alpha + X - 2r^2 \frac{q}{g} \omega^2 \sin \alpha = 0,$$

from which

$$N_0 = \frac{q}{g} \omega^2 r^2 - \frac{X}{2 \sin \alpha}. \tag{a}$$

The longitudinal force N at any cross section mn is

$$N = N_0 \cos \varphi + \frac{q\omega^2 r}{g} 2r \sin^2 \frac{\varphi}{2} = \frac{q\omega^2 r^2}{g} - \frac{X \cos \varphi}{2 \sin \alpha}. \tag{b}$$

The bending moment for the same cross section is

$$M = M_0 - N_0 r(1 - \cos \varphi) + \frac{q\omega^2 r^3}{g} 2 \sin^2 \frac{\varphi}{2}$$

$$= M_0 + \frac{Xr}{\sin \alpha} \sin^2 \frac{\varphi}{2}. \tag{c}$$

The force X and the moment M_0 cannot be determined from the equations of statics but are calculated by use of the theorem of least work. The strain energy of the portion AB of the rim is [25]

$$U_1 = 2 \int_0^\alpha \frac{M^2 r d\varphi}{2EI} + 2 \int_0^\alpha \frac{N^2 r d\varphi}{2EA}. \tag{d}$$

The tensile force N_1 at any cross section of the spoke at distance ρ from the center of the wheel is [26]

$$N_1 = X + \frac{q_1 \omega^2}{2g} (r^2 - \rho^2).$$

Hence, the strain energy of the spoke is

$$U_2 = \int_0^r \frac{N_1^2 d\rho}{2A_1 E}. \tag{e}$$

The equations for calculating M_0 and X are

$$\frac{\partial}{\partial M_0} (U_1 + U_2) = 0 \tag{f}$$

$$\frac{\partial}{\partial X} (U_1 + U_2) = 0. \tag{g}$$

---

[25] It is assumed that the thickness of the rim is small in comparison with r and only the energy of bending and tension is taken into account.

[26] The length of the spoke is taken equal to r. In practice it will be somewhat less than r.

Substituting (d) and (e), we obtain, from eqs. (f) and (g),

$$M_0 = -\frac{Xr}{2}\left(\frac{1}{\sin \alpha} - \frac{1}{\alpha}\right),$$ (233)

$$X = \frac{2}{3}\frac{q\omega^2 r^2}{g} \cdot \frac{1}{\dfrac{Ar^2}{I}f_2(\alpha) + f_1(\alpha) + \dfrac{A}{A_1}},$$ (234)

in which

$$f_1(\alpha) = \frac{1}{2\sin^2 \alpha}\left(\frac{\sin 2\alpha}{4} + \frac{\alpha}{2}\right),$$

$$f_2(\alpha) = \frac{1}{2\sin^2 \alpha}\left(\frac{\sin 2\alpha}{4} + \frac{\alpha}{2}\right) - \frac{1}{2\alpha}.$$

Several values of the functions f_1 and f_2, for various numbers of spokes, are given in Table 14 below.

TABLE 14

$n =$	4	6	8
$f_1(\alpha)$	0.643	0.957	1.274
$f_2(\alpha)$	0.00608	0.00169	0.00076

Using this table, the force X in the spoke is readily determined from eq. (234) and the bending moment M_0 from eq. (233). Then the longitudinal force and bending moment for any cross section mn of the rim may be found from eqs. (a), (b), and (c).[27]

Take, as an example, a steel flywheel rotating at 600 rpm, with radius $r = 60$ in., cross section of the rim a square 12×12 sq in. and with six spokes of cross sectional area $A_1 = 24$ sq in. If the rim

---

[27] The above theory was developed by R. Bredt, Z. Ver. deut. Ing., Vol. 45, p. 267, 1901; and H. Brauer, Dinglers Polytech. J., p. 353, 1908. See also J. G. Longbottom, Proc. Inst. Mech. Engrs. (London), p. 43, 1924; and K. Reinhardt, Forschungsarb., No. 226, 1920. A similar problem arises when calculating stresses in retaining rings of large turbogenerators; see E. Schwerin, Electrotech. Z., p. 40, 1931.

is first considered as a rotating ring which can expand freely, then the tensile stress due to centrifugal force is, from p. 32,

$$\sigma_0 = 0.106 \, v_1{}^2 = 0.106 \times \frac{2\pi}{60} \times 600 \times 5^2 = 10,450 \text{ lb per sq in.}$$

In the case of six spokes, $\alpha = 30°$, $f_1(\alpha) = 0.957$, $f_2(\alpha) = 0.00169$. Then the force in each spoke is, from eq. (234),

$$X = \frac{2}{3} \frac{q\omega^2 r^2}{g} \cdot \frac{1}{300 \times 0.00169 + 0.957 + 6} = 0.0893 \frac{q\omega^2 r^2}{g}.$$

The longitudinal force for the cross section bisecting the angle between the spokes is, from eq. (a),

$$N_0 = \frac{q\omega^2 r^2}{g} - 0.0893 \frac{q\omega^2 r^2}{g} = 0.911 \frac{q\omega^2 r^2}{g}.$$

The bending moment for the same cross section, from eq. (233), is

$$M_0 = -0.00402 \frac{q\omega^2 r^3}{g}.$$

The maximum stress at this cross section is

$$\sigma_{\max} = \frac{N_0}{A} - \frac{M_0}{Z} = 10,780 \text{ lb per sq in.}$$

For the cross section of the rim at the axis of the spoke, eqs. (b) and (c) give

$$(N)_{\varphi=\alpha} = 0.923 \frac{q\omega^2 r^2}{g}. \qquad (M)_{\varphi=\alpha} = 0.00794 \frac{q\omega^2 r^3}{g}.$$

The maximum stress at this cross section is

$$\sigma_{\max} = 12,100 \text{ lb per sq in.}$$

In this case the effect of the bending of the rim on the maximum stress is small and the calculation of the stresses in the rim as if it were a free rotating ring gives a satisfactory result.

83. Deflection Curve for a Bar with a Circular Center Line.
—In the case of a thin curved bar with a circular center line the differential equation for the deflection curve is analogous

to that for a straight bar (eq. 79, p. 139). Let *ABCD* (Fig. 334) represent the center line of a circular ring after deformation and let *u* denote the small radial displacements during this deformation. The variation in the curvature of the center line during bending can be studied by considering one element *mn* of the ring and the corresponding element m_1n_1 of the de-

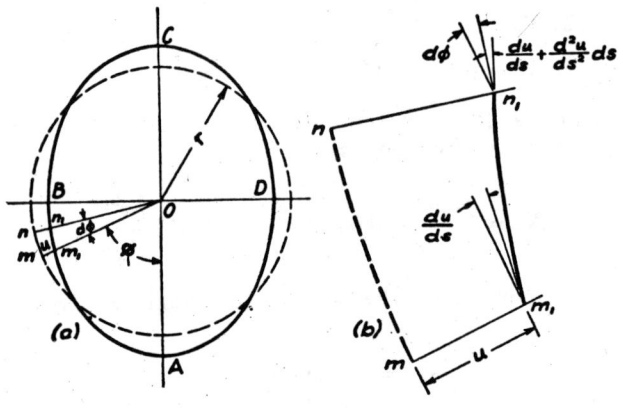

Fig. 334.

formed ring included between the same radii (Fig. 334*b*). The initial length of the element *mn* and its initial curvature are

$$ds = rd\varphi, \qquad \frac{d\varphi}{ds} = \frac{d\varphi}{rd\varphi} = \frac{1}{r}. \qquad (a)$$

For small deflections, the curvature of the same element after deformation can be taken equal to the curvature of the element m_1n_1. This latter is given by the equation

$$\frac{1}{r_1} = \frac{d\varphi + \Delta d\varphi}{ds + \Delta ds}, \qquad (b)$$

in which $d\varphi + \Delta d\varphi$ denotes the angle between the normal cross sections m_1 and n_1 of the deformed bar and $ds + \Delta ds$ the length of the element m_1n_1. The displacement *u* is considered positive if towards the center of the ring and is assumed to be very small in comparison with the radius of the ring. Then the angle between the tangent at m_1 to the center line and the normal to the radius m_1O is du/ds. The corresponding angle at the cross section n_1 is

$$\frac{du}{ds} + \frac{d^2u}{ds^2}\, ds.$$

Then

$$\Delta d\varphi = \frac{d^2u}{ds^2}\, ds. \qquad (c)$$

In comparing the length of the element m_1n_1 with that of the element mn, the small angle du/ds is neglected and the length m_1n_1 is taken equal to $(r - u)d\varphi$. Then

$$\Delta ds = -u\, d\varphi = -\frac{u\, ds}{r}. \qquad (d)$$

Substituting (c) and (d) into eq. (b), we obtain

$$\frac{1}{r_1} = \frac{d\varphi + \dfrac{d^2u}{ds^2}\, ds}{ds\left(1 - \dfrac{u}{r}\right)},$$

or neglecting the small quantities of higher order,

$$\frac{1}{r_1} = \frac{d\varphi}{ds}\left(1 + \frac{u}{r}\right) + \frac{d^2u}{ds^2} = \frac{1}{r}\left(1 + \frac{u}{r}\right) + \frac{d^2u}{ds^2},$$

from which

$$\frac{1}{r_1} - \frac{1}{r} = \frac{u}{r^2} + \frac{d^2u}{ds^2}. \qquad (e)$$

The relationship between the change in curvature and the magnitude of bending moment, from eq. (214), for thin bars is

$$\frac{1}{r_1} - \frac{1}{r} = -\frac{M}{EI}. \qquad (f)$$

The minus sign on the right side of the equation follows from the sign of the bending moment which is taken to be positive when it produces a decrease in the initial curvature of the bar (Fig. 308). From (e) and (f) it follows that

$$\frac{d^2u}{ds^2} + \frac{u}{r^2} = -\frac{M}{EI}. \tag{235}$$

This is the differential equation for the deflection curve fo a thin bar with circular center line. For an infinitely large r this equation coincides with eq. (79) for straight bars.

As an example of the application of eq. (235), let us consider the problem represented in Fig. 320. The bending moment at any cross section m_1n_1 is, from eq. (c), p. 380,

$$M = \frac{Pr}{2}\left(\cos\varphi - \frac{2}{\pi}\right),$$

and eq. (235) becomes

$$\frac{d^2u}{ds^2} + \frac{u}{r^2} = \frac{Pr}{2EI}\left(\frac{2}{\pi} - \cos\varphi\right)$$

or

$$\frac{d^2u}{d\varphi^2} + u = \frac{Pr^3}{2EI}\left(\frac{2}{\pi} - \cos\varphi\right).$$

The general solution of this equation is

$$u = A\cos\varphi + B\sin\varphi + \frac{Pr^3}{EI\pi} - \frac{Pr^3}{4EI}\varphi\sin\varphi.$$

The constants of integration A and B are determined from the condition of symmetry:

$$\frac{du}{d\varphi} = 0, \quad \text{at } \varphi = 0 \text{ and } \varphi = \frac{\pi}{2},$$

which are satisfied by taking

$$B = 0, \quad A = -\frac{Pr^3}{4EI}.$$

Then

$$u = \frac{Pr^3}{EI\pi} - \frac{Pr^3}{4EI}\varphi\sin\varphi - \frac{Pr^3}{4EI}\cos\varphi.$$

For $\wp = 0$ and $\varphi = \pi/2$, we obtain

$$(u)_{\varphi=0} = \frac{Pr^3}{EI}\left(\frac{1}{\pi} - \frac{1}{4}\right), \qquad (u)_{\varphi=\pi/2} = \frac{Pr^3}{EI}\left(\frac{1}{\pi} - \frac{\pi}{8}\right).$$

These results are in complete agreement with eqs. (227) and (226) obtained before by using the Castigliano theorem.[28]

84. Bending of Curved Tubes.—In discussing the distribution of bending stresses in curved bars (Art. 77), it was assumed that the shape of the cross section remains unchanged. Such an assumption is justifiable as long as we have a solid bar, because the very small displacements in the plane of the cross section due to lateral contraction and expansion have no substantial effect on the stress distribution. The condition is very different, however, in the case of a thin curved tube in bending. It is well known that curved tubes with comparatively thin walls prove to be more flexible during bending than would be expected from the usual theory of curved bars.[29] A consideration of the distortion of the cross section during bending is necessary in such cases.[30]

Consider an element between two adjacent cross sections of a curved round pipe (Fig. 335) which is bent by couples in

[28] Differential eq. (235) for the deflection of a circular ring was established by H. Résal. See his *Traité du mécanique géneral*, Vol. 5, p. 78, 1880. See also J. Boussinesq, *Compt. rend.*, Vol. 97, p. 843, 1883; and H. Lamb, *Proc. London Math. Soc.*, Vol. 19, p. 365, 1888. Various examples of applications of this equation are given in a paper by R. Mayer, *Z. angew. Math. u. Phys.*, Vol. 61, p. 246, 1913; see also K. Federhofer, *Wasserkraft u. Wasserwirtsch.*, Vol. 38, p. 237, 1943.

[29] Extensive experimental work on the flexibility of pipe bends was done by A. Bantlin, *Z. Ver. deut. Ing.*, Vol. 54, p. 45, 1910, and *Forschungsarb.*, *No. 96*. See also W. Hovgaard, *J. Math. and Phys.*, Vol. 7, 1928, and A. M. Wahl, *Trans. A.S.M.E.*, Vol. 49, 1927. The problem of flexibility of curved pipes is of practical importance in analyzing stresses in pipe lines. The recent literature on this subject is given in the paper by J. E. Brock, *J. Appl. Mech.*, Vol. 19, p. 501, 1952. See also the paper by N. Gross presented at a meeting of the Institution of Mechanical Engineers, London, Dec. 1952.

[30] This problem for pipes of circular section was discussed by Th. Kármán, *Z. Ver. deut. Ing.*, Vol. 55, p. 1889, 1911. Bending of pipes of elliptic cross section was investigated by M. T. Huber. See *Proc. 7th Internat. Congr. Appl. Math. Mech.*, Vol. 1, p. 322, 1948. The case of curved pipes of rectangular cross section was considered by the author; see *Trans. A.S.M.E.* Vol. 45, p. 135, 1923.

the direction indicated. Since both the tensile forces at the convex side of the tube and

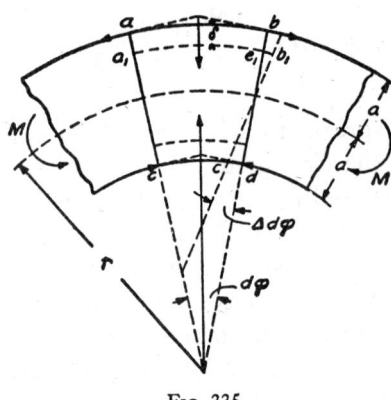

FIG. 335.

the compressive forces at the concave side have resultants towards the neutral axis, the previously circular cross sections are flattened and become elliptical. This flattening of the cross section affects the strain of longitudinal fibers of the tube. The outer fiber ab takes some position a_1b_1 after bending; denote its displacement towards the neutral axis by δ. The total elongation of the fiber is

$$a_1b_1 - ab = a_1b_1 - a_1e_1 - (ab - a_1e_1). \qquad (a)$$

The angle between the adjacent cross sections ac and bd is denoted by $d\varphi$, its variation during bending by $\Delta d\varphi$, the radius of the center line by r and the radius of the cross section of the tube by a. It is assumed that the ratio a/r is small enough so that the neutral axis can be taken as passing through the centroid of the cross section. Then, from the figure we obtain

$$a_1b_1 - a_1e_1 = (a - \delta)\Delta d\varphi \approx a\Delta d\varphi.^{31}$$

The total elongation of the fiber ab, as given by eq. (a), is

$$a\Delta d\varphi - \delta d\varphi$$

and the unit elongation is

$$\epsilon = \frac{a\Delta d\varphi - \delta d\varphi}{(r + a)d\varphi} = \frac{a}{r + a}\frac{\Delta d\varphi}{d\varphi} - \frac{\delta}{r + a}. \qquad (b)$$

The first term on the right side of this equation represents the strain in the fiber due to rotation of the cross section bd with respect to the cross section ac. This is the elongation which is considered in the bending of solid bars. The second term on

---

[31] The displacement δ is considered as very small in comparison with the radius a.

the right side of eq. (b) represents the effect of the flattening of the cross section. It is evident that this effect may be of considerable importance. Take, for instance, $r + a = 60$ in. and $\delta = 0.02$ in. Then $\delta/(r + a) = 1/3,000$ and the corresponding stress for a steel tube is 10,000 lb per sq in. Hence a very small flattening of the cross section produces a substantial decrease in the stress at the outermost fiber ab. A similar conclusion may be drawn for the fiber cd on the concave side of the bend.

A change in the direction of the bending moment causes a change of sign of the normal stresses and as a result, instead of a flattening of the tube in the radial direction, there is a flattening in the direction perpendicular to the plane of Fig. 335 and the fiber ab, due to this flattening, is displaced outward. From the same reasoning as above it may be shown that here again the flattening of the cross section produces a decrease in the stress at the most remote fibers. It may therefore be concluded that the fibers of the tube farthest from the neutral axis do not take the share in the stresses which the ordinary theory of bending indicates. This affects the bending of the tube in the same way as a decrease in its moment of inertia. Instead of eq. (214) which was derived for solid curved bars, the following equation must be used in calculating the deflections of thin tubes:

$$\Delta d\varphi = \frac{Mrd\varphi}{kEI_z}, \tag{236}$$

in which k is a numerical factor, less than unity, which takes account of the flattening. This factor depends upon the proportions of the bend and can be calculated from the following approximate formula: [32]

$$k = 1 - \frac{9}{10 + 12\left(\dfrac{tr}{a^2}\right)^2}, \tag{237}$$

in which t is the thickness of the tube. This indicates that the

---

[32] See paper by Th. Kármán, loc. cit., p. 405.

effect of the flattening depends only upon the magnitude of the ratio tr/a^2.

As for the effect of the flattening on the stress distribution, Kármán showed that instead of the simple equation for normal bending stresses [33] $\sigma = My/I_z$, in which y denotes the distance from the neutral axis, the following more complicated equation must be used:

$$\sigma = \frac{My}{kI_z}\left(1 - \beta\frac{y^2}{a^2}\right), \qquad (c)$$

in which

$$\beta = \frac{6}{5 + 6\left(\dfrac{tr}{a^2}\right)^2}.$$

The maximum stress, obtained from (c), is

$$\sigma_{max} = k_1\frac{Md}{2I_z}, \qquad (238)$$

in which d is the outer diameter of the tube and

$$k_1 = \frac{2}{3k\sqrt{3\beta}}$$

is a numerical factor which depends upon the proportions of the bend. Several values of k_1 are given in Table 15 below:

TABLE 15

$\dfrac{tr}{a^2} =$	0.3	0.5	1.0
$k_1 =$	1.98	1.30	0.88

It is seen that when tr/a^2 is small, the actual maximum stress is considerably greater than that given by the usual theory which neglects the flattening of the cross section.

---

[33] It is assumed that r is large in comparison with a and that a linear stress distribution is a sufficiently accurate assumption.

A theory analogous to the above may also be developed in the case of a tube of rectangular cross section.[34] For example, in the case of a thin tube of square cross section, the coefficient k in eq. (236) is found to depend upon the magnitude of the ratio

$$n = \frac{b^4}{r^2 t^2},$$

in which t is the thickness of the wall, r the radius of the center line of the bend and b the length of the side of the cross section. Then

$$k = \frac{1 + 0.0270n}{1 + 0.0656n}. \tag{239}$$

For instance, if $b/r = 0.1$ and $b/t = 50$, we obtain $n = 25$ and, from (239), $k = 0.63$. The maximum stress in tubes of rectangular section increases in the same proportion as the flexibility, i.e., in the above example the distortion of the cross section increases the maximum stress by approximately 60 per cent.[35]

If the cross section of a curved bar has flanges of considerable width, the question of distortion of the cross section again becomes of practical importance. We have such a problem, for example, when investigating bending stresses at a corner of a rigid frame of I section, Fig. 336a. Considering an element of the frame between the two consecutive cross

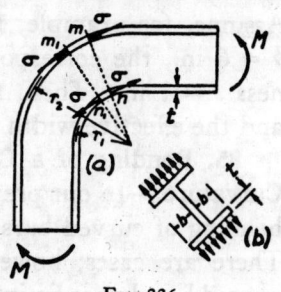

Fig. 336.

sections mn and m_1n_1, we see that the longitudinal bending stresses σ in the flanges have components in a radial direction which tend to produce bending of the flanges, Fig. 336b. This bending results in some diminishing of the longitudinal bending stress σ in portions of the flanges at a considerable distance

---

[34] Such a problem occurs, for instance, in the design of a Fairbairn crane. See footnote 30, p. 405.

[35] For further development of the theory of bending of curved tubes, see L. Beskin, *J. Appl. Mech.*, Vol. 12, 1945; E. Reissner, *Proc. Nat. Acad. Sci.*, Vol. 35, p. 204, 1949; R. A. Clark, T. I. Gilroy and E. Reissner, *J. Appl. Mech.*, Vol. 19, 1952.

from the web. To take this fact into account, an *effective width* αb of a flange must be used in applying formula (i), p. 371, for an I section. Naturally the magnitude of the factor α, defining the effective width of the flange, depends on the flexibility of the flanges, which is expressed by the quantity

$$\beta = 1.32 \frac{b}{\sqrt{tr}}, \qquad (d)$$

in which t is the thickness of the flange and r the radius of curvature of the flange. For the inner flange $r = r_1$, and for the outer flange $r = r_2$. The calculations show that if $\beta < 0.65$, the bending of the flanges can be neglected and we can directly apply the theory developed in Art. 79. For larger values of β the formula

$$\alpha b = \left(\frac{1}{\beta} - \frac{1}{2\pi\beta^3} \right) b \qquad (e)$$

can be used [36] for calculating the effective width of a flange. Assume, for example, that we have the width of the flange $b = 6$ in., the corresponding radius $r = 8$ in. and the thickness $t = 1$ in. Then, from formula (d), we obtain $\beta = 2.80$ and the effective width of the flange is $0.35 \times 6 = 2.1$ in.

85. Bending of a Curved Bar Out of Its Plane of Initial Curvature.—In our previous discussion we have dealt with the bending of curved bars in the plane of their initial curvature. There are cases, however, in which the forces acting on a curved bar do not lie in the plane of the center line of the bar.[37]

---

[36] For derivation of this formula see doctoral dissertation by Otto Steinhardt, Darmstadt, 1938. The experiments made by Steinhardt are in satisfactory agreement with the formula.

[37] Several problems of this kind have been discussed by I. Stutz, *Z. österr. Architekt.-u. Ing.-Ver.*, p. 682, 1904; H. Müller-Breslau, *Die neueren Methoden der Festigkeitslehre*, 2d Ed., p. 258, 1913, and 4th Ed., p. 265; and B. G. Kannenberg, *Eisenbau*, p. 329, 1913. The case of a circular ring supported at several points and loaded by forces perpendicular to the plane of the ring was discussed by F. Düsterbehn, *Eisenbau*, p. 73, 1920; and by G. Unold, *Forschungsarb.*, No. 255, 1922. The same problem was discussed by C. B. Biezeno by using the principle of least work, *Ingenieur* (*Utrecht*), 1927, and *Z. angew. Math. u. Mech.*, Vol. 8, p. 237, 1928. The application of trigonometric series in the same problem is shown by Biezeno and J. J. Koch, *ibid.*, Vol. 16, p. 321, 1936. The problem is of a practical importance

Then it is necessary to consider the deflection of the bar in two perpendicular planes and the twist of the bar. A simple problem of this kind is shown in Fig. 337a in which a portion of a horizontal circular ring, built in at A, is loaded by a vertical load P applied at the end B.[38] Considering a cross section D of the bar and taking the coordinate axes as shown in Figs. 337b and 337c [39] we find that the moments of the external load P with respect to these axes are

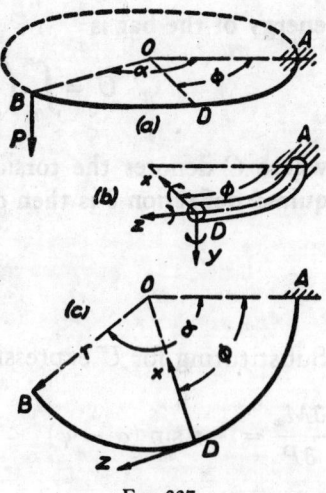

Fig. 337.

$$M_x = -Pr \sin (\alpha - \varphi), \qquad M_y = 0,$$
$$M_z = Pr[1 - \cos (\alpha - \varphi)]. \qquad (a)$$

By using these expressions the bending and torsional stresses can be calculated in any cross section of the bar. In calculating the deflection at the end B the Castigliano theorem will be used, for which purpose we need the expression for the strain energy of the bar. Assuming that the cross-sectional dimensions of the bar are small in comparison with the radius r, we apply the same formulas as previously derived for a straight bar (see eqs. 186 and 190). Thus the expression for the strain

in design of steam piping. The corresponding bibliography is given in the paper by H. E. Mayrose, *J. Appl. Mech.*, Vol. 4, p. 89, 1937. See also A. H. Gibson and E. G Ritchie, *A Study of the Circular-Arc Bow-Girder*, London, 1914. A complete study of the problem will be found in the book by Biezeno and R. Grammel, *loc. cit.*, p. 381. See also the publications of M. B. Hogan in *Bull. Univ. of Utah*, Vol. 34, 1943–44; Vol. 35, 1945, and 1947; Vol. 36, 1947; Vol. 38, 1948. In the last bulletin a list of numerous publications dealing with circular rings is given.

[38] This problem has been discussed by St.-Venant; see his papers in *Compt. rend.*, Vol. 17; 1843.

[39] It is assumed that the horizontal axis x and the vertical axis y are the axes of symmetry of the cross section and that the z axis is tangent to the center line of the ring at D.

energy of the bar is

$$U = \int_0^\alpha \left(\frac{M_x^2}{2EI_x} + \frac{M_s^2}{2C} \right) r d\varphi, \qquad (b)$$

where C denotes the torsional rigidity of the bar.[40] The required deflection δ is then obtained from the equation

$$\delta = \frac{\partial U}{\partial P}.$$

Substituting for U expression (b) and observing that

$$\frac{\partial M_x}{\partial P} = -r \sin (\alpha - \varphi) \quad \text{and} \quad \frac{\partial M_z}{\partial P} = r[1 - \cos (\alpha - \varphi)],$$

we obtain

$$\delta = \frac{Pr^3}{EI_x} \int_0^\alpha \left\{ \sin^2 (\alpha - \varphi) + \frac{EI_x}{C} [1 - \cos (\alpha - \varphi)]^2 \right\} d\varphi. \quad (240)$$

In the particular case when $\alpha = \pi/2$,

$$\delta = \frac{Pr^3}{EI_x} \left[\frac{\pi}{4} + \frac{EI_x}{C} \left(\frac{3\pi}{4} - 2 \right) \right]. \qquad (c)$$

If the cross section of the ring is circular, $C = GI_p = 2GI_x$; taking $E = 2.6G$, we obtain

$$\delta = \frac{Pr^3}{EI_x} \left[\frac{\pi}{4} + 1.3 \left(\frac{3\pi}{4} - 2 \right) \right] = 1.248 \frac{Pr^3}{EI_x}. \qquad (241)$$

As an example of a statically indeterminate problem, let us consider a horizontal semicircular bar with built-in ends, loaded at the middle, Fig. 338a. Considering only small

[40] The calculation of C for various shapes of cross section is discussed in Chap. VII, Part II. For a more rigorous discussion of stresses in a portion of a ring see *Theory of Elasticity*, p. 391, 1951.

vertical deflections of the bar, we can entirely neglect any displacements in the horizontal plane as small quantities of a higher order. Hence there will be no bending of the ring in its plane and no forces or moments in that plane at the ends A and B. Considering the built-in end B, we conclude from the equilibrium conditions that there will act a vertical reaction $P/2$ and the moment $M_{x_0} = Pr/2$. The moment M_{z_0} will also act, preventing the end section B from rotation with respect to the z_0 axis. The magnitude of this moment cannot be determined from statics. We shall find it by using the principle of least work, which requires that

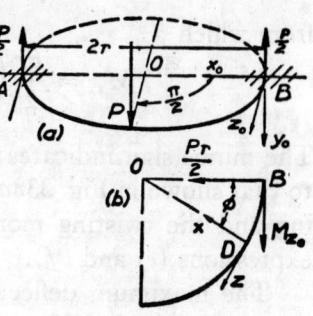

$$\frac{\partial U}{\partial M_{z_0}} = 0. \qquad (d)$$

In deriving the expression for strain energy of the bar we represent the moments applied at the

FIG. 338.

end B by the vectors $Pr/2$ and M_{z_0}, as shown in Fig. 338b. Then the moments M_x and M_z at any cross section D are

$$M_x = \frac{Pr}{2} \cos \varphi - M_{z_0} \sin \varphi - \frac{Pr}{2} \sin \varphi, \qquad (e)$$

$$M_z = \frac{Pr}{2} \sin \varphi + M_{z_0} \cos \varphi - \frac{Pr}{2} (1 - \cos \varphi), \qquad (f)$$

and the expression for strain energy is

$$U = 2 \int_0^{\pi/2} \left(\frac{M_x^2}{2EI_x} + \frac{M_z^2}{2C} \right) r d\varphi. \qquad (g)$$

Substituting this in eq. (d) and observing that

$$\frac{\partial M_x}{\partial M_z} = - \sin \varphi, \qquad \frac{\partial M_z}{\partial M_{z_0}} = \cos \varphi,$$

we obtain

$$\frac{1}{EI_x} \int_0^{\pi/2} \left(\frac{Pr}{2} \sin^2 \varphi + M_{z_0} \sin^2 \varphi \right.$$

$$\left. - \frac{Pr}{2} \sin \varphi \cos \varphi \right) d\varphi + \frac{1}{C} \int_0^{\pi/2} \left[\frac{Pr}{2} \sin \varphi \cos \varphi \right.$$

$$\left. + M_{z_0} \cos^2 \varphi - \frac{Pr}{2} (1 - \cos \varphi) \cos \varphi \right] d\varphi = 0,$$

from which

$$M_{z_0} = \frac{Pr}{2} \left(\frac{2}{\pi} - 1 \right) = -0.182 Pr. \tag{242}$$

The minus sign indicates that the direction of M_{z_0} is opposite to that shown in Fig. 338a. Knowing M_{z_0}, we obtain the bending and the twisting moments at any cross section from the expressions (e) and (f).

The maximum deflection is evidently under the load and we readily obtain it from Castigliano's equation:

$$\delta = \frac{\partial U}{\partial P}. \tag{h}$$

Substituting expression (g) for U and observing that

$$\frac{\partial M_x}{\partial P} = \frac{r}{2} (\cos \varphi - \sin \varphi),$$

$$\frac{\partial M_z}{\partial P} = \frac{r}{2} (\sin \varphi + \cos \varphi - 1), \tag{i}$$

we obtain

$$\delta = \frac{Pr^3}{2EI_x} \left\{ (2 - 0.363) \left(\frac{\pi}{4} - \frac{1}{2} \right) \right.$$

$$\left. + \frac{EI_x}{C} \left[(2 - 0.363) \left(\frac{\pi}{4} + \frac{1}{2} \right) + \frac{\pi}{2} - 4 + 0.363 \right] \right\}$$

$$= 0.514 \frac{Pr^3}{2EI_x}. \tag{j}$$

In the calculation of the partial derivatives (i) we disregarded the fact that the twisting moment M_{z_0} is not an independent quantity but is a function of P as defined by expression (242). If we take this into consideration, then the right-hand side of eq. (h) will be written in the following form:

$$\frac{\partial U}{\partial P} + \frac{\partial U}{\partial M_{z_0}} \cdot \frac{dM_{z_0}}{dP}. \qquad (k)$$

But the second term in this expression vanishes, by virtue of eq. (d). Hence our previous procedure of calculating deflection δ is justified.

Sometimes we have to consider curved bars for which the axis is not a curve in a plane, and we have a three-dimensional problem. Problems of such kind are encountered, for example, in analyzing edge reinforcements of cutouts in monocoques. Here again Castigliano's method can be used to advantage.[41]

Problems

1. A curved bar with circular axis and with $\alpha = \pi/2$ (Fig. 337) is loaded at end B by a twisting couple $M_t = T$. Find the deflection of the end B in the vertical direction.

Answer. Assuming $EI_x/C = 1.3$, $\delta = 0.506 \dfrac{Tr^2}{EI_x}$.

2. Solve the preceding problem, assuming that at end B a bending couple, $M_x = M_0$, is applied in the vertical plane tangent to the center line at B.

Answer. $\delta = 1.150 \dfrac{M_0 r^2}{EI_x}$.

3. A semicircular bar with the center line in a horizontal plane is built in at A and B and loaded symmetrically by two vertical loads P at C and D, Fig. 339. Find the twisting moments M_{z_0} at the built-in ends.

Answer.

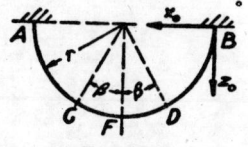

Fig. 339.

$$M_{z_0} = -\frac{2}{\pi} Pr \left(\frac{\pi}{2} - \cos \beta - \beta \sin \beta \right)$$

[41] See paper by K. Marguerre, *Luftfahrt-Forsch.*, Vol. 18, pp. 253–61, 1941; and *Nat. Advisory Comm. Aeronaut. Mem. No. 1005*, 1942. See also N. J. Hoff, *l.c. cit.*, p. 394.

4. Solve the preceding problem for the case of a uniform vertical load of intensity q distributed along the entire length of the bar.

Answer. $M_{z_0} = -qr^2\left(\dfrac{\pi}{2} - \dfrac{4}{\pi}\right) = -0.32qr^2$.

5. The horizontal semicircular bar, shown in Fig. 339 and uniformly loaded as in the preceding problem, is supported at the middle cross section F. Find the vertical reaction N at the support F.

Answer. $N = 2qr$.

APPENDIX A

MOMENTS OF INERTIA OF PLANE AREAS

I. The Moment of Inertia of a Plane Area with Respect to an Axis in Its Plane

In discussing the bending of beams, we encounter integrals of the type

$$I_z = \int_A y^2 dA, \tag{243}$$

in which each element of area dA is multiplied by the square of its distance from the z axis and integration is extended over the cross-sectional area A of the beam (Fig. 340). Such an integral is called the *moment of inertia* of the area A with respect to the z axis. In simple cases, moments of inertia can

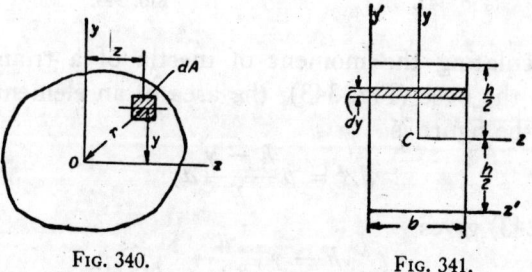

FIG. 340.   FIG. 341.

readily be calculated analytically. Take, for instance, a rectangle (Fig. 341). In calculating the moment of inertia of this rectangle with respect to the horizontal axis of symmetry z we can divide the rectangle into infinitesimal elements such as shown in the figure by the shaded area. Then

$$I_z = 2\int_0^{h/2} y^2 b \, dy = bh^3/12. \tag{244}$$

417

In the same manner, the moment of inertia of the rectangle with respect to the y axis is

$$I_y = 2\int_0^{b/2} z^2 h\, dz = hb^3/12.$$

Eq. (244) can also be used for calculating I_z for the parallelogram shown in Fig. 342, because this parallelogram can be obtained from the rectangle shown by dotted lines by a displacement parallel to the axis z of elements such as the one shown. The areas of the elements and their distances from the z axis remain unchanged during such displacement so that I_z is the same as for the rectangle.

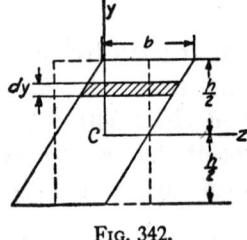

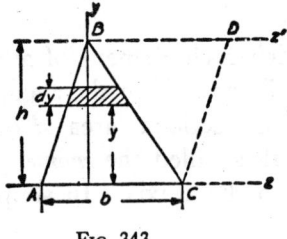

FIG. 342.   FIG. 343.

In calculating the moment of inertia of a triangle with respect to the base (Fig. 343), the area of an element such as shown in the figure is

$$dA = b\,\frac{h - y}{h}\,dy$$

and eq. (243) gives

$$I_z = \int_0^h b\,\frac{h - y}{h}\,y^2 dy = bh^3/12.$$

The method of calculation illustrated by the above examples can be used in the most general case. The moment of inertia is obtained by dividing the figure into infinitesimal strips parallel to the axis and then integrating as in eq. (243).

The calculation can often be simplified if the figure can be divided into portions whose moments of inertia about the axis are known. In such case, the total moment of inertia is the sum of the moments of inertia of all the parts.

From its definition, eq. (243), it follows that the moment of inertia of an area with respect to an axis has the dimensions of a length raised to the fourth power. Hence, by dividing the moment of inertia with respect to a certain axis by the cross-sectional area of the figure, the square of a certain length is obtained. This length k is called the *radius of gyration* with respect to that axis. For the y and z axes, the radii of gyration are

$$k_y = \sqrt{I_y/A}, \qquad k_z = \sqrt{I_z/A}. \tag{245}$$

Problems

1. Find the moment of inertia of the rectangle in Fig. 341 with respect to the base.

Answer. $I_{z'} = bh^3/3$.

2. Find the moment of inertia of the triangle ABC with respect to the axis z' (Fig. 343).

Solution. This moment of inertia is the difference between the moment of inertia of the parallelogram $ABDC$ and the triangle BDC. Hence,

$$I_{z'} = bh^3/3 - bh^3/12 = bh^3/4.$$

3. Find I_z for the cross sections shown in Fig. 344.

Answer. For (a), $I_z = a^4/12 - (a - 2h)^4/12$; for (b) and (c),

$$I_z = ba^3/12 - \frac{(b - h_1)(a - 2h)^3}{12}.$$

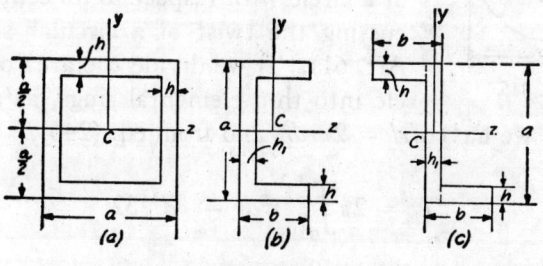

Fig. 344.

4. Find the moment of inertia of a square with sides a with respect to a diagonal of the square.

Answer. $I = a^4/12$.

5. Find k_y and k_z for the rectangle shown in Fig. 341.

Answer. $k_y = b/2\sqrt{3}$, $k_z = h/2\sqrt{3}$.

6. Find k_z for Figs. 344a and 344b.

II. Polar Moment of Inertia of a Plane Area

The moment of inertia of a plane area with respect to an axis perpendicular to the plane of the figure is called the *polar moment of inertia* with respect to the point where the axis intersects the plane (point O in Fig. 340). It is defined as the integral

$$I_p = \int_A r^2 dA, \qquad (246)$$

in which each element of area dA is multiplied by the square of its distance to the axis and integration is extended over the entire area of the figure.

Referring to Fig. 340, we have $r^2 = y^2 + z^2$, and from eq. (246)

$$I_p = \int_A (y^2 + z^2)dA = I_z + I_y. \qquad (247)$$

That is, the polar moment of inertia with respect to any point

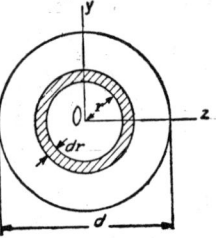

FIG. 345.

O is equal to the sum of the moments of inertia with respect to two perpendicular axes y and z through the same point.

Let us consider a *circular cross section*. We encounter the polar moment of inertia of a circle with respect to its center in discussing the twist of a circular shaft (see Art. 61). If we divide the area of the circle into thin elemental rings, as shown in Fig. 345, we have $dA = 2\pi r dr$, and from eq. (246),

$$I_p = 2\pi \int_0^{d/2} r^3 dr = \pi d^4/32. \qquad (248)$$

We know from symmetry that in this case $I_y = I_z$, hence, from eqs. (247) and (248),

$$I_y = I_z = \tfrac{1}{2} I_p = \pi d^4/64. \qquad (249)$$

The moment of inertia of an ellipse with respect to a principal axis z (Fig. 346) can be obtained by comparing the ellipse with the circle shown in the figure by the dotted line. The height y of any element of the ellipse, such as the one shown shaded, can be obtained by reducing the height y_1 of the corresponding element of the circle in the ratio b/a. From eq. (244), the moments of inertia of these two elements with respect to the z axis are in the ratio b^3/a^3. The moments of

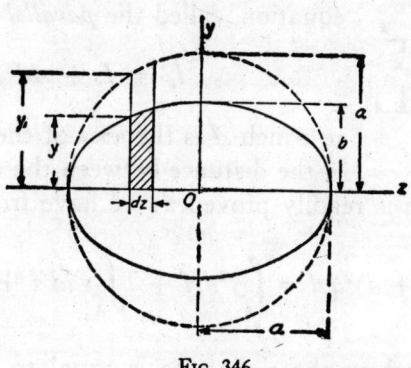

FIG. 346.

inertia of the ellipse and of the circle are evidently in the same ratio, hence, the moment of inertia of the ellipse is

$$I_z = \pi(2a)^4/64 \cdot b^3/a^3 = \pi ab^3/4. \qquad (250)$$

In the same manner, for the vertical axis

$$I_y = \pi ba^3/4,$$

and the polar moment of inertia of an ellipse is, from eq. (247),

$$I_p = I_y + I_z = \pi ab^3/4 + \pi ba^3/4. \qquad (251)$$

Problems

1. Find the polar moment of inertia of a rectangle with respect to the centroid (Fig. 341).

Answer. $I_p = bh^3/12 + hb^3/12.$

2. Find the polar moments of inertia with respect to their centroids of the areas shown in Fig. 344.

III. Parallel-Axis Theorem

If the moment of inertia of an area with respect to an axis z through the centroid (Fig. 347) is known, the moment of inertia with respect to any parallel axis z' can be calculated from the following equation, called the *parallel-axis theorem*:

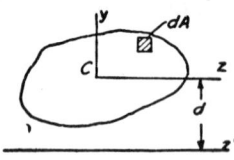

FIG. 347.

$$I_{z'} = I_z + Ad^2, \qquad (252)$$

in which A is the area of the figure and d is the distance between the axes.

The theorem is readily proved as we have from eq. (243):

$$I_{z'} = \int_A (y + d)^2 dA = \int_A y^2 dA + 2\int_A y\,d\,dA + \int_A d^2 dA.$$

The first integral on the right side is equal to I_z, the third integral is equal to Ad^2 and the second integral vanishes due to the fact that z passes through the centroid. Hence, this equation reduces to (252). Eq. (252) is especially useful in calculating moments of inertia of cross sections of built-up beams (Fig. 348). The positions of the centroids of standard angles and the moments of inertia of their cross sections with respect to axes through their centroids are given in handbooks. An abridged listing is also given in Appendix B. By use of the parallel-axis theorem, the moment of inertia of such a built-up section with respect to the z axis can readily be calculated.

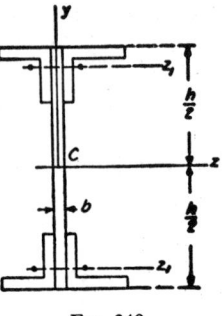

FIG. 348.

Problems

1. By the parallel-axis theorem, find the moment of inertia of a triangle (Fig. 343) with respect to the axis through the centroid and parallel to the base.

Answer. $I = bh^3/36$.

2. Find the moment of inertia I_z of the section shown in Fig. 348 if $h = 20$ in., $b = \frac{1}{2}$ in. and the angles have the dimensions $4 \times 4 \times \frac{3}{4}$ in.

Solution. $I_z = 20^3/(2 \times 12) + 4[7.70 + 5.44(10 - 1.27)^2] = 2,022$ in.^{4}

3. Find the moment of inertia with respect to the neutral axis of the cross section of the channel in Prob. 2, p. 104.

IV. Product of Inertia. Principal Axes

The integral

$$I_{yz} = \int_A yz\,dA, \tag{253}$$

in which each element of area dA is multiplied by the product of its coordinates and integration is extended over the entire area A of a plane figure, is called the *product of inertia* of the figure. If a figure has an axis of symmetry which is taken for the y or z axis (Fig. 349), the product of inertia is equal to zero. This follows from the fact that in this case for any element such as dA with a positive z there exists an equal and symmetrically situated element dA' with a negative z. The corresponding elementary products $yz\,dA$ cancel each other, hence integral (253) vanishes.

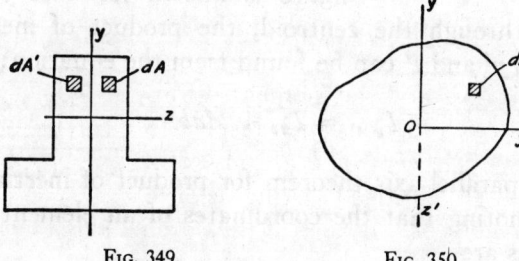

FIG. 349. FIG. 350.

In the general case, for any point of any plane figure, we can always find two perpendicular axes such that the product of inertia for these axes vanishes. Take, for instance, the axes y and z, Fig. 350. If the axes are rotated 90° about O in the clockwise direction, the new positions of the axes are y' and z as shown in the figure. There is then the following relation between the old coordinates of an element dA and its new coordinates:

$$y' = z, \qquad z' = -y.$$

Hence, the product of inertia for the new coordinates is

$$I_{y'z'} = \int_A y'z'dA = -\int_A yz\,dA = -I_{yz}.$$

Thus, during this rotation, the product of inertia changes its sign. As the product of inertia changes continuously with the angle of rotation, there must be certain directions for which this quantity becomes zero. The axes in these directions are called the *principal axes*. Usually the centroid is taken as the origin of coordinates and the corresponding principal axes are then called the *centroidal principal axes*. If a figure has an

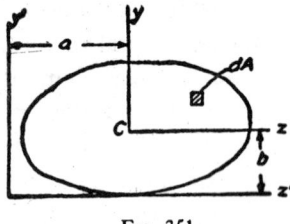

FIG. 351.

axis of symmetry, this axis and an axis perpendicular to it are principal axes of the figure, because the product of inertia with respect to these axes is equal to zero, as explained above.

If the product of inertia of a figure is known for axes y and z (Fig. 351) through the centroid, the product of inertia for parallel axes y' and z' can be found from the equation:

$$I_{y'z'} = I_{yz} + Aab. \qquad (254)$$

This is the parallel-axis theorem for product of inertia and is proved by noting that the coordinates of an element dA for the new axes are

$$y' = y + b, \qquad z' = z + a.$$

Hence,

$$I_{y'z'} = \int_A y'z'dA = \int_A (y + b)(z + a)dA$$

$$= \int_A yz\,dA + \int_A ab\,dA + \int_A ya\,dA + \int_A bz\,dA.$$

The last two integrals vanish because C is the centroid and the equation reduces to (254).

Problems

1. Find $I_{y'z'}$ for the rectangle in Fig. 341.
Answer. $I_{y'z'} = b^2h^2/4$.

2. Find the product of inertia of the angle section (Fig. 352) with respect to the y and z axes; also for the y_1 and z_1 axes.

Solution. Dividing the figure into two rectangles and using eq. (254) for each of these rectangles, we find

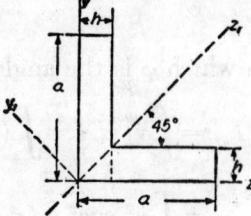

$$I_{yz} = a^2h^2/4 + h^2(a^2 - h^2)/4.$$

From the symmetry condition $I_{y_1z_1} = 0$.

3. Determine the products of inertia I_{yz} of the sections shown in Fig. 344 if C is the centroid.

FIG. 352.

Solution. For Figs. 344a and 344b, $I_{yz} = 0$ because of symmetry. In the case of Fig. 344c, dividing the section into three rectangles and using eq. (254) we find

$$I_{yz} = -2(b - h_1)h \left(\frac{a - h}{2}\right)\left(\frac{b}{2}\right).$$

V. Change of Direction of Axes.   Determination of the Principal Axes

Suppose that the moments of inertia

$$I_z = \int_A y^2 dA, \qquad I_y = \int_A z^2 dA \qquad (a)$$

and the product of inertia

$$I_{yz} = \int_A yz dA \qquad (b)$$

are known, and it is required to find the same quantities for the new axes y_1 and z_1 (Fig. 353). Considering an elementary area dA, the new coordinates from the figure are

$$z_1 = z \cos \varphi + y \sin \varphi, \qquad y_1 = y \cos \varphi - z \sin \varphi, \qquad (c)$$

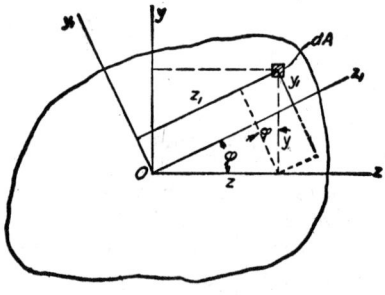

FIG. 353.

in which φ is the angle between z and z_1. Then,

$$I_{z_1} = \int_A y_1{}^2 dA = \int_A (y \cos \varphi - z \sin \varphi)^2 dA$$

$$= \int_A y^2 \cos^2 \varphi dA + \int_A z^2 \sin^2 \varphi dA - \int_A 2yz \sin \varphi \cos \varphi dA,$$

or by using (a) and (b)

$$I_{z_1} = I_z \cos^2 \varphi + I_y \sin^2 \varphi - I_{yz} \sin 2\varphi. \qquad (255)$$

In the same manner

$$I_{y_1} = I_z \sin^2 \varphi + I_y \cos^2 \varphi + I_{yz} \sin 2\varphi. \qquad (255')$$

By taking the sum and the difference of eqs. (255) and (255') we find

$$I_{z_1} + I_{y_1} = I_z + I_y, \qquad (256)$$

$$I_{z_1} - I_{y_1} = (I_z - I_y) \cos 2\varphi - 2I_{yz} \sin 2\varphi. \qquad (257)$$

These equations are very useful for calculating I_{z_1} and I_{y_1}. For calculating $I_{y_1 z_1}$, we find

$$I_{y_1 z_1} = \int_A y_1 z_1 dA = \int_A (y \cos \varphi - z \sin \varphi)(z \cos \varphi$$

$$+ y \sin \varphi) dA = \int_A y^2 \sin \varphi \cos \varphi dA$$

$$- \int_A z^2 \sin \varphi \cos \varphi dA + \int_A yz(\cos^2 \varphi - \sin^2 \varphi) dA,$$

or by using (a) and (b),

$$I_{y_1 z_1} = (I_z - I_y)\tfrac{1}{2} \sin 2\varphi + I_{yz} \cos 2\varphi. \qquad (258)$$

The product of inertia is zero for the principal axes. Thus the axes y_1 and z_1 in Fig. 353 are principal axes if the right-hand side of eq. (258) vanishes, or

$$(I_z - I_y)\tfrac{1}{2} \sin 2\varphi + I_{yz} \cos 2\varphi = 0.$$

This gives

$$\tan 2\varphi = 2I_{yz}/(I_y - I_z). \tag{259}$$

Eq. (259) may also be obtained by differentiating eq. (255) with respect to φ and then equating the resulting expression to zero. This shows that the moments of inertia about the principal axes are maximum and minimum.

Let us determine, as an example, the directions of the principal axes of a rectangle through a corner of the rectangle (Fig. 341). In this case,

$$I_{z'} = bh^3/3, \qquad I_{y'} = hb^3/3, \qquad I_{y'z'} = b^2h^2/4.$$

Hence,

$$\tan 2\varphi = \frac{b^2h^2}{2(hb^3/3 - bh^3/3)} = 3bh/2(b^2 - h^2). \tag{d}$$

The direction of φ is determined by noting that in the derivation of eq. (259), the angle φ was taken as positive in the counter-clockwise direction (Fig. 353). Eq. (d) gives two different values for φ differing by 90°. These are the two perpendicular directions of the principal axes. Knowing the directions of the principal axes, the corresponding moments of inertia can be found from eqs. (256) and (257).

The radii of gyration corresponding to the principal axes are called *principal radii of gyration*.

If y_1 and z_1 are the principal axes of inertia (Fig. 354) and k_{y_1} and k_{z_1} the principal radii of gyration, the ellipse with k_{y_1} and k_{z_1} as semi-axes, as shown in the figure, is called the *ellipse of inertia*. Having this ellipse, the radius of gyration k_z for any axis z can be obtained graphically by drawing a tangent to the ellipse parallel to z. The distance of the origin O from this tangent is the length of k_z. The ellipse of inertia gives a picture of how the moment of inertia changes as the axis z rotates in the plane of the figure about the point O, and shows that the maximum and minimum of the moments of inertia are the principal moments of inertia.

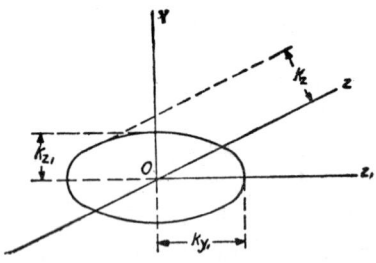

FIG. 354.

A Mohr's circle construction can also be used to find moments of inertia about inclined axes. The graphical procedure is entirely analogous to the procedure discussed in Chap. II for stresses. It is only necessary to replace σ_x, σ_y and τ by I_z, I_y and I_{yz}, respectively.

Problems

1. Determine the directions of the centroidal principal axes of the Z section (Fig. 344c) if $h = h_1 = 1$ in., $b = 5$ in., $a = 10$ in.

2. Find the directions of the centroidal principal axes and the corresponding principal moments of inertia for an angle section $5 \times 2\frac{1}{2} \times \frac{1}{2}$ in.

Answer. Tan $2\varphi = 0.547$; $I_{max} = 9.36$ in.^{4}; $I_{min} = 0.99$ in.^{4}

3. Determine the semi-axes of the ellipse of inertia for an elliptical cross section (Fig. 346).

Answer. $k_z = b/2$, $k_y = a/2$.

4. Under what conditions does the ellipse of inertia become a circle?

APPENDIX B

TABLES OF STRUCTURAL SHAPES

Note: In Tables 16–20 the properties of a few structural shapes are presented. These tables are very incomplete and are presented solely to enable the student to solve problems in the text. The data in the tables was taken from the Steel Construction Manual of the American Institute of Steel Construction, 1954.

TABLE 16: ELEMENTS OF WIDE FLANGE SECTIONS

(ABRIDGED LIST)

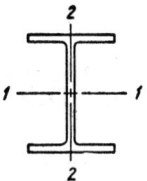

Section	Weight per Foot	Area	Depth	Flange		Web Thickness	Axis 1-1			Axis 2-2		
				Width	Thickness		I	Z	k	I	Z	k
	Lb	In.²	In.	In.	In.	In.	In.⁴	In.³	In.	In.⁴	In.³	In.
36 WF 300	300	88.17	36.72	16.655	1.680	0.945	20290	1105.1	15.17	1225.2	147.1	3.73
36 WF 245	245	72.03	36.06	16.512	1.350	0.802	16092	892.5	14.95	944.7	114.4	3.62
30 WF 210	210	61.78	30.38	15.105	1.315	0.775	9872.4	649.9	12.64	707.9	93.7	3.38
30 WF 172	172	50.65	29.88	14.985	1.065	0.655	7891.5	528.2	12.48	550.1	73.4	3.30
24 WF 160	160	47.04	24.72	14.091	1.135	0.656	5110.3	413.5	10.42	492.6	69.9	3.23
24 WF 120	120	35.29	24.31	12.088	0.930	0.556	3635.3	299.1	10.15	254.0	42.0	2.68
18 WF 114	114	33.51	18.48	11.833	0.991	0.595	2033.8	220.1	7.79	255.6	43.2	2.76
18 WF 85	85	24.97	18.32	8.838	0.911	0.526	1429.9	156.1	7.57	99.4	22.5	2.00
16 WF 96	96	28.22	16.32	11.533	0.875	0.535	1355.1	166.1	6.93	207.2	35.9	2.71
16 WF 50	50	14.70	16.25	7.073	0.628	0.380	655.4	80.7	6.68	34.8	9.8	1.54
14 WF 202	202	59.39	15.63	15.750	1.503	0.930	2538.8	324.9	6.54	979.7	124.4	4.06
14 WF 103	103	30.26	14.25	14.575	0.813	0.495	1165.8	163.6	6.21	419.7	57.6	3.72
14 WF 84	84	24.71	14.18	12.023	0.778	0.451	928.4	130.9	6.13	225.5	37.5	3.02
14 WF 38	38	11.17	14.12	6.776	0.513	0.313	385.3	54.6	5.87	24.6	7.3	1.49
12 WF 190	190	55.86	14.38	12.670	1.736	1.060	1892.5	263.2	5.82	589.7	93.1	3.25
12 WF 106	106	31.19	12.88	12.230	0.986	0.620	930.7	144.5	5.46	300.9	49.2	3.11
12 WF 72	72	21.16	12.25	12.040	0.671	0.430	597.4	97.5	5.31	195.3	32.4	3.04
12 WF 31	31	9.12	12.09	6.525	0.465	0.265	238.4	39.4	5.11	19.8	6.1	1.47
10 WF 112	112	32.92	11.38	10.415	1.248	0.755	718.7	126.3	4.67	235.4	45.2	2.67
10 WF 72	72	21.18	10.50	10.170	0.808	0.510	420.7	80.1	4.46	141.8	27.9	2.59
10 WF 45	45	13.24	10.12	8.022	0.618	0.350	248.6	49.1	4.33	53.2	13.3	2.00
10 WF 25	25	7.35	10.08	5.762	0.430	0.252	133.2	26.4	4.26	12.7	4.4	1.31
8 WF 67	67	19.70	9.00	8.287	0.933	0.575	271.8	60.4	3.71	88.6	21.4	2.12
8 WF 48	48	14.11	8.50	8.117	0.683	0.405	183.7	43.2	3.61	60.9	15.0	2.08
8 WF 40	40	11.76	8.25	8.077	0.558	0.365	146.3	35.5	3.53	49.0	12.1	2.04
8 WF 28	28	8.23	8.06	6.540	0.463	0.285	97.8	24.3	3.45	21.6	6.6	1.62

TABLE 17: ELEMENTS OF AMERICAN STANDARD I-BEAM SECTIONS

(ABRIDGED LIST)

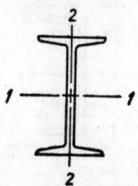

Section	Weight per Foot	Area	Depth	Flange		Web Thickness	Axis 1–1			Axis 2–2		
				Width	Mean Thickness		I	Z	k	I	Z	k
	Lb	In.²	In.	In.	In.	In.	In.⁴	In.³	In.	In.⁴	In.³	In.
24 I 120	120.0	35.13	24.00	8.048	1.102	0.798	3010.8	250.9	9.26	84.9	21.1	1.56
24 I 105.9	105.9	30.98	24.00	7.875	1.102	0.625	2811.5	234.3	9.53	78.9	20.0	1.60
24 I 79.9	79.9	23.33	24.00	7.000	0.871	0.500	2087.2	173.9	9.46	42.9	12.2	1.36
18 I 70.0	70.0	20.46	18.00	6.251	0.691	0.711	917.5	101.9	6.70	24.5	7.8	1.09
18 I 54.7	54.7	15.94	18.00	6.000	0.691	0.460	795.5	88.4	7.07	21.2	7.1	1.15
12 I 50	50.0	14.57	12.00	5.477	0.659	0.687	301.6	50.3	4.55	16.0	5.8	1.05
12 I 40.8	40.8	11.84	12.00	5.250	0.659	0.460	268.9	44.8	4.77	13.8	5.3	1.08
12 I 31.8	31.8	9.26	12.00	5.000	0.544	0.350	215.8	36.0	4.83	9.5	3.8	1.01
8 I 23	23.0	6.71	8.00	4.171	0.425	0.441	64.2	16.0	3.09	4.4	2.1	0.81
8 I 18.4	18.4	5.34	8.00	4.000	0.425	0.270	56.9	14.2	3.26	3.8	1.9	0.84
7 I 20	20.0	5.83	7.00	3.860	0.392	0.450	41.9	12.0	2.68	3.1	1.6	0.74
7 I 15.3	15.3	4.43	7.00	3.660	0.392	0.250	36.2	10.4	2.86	2.7	1.5	0.78
6 I 17.25	17.25	5.02	6.00	3.565	0.359	0.465	26.0	8.7	2.28	2.3	1.3	0.68
6 I 12.5	12.5	3.61	6.00	3.330	0.359	0.230	21.8	7.3	2.46	1.8	1.1	0.72
5 I 14.75	14.75	4.29	5.00	3.284	0.326	0.494	15.0	6.0	1.87	1.7	1.0	0.63
5 I 10	10.0	2.87	5.00	3.000	0.326	0.210	12.1	4.8	2.05	1.2	0.82	0.65
4 I 9.5	9.5	2.76	4.00	2.796	0.293	0.326	6.7	3.3	1.56	0.91	0.65	0.58
4 I 7.7	7.7	2.21	4.00	2.660	0.293	0.190	6.0	3.0	1.64	0.77	0.58	0.59

Table 18: Elements of American Standard Channel Sections

(Abridged List)

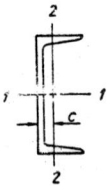

Section	Weight per Foot	Area	Depth	Flange		Web Thickness	Axis 1-1			Axis 2-2			
				Width	Mean Thickness		I	Z	k	I	Z	k	c
	Lb	In.²	In.	In.	In.	In.	In.⁴	In.³	In.	In.⁴	In.³	In.	In.
15 ⌐ 30	50.0	14.64	15.00	3.716	0.650	0.716	401.4	53.6	5.24	11.2	3.8	0.87	0.80
15 ⌐ 40	40.0	11.70	15.00	3.520	0.650	0.520	346.3	46.2	5.44	9.3	3.4	0.89	0.78
12 ⌐ 30	30.0	8.79	12.00	3.170	0.501	0.510	161.2	26.9	4.28	5.2	2.1	0.77	0.68
12 ⌐ 20.7	20.7	6.03	12.00	2.940	0.501	0.280	128.1	21.4	4.61	3.9	1.7	0.81	0.70
10 ⌐ 30	30.0	8.80	10.00	3.033	0.436	0.673	103.0	20.6	3.42	4.0	1.7	0.67	0.65
10 ⌐ 20	20.0	5.86	10.00	2.739	0.436	0.379	78.5	15.7	3.66	2.8	1.3	0.70	0.61
8 ⌐ 18.75	18.75	5.49	8.00	2.527	0.390	0.487	43.7	10.9	2.82	2.0	1.0	0.60	0.57
8 ⌐ 11.5	11.5	3.36	8.00	2.260	0.390	0.220	32.3	8.1	3.10	1.3	0.79	0.63	0.58
6 ⌐ 13.0	13.0	3.81	6.00	2.157	0.343	0.437	17.3	5.8	2.13	1.1	0.65	0.53	0.52
6 ⌐ 8.2	8.2	2.39	6.00	1.920	0.343	0.200	13.0	4.3	2.34	0.70	0.50	0.54	0.52
4 ⌐ 7.25	7.25	2.12	4.00	1.720	0.296	0.320	4.5	2.3	1.47	0.44	0.35	0.46	0.46
4 ⌐ 5.4	5.4	1.56	4.00	1.580	0.296	0.180	3.8	1.9	1.56	0.32	0.29	0.45	0.46

TABLE 19: ELEMENTS OF ANGLES—EQUAL LEGS

(ABRIDGED LIST)

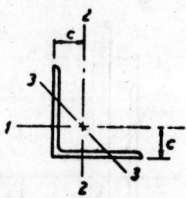

Size	Thick-ness	Weight per Foot	Area	Axis 1-1 and Axis 2-2				Axis 3-3
				I	Z	k	c	k_{min}
In.	In.	Lb	In.²	In.⁴	In.³	In.	In.	In.
8 × 8	1	51.0	15.00	89.0	15.8	2.44	2.37	1.56
8 × 8	½	26.4	7.75	48.6	8.4	2.50	2.19	1.59
6 × 6	1	37.4	11.00	35.5	8.6	1.80	1.86	1.17
6 × 6	½	19.6	5.75	19.9	4.6	1.86	1.68	1.18
5 × 5	⅞	27.2	7.98	17.8	5.2	1.49	1.57	0.97
5 × 5	½	16.2	4.75	11.3	3.2	1.54	1.43	0.98
4 × 4	¾	18.5	5.44	7.7	2.8	1.19	1.27	0.78
4 × 4	⅜	9.8	2.86	4.4	1.5	1.23	1.14	0.79
3½ × 3½	½	11.1	3.25	3.6	1.5	1.06	1.06	0.68
3½ × 3½	¼	5.8	1.69	2.0	0.79	1.09	0.97	0.69
3 × 3	½	9.4	2.75	2.2	1.1	0.90	0.93	0.58
3 × 3	¼	4.9	1.44	1.2	0.58	0.93	0.84	0.59

TABLE 20: ELEMENTS OF ANGLES—UNEQUAL LEGS
(ABRIDGED LIST)

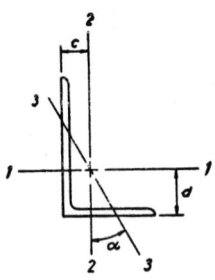

Size	Thickness	Weight per Foot	Area	Axis 1–1				Axis 2–2				Axis 3–3	
				I	Z	k	d	I	Z	k	c	k_{min}	$\text{Tan}\,\alpha$
In.	In.	Lb	In.^{2}	In.^{4}	In.^{3}	In.	In.	In.^{4}	In.^{3}	In.	In.	In.	
8 × 6	1	44.2	13.00	80.8	15.1	2.49	2.65	38.8	8.9	1.73	1.65	1.28	0.543
8 × 6	$\frac{1}{2}$	23.0	6.75	44.3	8.0	2.56	2.47	21.7	4.8	1.79	1.47	1.30	0.558
7 × 4	$\frac{7}{8}$	30.2	8.86	42.9	9.7	2.20	2.55	10.2	3.5	1.07	1.05	0.86	0.318
7 × 4	$\frac{1}{2}$	17.9	5.25	26.7	5.8	2.25	2.42	6.5	2.1	1.11	0.92	0.87	0.335
6 × 4	$\frac{7}{8}$	27.2	7.98	27.7	7.2	1.86	2.12	9.8	3.4	1.11	1.12	0.86	0.421
6 × 4	$\frac{1}{2}$	16.2	4.75	17.4	4.3	1.91	1.99	6.3	2.1	1.15	0.99	0.87	0.440
5 × 3½	$\frac{3}{4}$	19.8	5.81	13.9	4.3	1.55	1.75	5.6	2.2	0.98	1.00	0.75	0.464
5 × 3½	$\frac{3}{8}$	10.4	3.05	7.8	2.3	1.60	1.61	3.2	1.2	1.02	0.86	0.76	0.486
5 × 3	$\frac{1}{2}$	12.8	3.75	9.5	2.9	1.59	1.75	2.6	1.1	0.83	0.75	0.65	0.357
5 × 3	$\frac{1}{4}$	6.6	1.94	5.1	1.5	1.62	1.66	1.4	0.61	0.86	0.66	0.66	0.371
4 × 3	$\frac{5}{8}$	13.6	3.98	6.0	2.3	1.23	1.37	2.9	1.4	0.85	0.87	0.64	0.534
4 × 3	$\frac{3}{8}$	8.5	2.48	4.0	1.5	1.26	1.28	1.9	0.87	0.88	0.78	0.64	0.551

AUTHOR INDEX

435

SUBJECT INDEX

Numbers refer to pages

437